1/9/82 A.B. Brink

FAO SOILS BULLETIN 48

micronutrients
and
the nutrient status of soils:
a global study

by
mikko sillanpää

sponsored by
the government of finland
executed at
the institute of soil science
agricultural research centre
jokioinen, finland
and
soil resources, management
and conservation service
land and water development division
FAO

FOOD AND AGRICULTURE ORGANIZATION OF THE UNITED NATIONS
Rome 1982

The designations employed and the presentation of material in this publication do not imply the expression of any opinion whatsoever on the part of the Food and Agriculture Organization of the United Nations concerning the legal status of any country, territory, city or area or of its authorities, or concerning the delimitation of its frontiers or boundaries.

M-52
ISBN 92-5-101193-1

All rights reserved. No part of this publication may be reproduced, stored in a retrieval system, or transmitted in any form or by any means, electronic, mechanical, photocopying or otherwise, without the prior permission of the copyright owner. Applications for such permission, with a statement of the purpose and extent of the reproduction, should be addressed to the Director, Publications Division, Food and Agriculture Organization of the United Nations, Via delle Terme di Caracalla, 00100 Rome, Italy.

© **FAO** 1982

Printed in Finland by Werner Söderström Osakeyhtiö.

Foreword

During the last two decades, the increasing use of mineral fertilizers and organic manures of different types has led to impressive yield increases in developing countries. Major emphasis was given to the supply of the main macronutrients, nitrogen, phosphate and potash.

For a long time it was felt that under the existing farming systems and fertilizing practices the level of micronutrients was adequate and that the problem of micronutrient deficiencies was not a serious one. However, indications from developing countries show that micronutrient problems are becoming more and more frequent.

In the early 1970s the Government of Finland, through its Institute of Soil Science and FAO embarked on a system of investigation of microelement deficiencies in developing countries.

In 1974, the project "Trace Element Study, TF/129/FIN" started under the FAO/Finland Cooperative Programme, with the financial support of the Government of Finland, and with the cooperation of the staff of the Institute of Soil Science, under the direction of Prof. M. Sillanpää. The facilities of the Institute were made available to the project. This programme involved a worldwide study on micronutrients, in cooperation with 30 countries.

The study, which is presented in this document, shows that the undertaking was very timely. It appears that micronutrients are becoming deficient in developing countries and that remedial measuring will need to be taken to ensure full soil and plant productivity. The study also gives guidance for the investigation of micronutrient deficiencies, which should be dealt with at the country level.

FAO expresses its appreciation to the Government of Finland for its generous support of the study and for its excellent cooperation throughout.

F.W. Hauck
Chief, Soil Resources, Management
and Conservation Service

CONTENTS

	page
PREFACE	
PART I	1
PART I SOIL AND PLANT DATA, METHODOLOGY AND INTERPRETATION	3
1 MATERIALS AND METHODS	3
1.1 Collection of original plant and soil samples	3
1.2 Indicator plants	3
1.2.1 General	3
1.2.2 Sampling time	4
1.2.3 Contamination	4
1.2.4 Plant varieties	6
1.2.5 Need for better grounds for comparing plant data	6
1.2.6 Growing new indicator plants in pots	7
1.3 Analytical methods	8
1.3.1 Methods of soil analyses	9
1.3.2 Methods of plant analyses	11
1.3.3 Statistical methods	12
1.4 Expression of analytical data	13
1.4.1 Soils	13
1.4.2 Plants	14
1.4.3 Regression graphs	14
2 RESULTS AND DISCUSSION	17
2.1 General properties of soils and their mutual relations	17
2.2 Macronutrients	21
2.2.1 General aspects	21
2.2.2 Comparison of the two original indicator plants	21
2.2.3 Comparison of macronutrient status in different countries	21
2.2.4 Macronutrient contents of plants and soils in relation to four soil characteristics	28
2.3 Micronutrients	33
2.3.1 Plant analysis versus soil analysis	33
2.3.2 Molybdenum	35
2.3.2.1 General aspects	35
2.3.2.2 Soil factors affecting the molybdenum contents of plants and soils	36
2.3.3 Boron	46
2.3.3.1 General aspects	46
2.3.3.2 Soil factors affecting the boron contents of plants and soils	48
2.3.4 Copper	54
2.3.4.1 General aspects	54
2.3.4.2 Soils factors affecting the copper contents of plants and soils	55
2.3.5 Iron	61
2.3.5.1 General aspects	61
2.3.5.2 Soil factors affecting the iron contents of plants and soils	61
2.3.6 Manganese	64
2.3.6.1 General aspects	64
2.3.6.2 Soil factors affecting the manganese contents of plants and soils	66
2.3.7 Zinc	75
2.3.7.1 General aspects	75
2.3.7.2 Soil factors affecting the zinc contents of plants and soils	75

		page
2.4	Mutual relations between micronutrients	83
2.5	Micronutrient contents of plants and soils in relation to yields with special reference to the "concentration-dilution" phenomenon	86
2.6	Suitability of plant and soil analyses for large scale operations	90
2.7	Estimation of critical micronutrient limits	90
2.8	Plant and soil micronutrieents and other soil data in relation to FAO/Unesco soil units	96

3 SUMMARY FOR PART I 97

PART II 101

PART II NUTRIENT STATUS BY COUNTRIES 103

4 INTRODUCTION 103

5 EUROPE AND OCEANIA 105

5.1 Belgium 105
5.2 Finland 117
5.3 Hungary 128
5.4 Italy 139
5.5 Malta 150
5.6 New Zealand 158

6 LATIN AMERICA 169

6.1 Argentina 169
6.2 Brazil 180
6.3 Ecuador 190
6.4 Mexico 194
6.5 Peru 207

7 FAR EAST 219

7.1 India 219
7.2 Korea, Republic of 232
7.3 Nepal 242
7.4 Pakistan 251
7.5 Philippines 262
7.6 Sri Lanka 271
7.7 Thailand 280

8 NEAR EAST 289

8.1 Egypt 289
8.2 Iraq 300
8.3 Lebanon 313
8.4 Syria 322
8.5 Turkey 332

9 AFRICA 343

9.1 Ethiopia 343
9.2 Ghana 355
9.3 Malawi 363
9.4 Nigeria 372
9.5 Sierra Leone 383
9.6 Tanzania 392
9.7 Zambia 401

VII

	page
10 SUMMARY FOR PART II	411
REFERENCES	417

APPENDIXES

1	Instructions for taking plant and soil samples	421
2	National mean values of basic analytical data on original wheat plants and respective soils	422
3	National mean values of basic analytical data on original maize plants and respective soils	428
4	National mean values of basic analytical data on pot-grown wheat plants and respective soils	434
5	Indicative data on fertilizer application to the original indicator crops	440
6	Average soil properties and macronutrient contents of soils classified by FAO/Unesco soil units	441
7	Average micronutrient contents of soils and respective pot-grown wheats for FAO/Unesco soil units	442

LIST OF TABLES

		page
Table 1.	Comparison of nutrient content of two soils expressed on weight and volume bases.	13
Table 2.	Numerical products of plant content x soil content along the lines dividing the regression graphs into five zones.	15
Table 3.	Mean macronutrient contents of the two original indicator plants, of the respective soils and the mean amounts of N, P, and K applied in fertilizers to each crop.	21
Table 4.	Regressions of pooled plant N, P, K, Ca, and Mg contents on respective soil macronutrients.	28
Table 5.	Plant and soil macronutrients as functions of four soil characteristics in the whole international material.	28
Table 6.	Directions (+ or — symbols) of regressions of plant and soil macronutrient contents on four soil characteristics.	32
Table 7.	Equations and correlation coefficients for regressions of AO-OA extractable soil Mo (uncorrected and pH-corrected) and plant Mo on six soil characteristics.	39
Table 8.	Correlations between plant Mo and AO-OA extractable soil Mo in various countries. Soil Mo analyses are both uncorrected and corrected for pH and texture.	45
Table 9.	Correlations between plant B and hot water extractable soil B in various countries. Soil B analyses are both uncorrected and corrected for cation exchange capacity.	51
Table 10.	Equations and correlation coefficients for regressions of hot water soluble soil B (uncorrected and CEC-corrected) and plant B on various soil factors.	53
Table 11.	Correlations between plant Cu and AAAc-EDTA extractable soil Cu in various countries. Soil Cu analyses are both uncorrected and corrected for soil organic carbon contents.	58

		page
Table 12.	Equations and correlation coefficients for regressions of AAAc-EDTA extractable soil Cu (uncorrected and org. C-corrected) and plant Cu on various soil factors.	60
Table 13.	Equations and correlation coefficients for regressions of AAAc-EDTA extractable soil Fe and plant Fe on various soil factors.	64
Table 14.	Correlations between plant Mn and DTPA extractable soil Mn in various countries. Soil Mn analyses are both uncorrected and corrected for pH($CaCl_2$).	69
Table 15.	Equations and correlation coefficients for regressions of DTPA extractable soil Mn (uncorrected and pH-corrected) and plant Mn on various soil factors.	71
Table 16.	Correlations between plant Mn content and AAAc-EDTA extractable soil Mn in various countries. Soil Mn analyses are both uncorrected and corrected for pH.	73
Table 17.	Mutual correlations between extractable Mn contents of soils in the whole material as determined by four methods.	74
Table 18.	Correlations in various countries between plant Zn and soil Zn determined by three different methods.	79
Table 19.	Equations and correlation coefficients for regressions of AAAc-EDTA extractable soil Zn (uncorrected and pH-corrected), plant Zn, and DTPA extractable soil Zn on various soil factors.	81
Table 20.	Mutual correlations between extractable soil micronutrients in the whole material.	86
Table 21.	Two-year averages of micronutrient contents of 17 crops grown side by side at nine sites.	91
Table 22.	Tentative critical levels of micronutrients determined by various methods.	96
Table 23.	Correlations between micronutrient contents of the three types of indicator plants and respective soils.	97
Table 24.	Comparison of micronutrient contents between high yielding maize varieties and local variety 'Criollo', Mexico.	206
Table 25.	Comparison of macronutrient contents of high yielding and local varieties of original Indian wheats.	231
Table 26.	Comparison of micronutrient contents of high yielding and local varieties of original Indian wheats.	231
Table 27.	Comparison of analytical data from non-irrigated and irrigated soils of Iraq and respective plants.	311
Table 28.	Comparison of macronutrient contents of high yielding and local varieties of original Nigerian wheats.	376
Table 29.	Relative frequencies of low, medium and high values of soil texture (TI), pH, organic carbon content and cation exchange capacity in 30 countries.	412
Table 30.	Relative frequencies of low, medium and high values of nitrogen, phosphorus, potassium, calcium and magnesium in soils of 30 countries.	413
Table 31.	Relative distributions of six micronutrients in five plant x soil content zones in various countries.	414

LIST OF FIGURES

		page
Fig. 1.	Variations in B, Fe, Cu, Mn, Mo, and Zn contents of wheat (ppm in DM) during the growing season.	5
Fig. 2.	Growing new indicator plants in pots (photo).	8
Figs 3 and 4.	Mutual correlations between six soil characteristics.	18, 19
Fig. 5.	Relationship between $pH(H_2O)$ and $pH(CaCl_2)$.	20
Fig. 6.	Regressions of nitrogen content of original wheat and maize (pooled) on total nitrogen content of soil a) for the whole international material, b) for crops fertilized with $>$ 100 kg N/ha, and c) for crops fertilized with $<$ 50 kg N/ha. The national mean values are given.	23
Fig. 7.	Regressions of phosphorus content of original wheat and maize (pooled) on $NaHCO_3$ extractable phosphorus content of soil a) for the whole international material, b) for crops fertilized with $>$ 30 kg P/ha, and c) for crops fertilized with $<$ 10 kg P/ha. The national mean values are given.	24
Fig. 8.	Regressions of potassium content of original wheat and maize (pooled) on CH_3COONH_4 exchangeable potassium content of soil a) for the whole international material, b) for crops fertilized with $>$ 30 kg K/ha, and c) for crops fertilized with $<$ 10 kg K/ha. The national mean values are given.	25
Fig. 9.	Regression of calcium content of original wheat and maize (pooled) on CH_3COONH_4 exchangeable calcium content of soil for the whole international material. The national mean values are given.	26
Fig. 10.	Regression of magnesium content of original wheat and maize (pooled) on CH_3COONH_4 exchangeable magnesium content of soil for the whole international material. The national mean values are given.	27
Fig. 11.	Plant and soil macronutrients as functions of soil pH in the whole international material.	29
Fig. 12.	Plant and soil macronutrients as functions of soil texture in the whole international material.	30
Fig. 13.	Plant and soil macronutrients as functions of soil organic carbon content in the whole international material.	31
Fig. 14.	Plant and soil macronutrients as functions of cation exchange capacity in the whole international material.	32
Fig. 15.	Regression of Mo content of pot-grown wheat on ammonium oxalate-oxalic acid extractable soil Mo for the whole international material. The national mean values are given.	37
Fig. 16.	Relationships of AO-OA extractable soil Mo (uncorrected and pH-corrected) and plant Mo to various soil characteristics in the whole international material.	38
Fig. 17.	Relationships of AO-OA extractable soil Mo (uncorrected and pH-corrected) and plant Mo to soil pH.	40
Fig. 18.	Relationship of plant Mo to AO-OA extractable soil Mo in the whole material classified for pH.	42

		page
Fig. 19.	Relationship of plant Mo to pH-corrected soil Mo (AO-OA) in the whole material classified for pH.	42
Fig. 20.	Regression of Mo content of the pot-grown wheat on the pH-corrected AO-OA extractable soil Mo for the whole international material. National mean values are given.	43
Fig. 21.	Regression of Mo content of the pot-grown wheat on the pH- and texture-corrected soil Mo for the whole international material. National mean values are given.	45
Fig. 22.	Regression of B content of pot-grown wheat on hot water extractable soil B for all the international material. National mean values are given.	47
Fig. 23.	Regression of B uptake by pot-grown wheat on hot water extractable soil B for all the international material. The national mean values are given.	48
Fig. 24.	Relationships of plant B and soil B (uncorrected and CEC-corrected) to cation exchange capacity.	49
Fig. 25.	Regression of B content of the pot-grown wheat on CEC-corrected hot water soluble B for the whole international material. National mean values of plant and soil B are plotted in the graph.	50
Fig. 26.	Relationships of soil B and plant B to various soil factors.	52
Fig. 27.	Regression of Cu content of pot-grown wheat on acid ammonium acetate-EDTA extractable soil Cu for the whole international material. National mean values are given.	55
Fig. 28.	Relationships of soil Cu (uncorrected and org.C-corrected) and plant Cu to organic carbon content of soil.	56
Fig. 29.	Regression of Cu content of pot-grown wheat on k(org. C)-corrected AAAc-EDTA extractable soil Cu for the whole international material. National mean values are plotted in the graph.	57
Fig. 30.	Relationships of AAAc-EDTA extractable soil Cu (uncorrected and org- C-corrected) and plant Cu to various soil factors.	59
Fig. 31.	Regression of Fe content of pot-grown wheat on acid ammonium acetate-EDTA extractable soil Fe for all the international material. National mean values are given.	62
Fig. 32.	AAAc-EDTA extractable soil Fe and plant Fe as a function of six soil characteristics.	63
Fig. 33.	Regression of Mn content of pot-grown wheat on DTPA extractable soil Mn for the whole international material. National mean values are given.	65
Fig. 34.	Relationships of DTPA extractable soil Mn (uncorrected and pH-corrected) and plant Mn to soil pH.	67
Fig. 35.	Relationships of Mn content of pot-grown wheat on DTPA extractable soil Mn in the whole material classified according to pH.	68
Fig. 36.	Relationships of plant Mn to pH-corrected soil Mn (DTPA) on the whole material classified according to pH.	68
Fig. 37.	Regression of Mn content of pot-grown wheat on pH-corrected DTPA extractable soil Mn for the whole international material. National mean values of plant and soil Mn are plotted in the graph.	69
Fig. 38.	Relationships of DTPA extractable soil Mn (uncorrected and pH-corrected) and plant Mn to various soil factors.	70

page

Fig. 39. Relationships of AAAc-EDTA extractable soil Mn (uncorrected and pH-corrected) and plant Mn to soil pH. 72

Fig. 40. Regression of Zn content of pot-grown wheat on acid ammonium acetate-EDTA extractable soil Zn for the whole international material. National mean values of plant and soil Zn are also given. 76

Fig. 41. Regression of Zn content of the pot-grown wheat on DTPA extractable soil Zn for the whole international material. The national mean values are given. 77

Fig. 42. Relationships of extractable soil Zn (AAAc-EDTA, AAAc-EDTA + pH-correction, DTPA) and plant Zn to soil pH. 78

Fig. 43. Regression of Zn content of the pot-grown wheat on pH-corrected AAAc-EDTA extractable soil Zn for the whole material. The national mean values are given. 79

Fig. 44. Relationships of plant Zn and soil Zn (determined by three methods) to various soil factors. 80

Figs 45 and 46. Mutual correlations between B, Cu, Fe, Mn, Mo, and Zn contents of pot-grown wheat in the whole international material. 84—85

Fig. 47. B, Cu, Fe, Mn, Mo, and Zn contents of pot-grown wheat and respective extractable soil micronutrients as a function of yield. 87

Fig. 48. Regressions of B content of pot-grown wheat on hot water soluble soil B corrected for CEC, (a) all samples, (b) low yielding samples, (c) high yielding samples. 89

Fig. 49. Coefficients for correcting AAAc-EDTA extractable soil Cu for soil organic carbon content, k(org. C), and hot water soluble B for cation exchange capacity, k(CEC). 99

Fig. 50. Coefficients for correcting DTPA extractable soil Mn, AO-OA extractable Mo and AAAc-EDTA extractable Zn for soil pH. 99

Figure Nos	National data for	page	Figure Nos	National data for	page
51—68	Belgium	105—116	69—86	Finland	117—126
87—104	Hungary	128—138	105—122	Italy	139—149
123—139	Malta	150—157	140—157	New Zealand	158—168
158—175	Argentina	169—179	176—193	Brazil	180—188
194—199	Ecuador	190—191	200—217	Mexico	194—205
218—235	Peru	207—217	236—253	India	219—229
254—271	Korea, Rep.	232—241	272—288	Nepal	242—250
289—306	Pakistan	251—261	307—324	Philippines	262—270
325—342	Sri Lanka	271—279	343—360	Thailand	280—288
361—378	Egypt	289—299	379—396	Iraq	300—310
397—414	Lebanon	313—321	415—432	Syria	322—331
433—450	Turkey	332—342	451—468	Ethiopia	343—353
469—485	Ghana	355—362	486—503	Malawi	363—371
504—521	Nigeria	372—382	522—539	Sierra Leone	383—391
540—557	Tanzania	392—400	558—575	Zambia	401—409

Preface

In spite of increased use of fertilizers, substantially more macronutrients are still being removed annually from the soil than are applied to it as mineral fertilizers. Some of the nutrients removed are replaced by those in straw, farmyard manure, etc., but on the average the nutrient balance is likely to remain negative. This applies especially to less developed parts of the world.

Micronutrients have not generally been applied regularly to soil in conjunction with common fertilizers, and fertilizing soils with macronutrients only is likely to promote imbalances between these nutrient groups as well as between individual nutrients. Furthermore, increased yields, losses of micronutrients through leaching, liming, a decreasing proportion of farmyard manure compared with chemical fertilizers and the increasing purity of chemical fertilizers are among the factors contributing toward accelerated exhaustion of the supply of available micronutrients in soils.

Hidden micronutrient deficiencies are far more widespread than is generally suspected. Micronutrient problems, which are only local today, may well become more serious and widespread in the relatively near future. They must be diagnosed and studied promptly to avoid production problems related to the quantity and quality of foods and feedingstuffs, keeping in mind also the geomedical aspects concerning both humans and animals. Even though much is known of the nature of micronutrient functions, extensive research and experimentation is needed to avoid mistaken use of this knowledge.

One of the greatest difficulties encountered in the study of microelements is that different methodologies have been adopted for analyses in different countries and laboratories. Sometimes even slight variations in procedure may cause quite substantial differences in analytical results. Since these vary from case to case, it is difficult to compare the results obtained by different scientists. It could be said that scientists using different analytical methods are speaking "different languages". Consequently, knowledge of microelement status is rather fragmentary and there is very little basis for comparison. Discussions on this subject between FAO and Finland in 1973 led to the establishment of an international project called the "Trace Element Study, TF 129 FIN".

The purpose of the study, as initially stated, "is to produce fresh information on a world-wide basis on the problems of a number of micronutrients under different soil, climatic and cultural conditions. The results to be obtained will be comparable because all analytical work will be done in one laboratory and because there will be uniformity of the analytical methods to be used. These results should then be used to make interested countries better acquainted with the problems of micronutrients when developing their agriculture and to provide them with guidelines for future research and practical work on a national basis. In other words, the aim is 1) to obtain a general picture of micronutrient status on a world-wide basis, 2) to locate and limit the problematic areas, soils and conditions where one or more of the micronutrients are likely to be deficient and where more detailed future research and field experimentation is needed and 3) to give guidelines to solve the problems in practice."

The project was started at the end of 1974 in cooperation with FAO and was financed by the government of Finland. Originally, it was hoped that 25 countries would participate in the project and invitations were sent to 40 countries. Interest in the study was, however, greater than anticipated, and acceptances were received from 35 countries,

30 of which took an active part in the project by collecting and sending soil and plant samples to Finland for analysis in one laboratory (the Institute of Soil Science, Finland) using the same procedures for all samples. Most of the participants were developing countries, but some highly developed countries were invited for purposes of comparison.

The following persons acted as Head Cooperators for the 30 countries participating in the study and were responsible for organizing the collection of representative plant and soil samples for the study: *Argentina:* Dr. P.H. Etchevehere, Prof. A. Berardo; *Belgium:* Prof. Dr. A. Cottenie; *Brazil:* Mr. A. F. Castro Bahia Filho, Mr. S. Wiethölter; *Ecuador:* Ing. W. Bejarano, Ing. J. Villavicencio; *Egypt:* Dr. Abdel Hamit Fathi; *Ethiopia:* Ato Desta Beyene; *Finland:* Prof. Dr. P. Elonen; *Ghana:* Dr. H.B. Obeng; *Hungary:* Prof. Dr. I. Szabolcs, Dr. Eva Elek; *India:* Dr. N.S. Randhawa; *Iraq:* Dr. Khalil Mosleh, Dr. Nouri A.K. Hassan; *Italy:* Prof. Dr. F. Mancini; *Korea, Rep.:* Dr. Tai-Soon Kim; *Lebanon:* Dr. A. Sayegh; *Malawi:* Mr. D.R.B. Manda; *Malta:* Mr. A. Scicluna-Spiteri; *Mexico:* Ing. G. Flores Mata, Dr. L. Lopez Martinez de Alva; *Nepal:* Mr. Manik L. Pradhan; *New Zealand:* Dr. R.B. Miller; *Nigeria:* Dr. B.O.E. Amon, Dr. S.A. Adetunji; *Pakistan:* Dr. M.B. Choudhri; *Peru:* Mr. J.H. Christensen; *Philippines:* Dr. G.N. Alcasid; Dr. J. Mariano; *Sierra Leone:* Dr. I. Haque; *Sri Lanka:* Dr. S. Nagarajah; *Syria:* Mr. S. Kourdi; *Tanzania:* Mr. J.K. Samki; *Thailand:* Mr. Samrit Chaiwanakupt; *Turkey:* Dr. M. Özuygur; *Zambia:* Mr. Kalaluka Munyinda.

I wish to express my gratitude to the above Head Cooperators, their colleagues and subordinates for their excellent cooperation in furnishing me with the sample material and other relevant data requested. A total of 7 488 samples (half of them soil and the other half plant) were received, amounting to 94 per cent of the original target of 8 000.

I am greatly indebted to the professional and technical staff of the Institute of Soil Science, Finland, for carrying out the demanding analytical task consisting of about 170 000 chemical and physical analyses and I wish to acknowledge the stimulating discussions with Dr. J. Sippola and Mr. T. Yläranta and their practical help during various stages of the study. I would also like to thank Mr. S. Hyvärinen, who sparing no trouble has performed the statistical analyses and computer work involved. My special thanks are due to Mr. H. Jansson, my assistant, who with unfailing enthusiasm and interest in this work has been of excellent help to me during the long course of this study and to Dr. F. W. Hauck, Chief, AGLS, FAO, for his keen interest in this investigation and for his support and cooperation from the very beginning of the project.

I wish to extend my thanks to Dr. P. Arens, Prof. Dr. A. Cottenie and Dr. K. Harmsen who read parts of the first manuscript and made several valuable comments and suggestions for clarification, and to Dr. N.E. Borlaug and Dr. R.L. Paliwal who were kind enough to check the lists of wheat and maize cultivars included in this study and furnished me with useful information on this subject.

My thanks are also due to Mr. H.K. Ashby for his assistance with the final editing of the text.

Mikko Sillanpää

PART I

PART I

SOIL AND PLANT DATA, METHODOLOGY AND INTERPRETATION

1. Materials and methods

1.1 Collection of original plant and soil samples

Detailed instructions for taking the plant and soil samples were sent to all the field cooperators in various countries in order to standardize the sampling techniques and procedures and to obtain material of good quality for analytical comparisons. The instructions (condensed) are given in Appendix 1.

The collectors of samples were also asked to provide additional background information on the sampled fields including: plant cultivars, estimated yields, use of chemical NPK fertilizers during the past three years, applications of farmyard manure (FYM) and lime, pest control measures, soil units (national and FAO/Unesco classification), known or suspected micronutrient deficiencies or toxicities, etc. These data were collected on "Field information forms" provided for the purpose.

1.2 Indicator plants

1.2.1 General

Wheat and maize were selected as original indicator plants for this study, not because they would be the optimum indicators of soil micronutrients, but because they are the most widely grown crops all over the world. They also complement each other territorially because maize is often grown in more humid areas while wheat grows also in more arid soils.

However, the original plant samples received from different countries proved to be too heterogeneous to reflect reliably the micronutrient status of the soils where the plants had grown. It was already known that the different micronutrient contents of wheat and maize would alone cause heterogeneity among the plant samples as a whole and make comparison of the findings difficult. In addition, among other factors causing uncontrolled variation in the results of plant analyses were:
— variation in the physiological age of plants at sampling. The age of maize and

spring wheat samples at sampling varied from less than 20 to over 60 days and similarly for winter wheat;
— contamination of plants by soil in the field or during sampling, sample pre-treatment or transportation;
— differences in main nutrient fertilization, yield levels, irrigation, use of herbicides and pesticides possibly containing micronutrients. Some on these are referred to in Sections 2.2.1 to 2.2.3 and 8.2.4;
— variation arising from different plant cultivars.

The magnitude of the effects of the above factors is variable from one sample to another and cannot be reliably estimated, but the following attempts were made to quantify some of them.

1.2.2 Sampling time

The effect of sampling time on micronutrient contents of plants was studied by planting two fields each of spring and winter wheat. Sampling, sample pre-treatment and analytical methods were the same as applied to the international study (Appendix 1) except that sampling was carried out twice a week during the whole growing season. There were no clear differences in the concentration or behaviour of micronutrients between the spring and winter wheats. The data from all fields and both cultivars were therefore combined and are illustrated in Fig. 1. They were reported in greater detail by Yläranta et al. (1979). The highest concentrations of B, Cu, Fe, Mn, and Mo were recorded at an early stage of plant growth and followed by substantial decreases toward the end of the growing season. The decreases in B were from about 7 to less than 2 ppm, in Cu from about 7 to 3—4 ppm, and in Fe from 120 to 40—50 ppm. The changes in the Mn and Mo concentrations were less regular, and in the case of Mn a tendency to increase could be observed toward harvesting time. The behaviour of Zn differed from that of other micronutrients. The changes in Zn concentrations were relatively small, consisting mainly of a slight tendency to increase toward the later stages of growth. Clearly decreasing trends in the macronutrient (P, K, Ca, and Mg) contents of wheat were also reported by Sippola et al. (1978). Similar results have been published elsewhere with other plants.

Since the concentration of micronutrients in a plant is dependent on the sampling time (physiological age of the plant), it is evident that concentrations in samples collected at different stages of growth are not strictly comparable and may give a misleading estimate of the micronutrient status of soils. See also Sections 2.5 and 7.1.4.

1.2.3 Contamination

Soil is likely to be the most contaminant of plant samples. It may arise from soil dust raised by wind, the spattering of rain or some other plant-soil contacts. The effect of soil contamination on plant micronutrient concentration depends on the degree of contamination, the chemical element, the total micronutrient content of the contaminating soil, etc. For example, if a plant sample of 2 g dry matter containing 6 ppm B, 7 ppm Cu, 61 ppm Fe, 112 ppm Mn, 0.32 ppm Mo, and 18 ppm Zn (average contents of pot-grown wheat in this study) is contaminated with 10 mg of soil containing 10 ppm B, 50 ppm Cu, 40 000 ppm Fe, 1 000 ppm Mn, 2 ppm Mo and 80 ppm Zn (total contents commonly

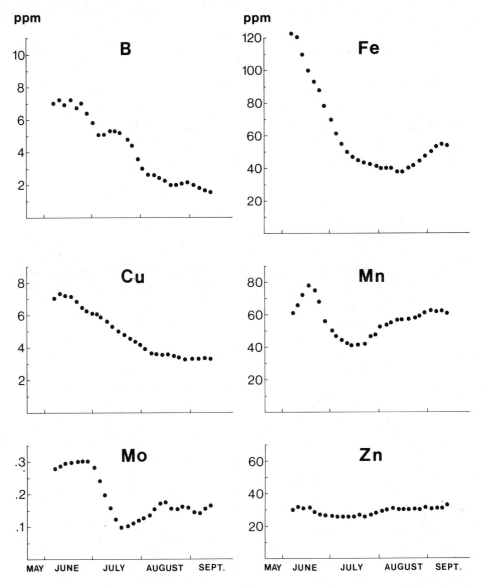

Fig. 1. Variations in B, Fe, Cu, Mn, Mo, and Zn contents of wheat (ppm in DM) during the growing season.

found in mineral soils) the result of the micronutrient analysis of the plant sample would theoretically be affected by the contamination as follows: The B content would be increased from 6.00 to about 6.02 ppm, i.e. an error of + 0.3 per cent. Similarly, the error for Cu would be 3.1 per cent, for Fe 327 per cent, for Mn 4.0 per cent, for Mo 2.8 per cent and for Zn 1.7 per cent.

According to this theoretical calculation, the risk of soil contamination is relatively small in the case of B, Cu, Mn, Mo and Zn if compared to that of Fe. Comparison of the Fe concentration in individual original plants led to the assumption that the samples were liable to be severely contaminated with soil. Not uncommonly the plant Fe concentrations were increased by obvious contamination to a much greater extent than in the above example. Therefore, all results of Fe analyses from original plants were discarded.

1.2.4 Plant varieties

The ability of different plant varieties (cultivars) to absorb micronutrients from soil as well as their micronutrient requirements may vary considerably. For example, Randhawa and Takkar (1975) have drawn special attention to differences in the susceptibility of various species and varieties of crops to Zn deficiency. Because of the large number of plant varieties, data on variety effects are fragmentary and may only be used in special cases for interpreting the results of micronutrient studies.

The original plant material received from various countries for this study included over 200 wheat and over 200 maize cultivars, so preventing the study of differences between individual cultivars. An attempt was therefore made to classify both species in two categories each, high yielding varieties (HYV) and local varieties, and to compare the nutrient contents of the two groups. These comparisons covering all plant samples failed for a number of reasons. For example:
— Neither of the two variety groups was homogeneous. Different high yielding varieties were grown in different countries and the group of local varieties especially was very heterogeneous, consisting as it did of native varieties in different developing countries as well as highly improved wheat and maize cultivars, the products of plant breeding work in developed countries.
— Great differences between the variety groups in the rates of application of N, P and K fertilizers in different countries affected the nutrient uptake.
— The estimated yields of the HYV plants were generally higher than those of other varieties; a factor which should be taken into account when nutrient concentrations of plants are to be compared.
— The nutrient contents of soils where the plants of the two variety groups had grown varied so much that often the grounds of comparison were not justified.
— In some countries plants of one variety group were mainly grown in irrigated and those of the other mainly in rainfed conditions.

However, in a few countries, in spite of some differences in soils, yields or fertilization, the differences in the plant nutrient concentrations between the two variety groups were so clear that some indicative conclusions on their genetic diversities may be drawn. Some of these are pointed out in Part II, see e.g. Mexico, B, Mn and Zn; India, N, P, K, Ca, Mg, B, Cu, Mn and Mo; Ethiopia, Mg, Mn and Zn; and Nigeria, N and P. It must be kept in mind, however, that these comparisons are local in nature and cannot be generalized or applied to other conditions. They serve to draw attention to one of the factors partly responsible for the heterogeneity of the original plant sample material and consequently, for the discrepancy between the results of plant and soil analyses.

1.2.5 Need for better grounds for comparing plant data

In order to reduce the causes of uncontrolled variation in the micronutrient contents as found in the original indicator plants, and to establish sounder grounds for comparison of the results of plant analyses, fresh samples of indicator plants (wheat, cv. 'Apu') were grown in pots on the soils received from the participating countries.

It must be understood, however, that environmental factors affecting the availability of micronutrients to plants in the field, such as temperature, soil moisture and aeration (redox potential), were also standardized when growing the indicator plants in pots. These effects must therefore be taken into account when interpreting the results obtained from widely varying environmental conditions. Relatively few quantitative data on the effects of temperature on the availability of micronutrients are available, but it is generally known that reduction-oxidation processes affect the availability of Fe and Mn more than that of the other four micronutrients included in the study. The reduced forms, Fe^{2+} and Mn^{2+}, are more soluble and available to plants than are the oxidiced Fe^{3+} and Mn^{3+} or Mn^{4+}. For example, in this study the data given in Table 27 in Part II (comparison of original and pot-grown plants) indicate that the plant uptake of Mn is substantially more affected by soil moisture conditions than that of B, Cu, Mo or Zn. Another shortcoming of pot-grown plants is the effect of subsoils. Differences in the micronutrient content and/or availability in the subsoil and top soil may affect the micronutrient uptake by plants, especially during later stages of growth when the root system is more highly developed. It is believed, however, that the foregoing disadvantages are overshadowed by the benefits gained by growing fresh indicator plants in pots.

Although the results of micronutrient analyses of the original plants are given in Appendixes 2 and 3, all further discussion (if not specifically stated otherwise) concerning the behaviour of the six micronutrients B, Cu, Fe, Mn, Mo and Zn, is based on results obtained from analyses of the wheat samples grown in pots under controlled and uniform conditions.

1.2.6 Growing new indicator plants in pots

The small quantity of soil available from samples received from participating countries necessitated the use of small pots (200 ml) partly (75 ml) filled with quartz sand (Fig. 2). The quartz sand was washed with 6 N HCl and thereafter repeatedly with deionized water until no reaction to $AgNO_3$ was observed. Two-thirds (50 ml) of the quartz sand was placed below the soil (100 ml) and one-third (25 ml) above the soil to prevent destruction of the soil surface structure and prevent crust formation due to repeated irrigations as well as to minimize contamination of plants by soil. Holes in the bottoms of the pots were covered with filter paper to prevent leakage of quartz sand but to allow excess irrigation water to drain off. Any excess water was collected in a dish beneath the pot and re-used for irrigating the plants in that pot. Irrigation was carried out once a day using deionized water in quantities sufficient to keep the soil moisture content close to field capacity.

To prevent deficiencies of macronutrients, fertilizers corresponding to about 60 kg N, 30 kg P, 30 kg K and 15 kg Mg per hectare were applied in liquid form in irrigation water in two equal dressings, one at the end of the second and the other in the middle of the fourth week from planting.

Spring wheat (cv. 'Apu') was used as an indicator plant. Ten seeds were placed in each pot on the surface of the soil and covered with the 25 ml of quartz sand. After sprouting, the number of plants was reduced to eight. The plant tissues from 1 cm above the surface of quartz sand were harvested on the 36th day from planting, dried, weighed and analysed as explained in Section 1.3.2. About five percent of the soil samples were too small to enable new indicator plants to be grown in pots.

Fig. 2. Growing new indicator plants (spring wheat cv. 'Apu') in pots.

1.3 Analytical methods

In order to obtain background information on factors affecting the micronutrient contents of soils and plants, a number of additional analyses were made both on soils and plants. These data include soil texture, cation exchange capacity, pH, electrical conductivity, $CaCO_3$ equivalent, organic carbon content, volume weight of soils and macronutrient contents of both soils and plants.

At an early stage of the study when selecting the methods for soil micronutrient analyses, preliminary studies with different extraction methods were carried out on a limited number of samples. For four micronutrients (Cu, Fe, Mn, Zn) each of the international samples was analysed by two different methods as described below.

1.3.1 Methods of soil analyses

Particle size distribution was determined by dry and wet sieving and for the finer fractions by a pipette method (Elonen, 1971). Four fractions (< 0.002, 0.002—0.06, 0.06—2.0 and >2.0 mm) were determined. For the expression of texture as a single figure a texture index (TI) was calculated (TI = 1.0 x % of fraction < 0.002 + 0.3 x % of fraction 0.002—0.06 + 0.1 x % of fraction 0.06—2.0 mm).

Cation exchange capacity (CEC) of soil was determined by the Bascomb (1964) method except that the Mg concentration was measured with an atomic absorption spectrophotometer (Varian Techtron 1200; air-acetylene flame) after diluting the final filtrate by a factor of ten with a 0.24 M HCl + 0.28 % La solution.

Electrical conductivity was measured from a soil:water (1:2.5) suspension after letting the suspension settle overnight.

Soil pH(H_2O) was determined on the above suspension after stirring. For measuring pH($CaCl_2$) the suspension was made 0.01 M with respect to $CaCl_2$ and stirred. The pH ($CaCl_2$) was measured after two hours from a restirred suspension. The relationship between the two pH values is given in Section 2.1 and Fig. 5.

$CaCO_3$ equivalent of soil was determined by a method described by Trierweiler and Lindsay (1969).

Organic carbon was determined by a modification of Alten's (1935) method (Tares and Sippola, 1978). An amount of soil containing 30—100 mg of humus, 25 ml 0.25 M $K_2Cr_2O_7$ and 40 ml conc. H_2SO_4 were put into a 400 ml flask. The mixture was kept for 1.5 h on a hot water bath, allowed to cool for 30 min. and 175 ml of water were added. After standing overnight, the solution was measured colorimetrically and the reading compared to standards (red filter, 620—645 nm).

Volume weight of soil was determined by weighing a 25 ml sample of air dried soil passed through a 2 mm sieve (Tares and Sippola, 1978).

Soil total nitrogen was determined by the Kjeldahl method using Tecator equipment. A 1—3 g sample of air dry soil was weighed into a digestion tube containing 8 g of K_2SO_4 and 1 g of $CuSO_4$. Conc. H_2SO_4 (20 ml) was added. The mixture was maintained at about 420 °C (for 1—2 h) until clear. To the cooled digest 50 ml of water were added, made alkaline with NaOH and nitrogen distilled into 4 % boric acid and titrated with 0.01 M HCl using bromocresol green—methyl red as an indicator.

Phosphorus was extracted with 0.5 M $NaHCO_3$, pH 8.5 (Olsen *et.al.*, 1954). The volumetric soil:extractant ratio was 1:20 and shaking time 1 h (27 r.p.m.). Phosphorus was measured colorimetrically with ammonium molybdate-ascorbic acid reagent (Watanabe and Olsen, 1965).

Potassium, calcium, magnesium and sodium were extracted from soil with 1 M CH_3COONH_4, pH 7.0 (Cahoon, 1974). The volumetric soil:extractant ratio was 1:10 and shaking (end over end) time 1 h (27 r.p.m.). Conc. HCl and La were added to the extract to make the extract 0.2 M with respect to HCl and contain 0.25 % La. The cations were determined with an atomic absorption spectrophotometer (Techtron AA-4 or Varian Techtron 1200) using an air-acetylene flame for Ca and Mg and air-propane for K and Na.

Boron. Hot water soluble soil B was determined by a modified Berger and Truog (1944) method (Sippola and Erviö, 1977). A 25 ml soil sample, 50 ml of water and about 0.5 ml of activated charcoal were boiled for 5 min. in a quartz flask and filtered immediately. Two ml of the extract and 4 ml of buffer masking agent (250 g CH_3COONH_4 and 15 g Na_2EDTA dissolved in 400 ml water, and 125 ml 100 % CH_3COOH added) were mixed and 4 ml of azomethine reagent (0.9 g azomethine-H and 2 g ascorbic acid dissolved in 200 ml water, prepared daily or weekly if refrigerated) added. The colour was allowed to develop for 1 h, intensity measured spectrophotometrically at 420 nm and compared to standards varying from 0 to 2 mg boron per litre.

To correct soil B values for CEC, see Section 2.3.3.

Copper, iron, manganese and zinc were extracted from all soils by two extraction methods, AAAc-EDTA and DTPA as extractants.

The acid ammonium acetate-EDTA extraction solution, 0.5 M CH_3COONH_4, 0.5 M CH_3COOH, 0.02 M Na_2EDTA, was made by diluting 571 ml 100 % CH_3COOH, 373 ml 25 % NH_4OH and 74.4 g Na_2EDTA (EDTA — ethylenediaminetetracetic acid) to 10 litres with water. The pH was adjusted to 4.65 with acetic acid or ammonium hydroxide (Lakanen and Erviö, 1971). Soil (25 ml) and extracting solution (250 ml) were shaken for 1 h (end over end, 27 r.p.m.). The suspension was filtered using Whatman No. 42 filter paper. The concentrations of Cu, Fe, Mn and Zn were determined with an atomic absorption spectrophotometer (Varian Techtron 1200, air-acetylene flame) and appropriate standards.

The DTPA extracting solution was prepared to contain 0.005 M DTPA (diethylenetriamine pentacetic acid), 0.01 M $CaCl_2$, 0.1 M TEA (triethanolamine) and was adjusted to pH 7.3 with HCl (Lindsay and Norvell, 1969, 1978). Soil (25 ml) and extracting solution (50 ml) were shaken for 2 h (end over end, 27 r.p.m.). The suspension was filtered using Schleicher & Schüll selecta 589/3 filter paper. The concentrations of Cu, Fe, Mn and Zn were determined with an atomic absorption spectrophotometer (Varian Techtron 1200, air-acetylene flame) and appropriate standards.

The results presented in this paper are based an AAAc-EDTA extraction of Cu and Fe and on DTPA extraction of Mn. For Zn the results obtained by both extraction methods are given. For details, see Sections 2.3.4 to 2.3.7.

Molybdenum was extracted with AO-OA solution. The ammonium oxalate-oxalic acid extracting solution, a modified Tamm's (1922) solution, was prepared by dissolving 249 g ammonium oxalate and 126 g oxalic acid in 10 litres of water. The pH was adjusted to 3.3 with ammonium hydroxide or oxalic acid. Soil (10 ml) and extracting solution (100 ml) were shaken for 16 h (end over end, 27 r.p.m.) and the suspension was filtered.

Mo was determined by the zinc dithiol method (Stanton and Hardwick, 1967). A 50 ml aliquot of the filtrate was evaporated to dryness in a quartz dish on a water bath, oven dried at 105 °C overnight and ashed by raising the temperature slowly to 600 °C. The ash was moistened with 1 ml of water and dissolved in 15 ml 4 M HCl on a water bath for 1 h. If the residue had dissolved completely the solution was washed (4 times with 4 ml of warm 4 M HCl) into a separating funnel. If there was an insoluble residue in the quartz dish, the solution was filtered into a separating funnel and the dried (105 °C) filter paper with the residue ashed at 600 °C for 1 h. The ash was treated with 0.5 ml 40 % HF and 0.1 ml 70 % $HClO_4$ in a teflon dish, dissolved in 1 ml 4 M HCl and combined with the main solution in the separating funnel. The following reagents were added to the main solution:
— 2 ml of Fe reagent (10 g $FeNH_4(SO_4)_2$ in 1 litre of 6 M HCl),
— 2 ml of reducing solution (15 g ascorbic acid + 7.5 g citric acid/100 ml H_2O) shaken and left to stand for 2 min.,
— 2 ml of 50 % potassium iodide, shaken,
— 2 ml of zinc dithiol solution (0.3 g Zn dithiol, 2 ml ethanol, 4 ml water, 2 g NaOH and 1 ml of thioglycolic acid were mixed and when clear, diluted with water to 50 ml, and 50

ml of fresh 50 per cent potassium iodide solution added) mixed and left to stand for 2 min.,
— 5 ml of chloroform.

The separating funnel was shaken vigorously for 2 minutes. The chloroform phase was separated. The intensity of colour in the chloroform was compared to molybdenum standards using a spectrophotometer at 680 nm wavelength. The standards (0.2—10.0 mg Mo/30 ml 4 M HCl) also contained the same reagents as the sample solution in the separating funnel and received a similar treatment.

For correcting the results of AO-OA extractable soil Mo values for pH, see Section 2.3.2.

1.3.2 Methods of plant analyses

Preparation of samples. Air dried plant material was ground with a hammer mill of pure carbon steel to pass a 2 mm sieve (pure carbon steel). To avoid contamination, the samples were stored in plastic (pure polyethylene) bags.

Since the mineral element contents were expressed on a dry matter basis, the DM contents of air dry samples (5.0 g) were determined by drying the samples at 105 °C for 4 h. The samples were cooled in a desiccator for 2 h and weighed before ashing.

The **nitrogen** content of plant material was determined by the Kjeldahl method using Tecator equipment. The procedure differed from that of soil total nitrogen determination only as regards the size of sample which was 0.50 g of air dry plant material. Results were expressed on a DM (105 °C) basis.

The **boron** content of plant material was determined by a modified azomethine-H method (Basson et al., 1969, John et al., 1975, Sippola and Erviö, 1977). A 100 mg sample of air dry plant material was weighed into a quartz dish, ashed at 450 °C overnight and cooled. A 10 ml aliquot of 0.1 M HCl was added and covered with a watch glass. The solution was allowed to stand for 4 h and filtered into a test tube.

Two ml azomethine reagent were added and the spectrophotometric measurement was made according to the procedure described for soil boron analysis. Results were expressed on a DM (105 °C) basis.

Ashing and determination of Ca, K, Mg, P, Cu, Fe, Mn, Mo, and Zn. The oven dry samples described above were ashed by raising the temperature slowly (about 2.5 h) to 450 °C and maintaining this temperature overnight. The ash was moistened with a few drops of water, covered with a watch glass and 10 ml of 6 M HCl slowly added. When the reaction had ceased, the watch glass was removed and washed with 10 ml of 6 M HCl. The solution was evaporated to dryness on a water bath, 40 ml of 0.2 M HCl were added, covered with a watch glass and kept on the water bath for 30 minutes. The hot solution was filtered (Schleicher—Schüll selecta 589/2) into a 100 ml volumetric flask. The dish, filter, and undissolved residue were washed with about 40 ml of hot 0.2 M HCl.

The filter paper and residue were ashed in the original quartz dish by raising the oven temperature (over a period of about 2 h) to 600 °C, thereafter cooled and transferred to a teflon dish to which 0.2 ml $HClO_4$ and 5 ml HF were added. The dish was kept on a hot plate (200 °C) until all the HF had been removed, cooled, and 0.5 ml 6 M HCl was added. The dish was kept on the hot plate until nearly dry, the residue dissolved in 5 ml of 0.2 M HCl, filtered, and the filtrate combined with the main sample solution in the 100 ml volumetric flask. The teflon dish and filter paper were rinsed and the flask was filled to volume with 0.2 M HCl. For storage, polyethylene bottles were used.

For Ca, K, Mg, and P determinations, 8 ml of the main sample solution were diluted to 40 ml with lanthanum-HCl solution (0.2 M HCl containing 0.31 % La). Ca, K, and Mg were determined with an atomic absorption spectrophotometer (Techtron AA-4 or Varian Techtron 1200) using an air-acetylene flame for Ca and Mg and air-propane for K. For determining P colorimetrically with a modified ammonium vanadate-molybdate method (Gericke and Kurmies, 1952), 5 ml of the dilute sample solution and 10 ml of reagent (7.5 g $(NH_4)_6Mo_7O_{24} \cdot 4 H_2O$ and 0.38 g NH_4VO_3 in 1 litre of 0.13 M H_2SO_4) were mixed. The colour intensity was measured and compared with appropriate standards (0—70 mg P/l) one hour after mixing.

Cu, Fe, Mn, and Zn were measured directly on the main sample solution with an atomic absorption spectrophotometer (Varian Techtron 1200) as on soil extracts.

Mo determination was made by the zinc dithiol method (Stanton and Hardwick, 1967). To make the solution 4 M with respect to HCl, 20 ml of the main sample solution and 10 ml of conc. HCl were pipetted into a separating funnel and shaken. Fe reagent, reducing solution, potassium iodide, zinc dithiol and chloroform were added and Mo was determined according to the procedure described for soil Mo analysis. The same Mo standards were used for plant and soil analyses.

Micronutrient contents of pot-grown wheat. Owing to small pot yields, the contents of the six micronutrients (B, Cu, Fe, Mn, Mo and Zn) only were determined. The analytical procedures and methods deviated from those described above only in that smaller quantities of sample material were analysed.

1.3.3 Statistical methods

The evaluation of the micronutrient status of each soil was based on two analyses, i.e. one of the soil and one of the plant grown on that soil. Regression graphs are used to present the results of both analyses simultaneously. Furthermore, the correlation between the results of the two analyses can be considered a good measure for judging the reliability of the results.

In general, for statistical evaluation of the results of chemical micronutrient analyses by countries (Part II) four simple regression equations were computed: the linear, logarithmic and two semi-logarithmic models. The best fitting equations are graphically presented. In addition to the above regression models, parabolas, cubic parabolas and their respective logarithmic modifications were adopted when studying the relations between micronutrients and various soil factors (Part I). In special cases multiple regression and linear stepwise multiple regression analyses were used.

In respect of macronutrients and general soil properties the data in Part II are presented only as arithmetic means and their standard deviations and frequency distribution graphs of classified materials.

1.4 Expression of analytical data

1.4.1 Soils

The results of nutrient analyses of soils have traditionally been expressed on a weight basis, i.e. in ppm or mg/kg soil, but expressions on a volume basis, e.g. in mg/dm^3 or mg/litre of soil, are becoming increasingly frequent. The weights of a certain volume of soil may vary considerably depending on soil texture, mineral composition etc., but especially on the organic matter content of the soils. In the soil samples studied, which included only a few organic soils, the volume weights varied from 0.51 g/cm^3 (a New Zealand peat soil) to 1.77 g/cm^3 (a coarse sand from Nigeria). However, much lower volume weights have been recorded, e.g. Sippola and Tares (1978) reported an average volume weight of 0.15 for cultivated *Sphagnum* peat soils and a minimum value of 0.06 g/cm^3. Although the volume of soil where the roots of a plant are distributed may vary from one plant to another, it can be assumed that the density of roots and nutrient absorption by plants are much better related to soil volume than to the weight of the soil, since in the latter ten- or twenty-fold variations often exist.

Often, in order to understand or to interpret conflicting results, the importance of the dimension or unit in which the results are given must be realized. For example, if two soils, one a mineral soil with a volume weight of 1.5 and the other a peat soil with a volume weight of 0.1 g/cm^3 are analysed and both show an equal content of 100 ppm of nutrient X when expressed on a weight basis, the result is completely different if expressed on a volume basis, as shown below:

Table 1. Comparison of nutrient content of two soils expressed on weight and volume bases.

Soil	Volume weight	Content of nutrient X as expressed	
		on weight basis ppm	on volume basis mg/litre of soil
Mineral soil	1.5	100	150
Peat soil	0.1	100	10

Thus, in this example, the 100 ppm of nutrient X in the mineral soil is 1400 per cent higher than the 100 ppm in the peat soil when the same results are expressed on a volume basis. The above is directly applicable when the total nutrient contents of soils are determined. In the case of a determination of extractable contents, the difference is not as pronounced because of the effect of changing extraction ratio. Even so, misleading conclusions have too often been drawn because insufficient thought has been given to the dimension of the results presented. In areas where peat soils do not exist and volume weights of soils vary less, the difference between the methods of expressing the results is, however, relatively small. On the other hand, even within the same soil profile the volume weight of the topsoil may be markedly different to that of the lower horizons.

The relative error due to inaccuracies in measuring a certain volume of soil for analysis, or due to variation in extraction ratio, is negligible compared to the effects of variation of volume weight among samples including both mineral and organic soils. The advantages of volume-based expressions have been pointed out, for example, by Mehlich (1972) and Sillanpää (1962a, b, 1972a, b).

In accordance with the above principles, the contents of all soil nutrients in this study are expressed on a volume basis, i.e. in mg/l. However, the total soil N content is also given as a percentage (Appendixes 2, 3 and 4) but for some of the soil characteristics such as organic carbon content and CEC the traditional dimensions are used. The quantities of various soil nutrients are expressed in terms of the element rather than oxide. The same applies also to the nutrient contents of plants and fertilizers.

1.4.2 Plants

Plant analytical data are expressed on a concentration (% or ppm) basis rather than as uptake per hectare. Both methods have certain advantages as well as disadvantages, as discussed in Section 2.6. The plant/soil micronutrient correlations calculated for B, Cu, Fe, Mn, Mo and Zn in all samples (Section 2.3) showed no indication of the superiority of one method over the other. The decision to express as concentration was due to the practical fact that in order to be able to calculate uptake, yield must also be known and, generally, at the time of plant sampling only rough estimates of yield can be made.

1.4.3 Regression graphs

To facilitate the comparison of results (e.g. between different countries), the corresponding values of the soil and plant analyses are presented in regression graphs in order to combine the results of the two techniques. The higher the correlation coefficient, the better the link between the soil and plant analyses. The regression line and equation as well as the mean values and standard deviations for all the international material are usually given as background information and also in regression graphs concerning only individual countries or other groups of data. Because of widely varying micronutrient contents and their abnormal (linear) distributions, logarithmic scales were adopted for regression graphs.

Each regression graph has been divided into five Zones (I—V) in such a way that the numerical product of plant micronutrient content multiplied by soil micronutrient content is constant at every point on the line separating two zones. The numerical values along the lines demarcating the zones are given in Table 2. The locations of the zone limits have been determined so that five per cent of sample pairs (plant micronutrient content x soil micronutrient content) of the whole international material falls below the line separating Zones I and II, i.e. into Zone I. Accordingly, ten per cent of the material falls below the line between Zones II and III, i.e. five per cent into Zone II. The next 80 per cent of the material falls in Zone III and the highest ten per cent of the values in Zones IV and V, i.e. five per cent in each.

Even though the most likely cases of deficiency are to be found among samples falling in Zone I and the most probable cases of toxicity in Zone V, there is no proof that the above zone limits would also be critical limits for deficiency or toxicity. Furthermore, critical deficiency and toxicity limits vary from one plant species to another.

Table 2. Numerical products of plant content x soil content along the lines dividing the regression graphs into five zones.

Micronutrient and soil extraction method[1]		Lower 5 % — limit between Zones I and II	Lower 10 % — limit between Zones II and III	Higher 10 % — limit between Zones III and IV	Higher 5 % — limit between Zones IV and V
B	(hot water sol.)	0.67	0.87	9.4	18.7
B	(CEC-corr. hot w.s.)	0.72	0.98	9.2	17.5
Cu	(AAAc-EDTA)	3.7	6.4	107	160
Cu	(Org.C-corr. AAAc-EDTA)	3.9	6.7	102	144
Fe	(AAAc-EDTA)	1680	2300	22500	31900
Mn	(DTPA)	206	295	14100	25100
Mn	(pH-corr. DTPA)	166	201	14100	37100
Mo	(AO-OA)	0.00260	0.0040	0.166	0.280
Mo	(pH-corr. AO-OA)	0.00070	0.0017	0.275	0.441
Zn	(DTPA)	2.0	2.8	86	183
Zn	(AAAc-EDTA)	7.2	9.5	169	336
Zn	(pH-corr. AAAc-EDTA)	6.4	8.3	180	370

[1] For CEC-, org. C-, and pH-corrections, see Sections 2.3.2—2.3.7.

2. Results and discussion

2.1 General properties of soils and their mutual relations

The amounts of micronutrients removed yearly with normal crop yields represent only a very small proportion, generally less than one per cent of the total amount present in soils. The total amounts, even in serious deficiency cases, therefore far exceed the crop requirements. Cases of **primary deficiency** that are mainly caused by a low total content of micronutrients, are therefore very rare in normal agricultural soils, but may occur in severely leached sands or in certain peat soils.

Secondary micronutrient deficiencies are the most common and are caused by soil factors reducing the availability to plants of otherwise ample supplies of micronutrients. For the effective correction in the field of a micronutrient deficiency, it is necessary to know which element is deficient and the reason why it is deficient.

While primary deficiency can usually be remedied by applying salts containing the deficient micronutrient to the soils, the correction of secondary deficiencies is more complicated. Direct salt applications to the soil may not lead to a cure because the original cause of deficiency persists and renders the element added unavailable. Should it be impossible to lessen the effect of the causative factor, other measures must be taken, such as foliar applications or the use of chelates. It is most important to know the main reasons for a deficiency.

Where micronutrient disorders occur, or are suspected, a general background knowledge of soil characteristics is essential. Therefore, in this study emphasis has been laid on investigations concerning the relationships of plant and soil micronutrients to various soil characteristics. For such work, the present abundant material with exceptionally wide variations in respect of all soil characteristics was well suited. Relationships of plant and soil micronutrients to various soil characteristics are presented in detail in Section 2.3, separately for each of the six micronutrients under study. General information on soil characteristics by countries is given in Appendixes 2, 3 and 4. The frequency distributions of texture, pH, organic matter content and cation exchange capacity of soils in each country against the background of corresponding distributions in the whole international material are presented in Part II.

The effects of two or more different soil factors on the availability of certain micronutrients are often very similar (e.g. effects of texture and CEC on Cu, Fig. 30). In other cases the same soil factor affects the availability of two or more micronutrients very differently (e.g. effects of pH on Mo and Mn, Figs 16 and 34). Furthermore, it is often difficult to define which soil factor affects directly the availability of a certain micronutrient and which factor is only in "pseudocorrelation" with the micronutrient, owing to the mutual correlation between the two soil factors concerned. From the viewpoint of soil chemistry these questions are of importance, but in practical micronutrient studies a "pseudocorrelating" soil factor may often be as informative as a factor of direct effect.

To understand better the relations between various micronutrients and other soil factors to be given in Sections 2.3 and 2.4 the mutual relations between the six soil characteristics are presented in Figs 3 and 4.

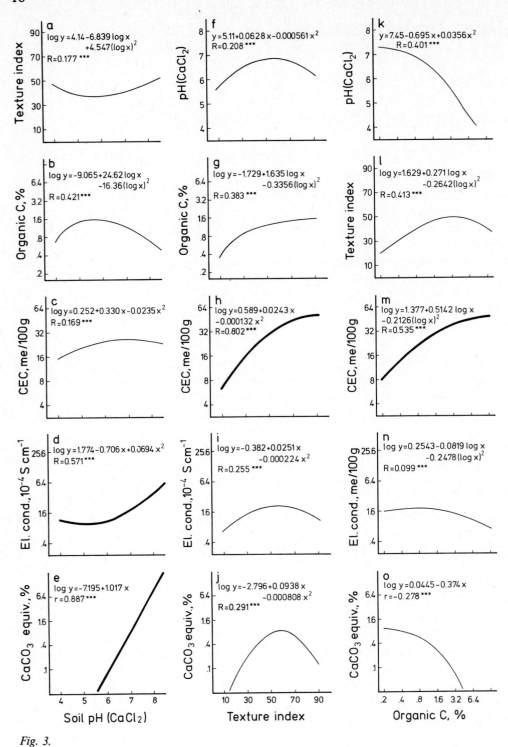

Fig. 3.

Figs 3 and 4. Mutual correlations between soil pH(CaCl$_2$), texture (TI), cation exchange capacity (me/100 g), organic carbon (%), electric conductivity (10^{-4} S cm^{-1}), and CaCO$_3$ equivalent (%), in the whole international material (n = 3536). The best fitting regressions are given.

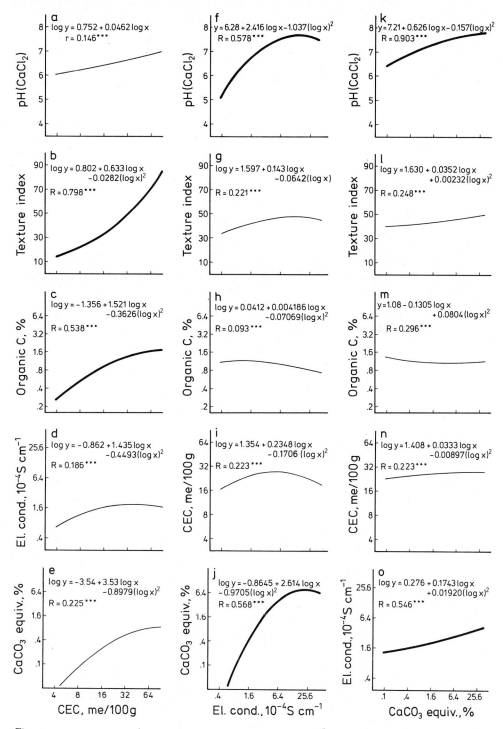

Fig. 4.

Although all six soil factors are statistically highly significantly (0.1 per cent level) correlated with each other, the correlation coefficients vary from 0.093*** to 0.903***. (Regression curves between the best correlating soil factors, R>0.5, are drawn with thick lines).

The closest correlations were found for pH—$CaCO_3$ equivalent, CEC—texture, and pH—electrical conductivity, but those for $CaCO_3$ equivalent—electrical conductivity and CEC—organic C were also high. The relations between electrical conductivity and organic C, pH and CEC, and pH and texture were the weakest but still highly significant.

Three of these soil factors (pH, CEC and organic carbon content) are of special importance in connection with micronutrients because (a) they are highly responsible for the degree of availability of micronutrients to plants, and (b) they influence the extractability of micronutrients by the extractants used, and (c) because their effects during the above processes (a and b) are dissimilar. With regard to soil pH the results given in this study are based on determination of $pH(CaCl_2)$. The advantages of this method have been pointed out, e.g. by Peech (1965). In addition $pH(H_2O)$ was determined from all soil samples. The relationship between the two pH values is given in Fig. 5. In general, the $pH(H_2O)$ is higher than the $pH(CaCl_2)$ and the difference was somewhat greater in acid than in alkaline soils. To convert $pH(CaCl_2)$ to $pH(H_2O)$ or *vice versa* the regression lines in Fig. 5 can be used.

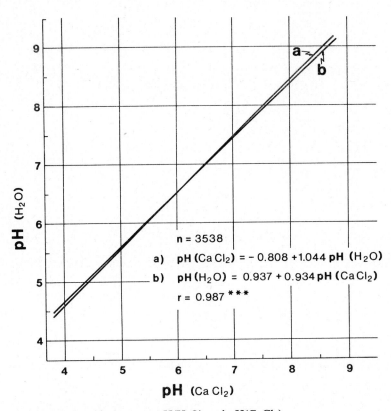

Fig. 5. Relationship between $pH(H_2O)$ and $pH(CaCl_2)$.

2.2 Macronutrients

2.2.1 General aspects

The data on macronutrients are based on analyses of soils and of the **two original indicator plants,** wheat and maize, grown in those countries participating in the study. This plant sample material was quite heterogeneous because, in addition to two indicator plants, it consisted of numerous plant varieties, sampled at considerably varying stages of growth and fertilized with widely different dressings of N, P and K fertilizers. Unlike the micronutrient data of this study, macronutrients were not analysed from the more homogeneous plant sample material obtained from the pot experiment because these samples were too small in size for determining macronutrients in addition to micronutrients. Furthermore, the pot-grown wheats were fertilized with NPK.

2.2.2 Comparison of the two original indicator plants

In general the sampled wheats contained more N, P and K, but less Ca and Mg than maize (Table 3). Despite the wide variations in the soil macronutrient contents, the mean values for wheat and maize soils differed relatively little. More N, P and K had been applied to the wheat than to the maize crops, so partly accounting for their higher N, P, and K contents.

Table 3. Mean macronutrient contents of the two original indicator plants (%), of the respective soils (total N %, others mg/l), and the mean amounts (kg/ha) of N, P and K applied in fertilizers to each crop.

		Wheat fields (n = 1768)					Maize fields (n = 1976)				
		N	P	K	Ca	Mg	N	P	K	Ca	Mg
Plants[1],	mean	4.28	0.375	4.03	0.428	0.172	3.14	0.330	3.13	0.470	0.251
	± s	1.15	0.125	0.97	0.166	0.060	0.87	0.104	0.96	0.205	0.119
Soils[1],	mean	0.133	20.2	365	4671	489	0.135	22.5	330	3450	446
	± s	0.084	24.7	283	3076	437	0.088	33.0	356	2815	462
Fert. appl.,	mean	66	21	21			27	6	7		
	± s	61	27	44			37	10	18		

[1] For analytical methods, see Section 1.3.

2.2.3 Comparison of macronutrient status in different countries

Since plant samples received from some of the participating countries consisted of wheat only and others maize only, their plant macronutrient data were not directly comparable. An attempt was therefore made to eliminate the difference in the uptake of macronutrients between the two plant species by raising (pooling) the average macronutrient contents of maize to the same level as those of wheat. For example, the N contents of individual maize samples were multiplied by a factor of 1.36 to raise the average N content of maize (3.14 %) to the same average level as that of wheat (4.28 %). The national mean data for soil macronutrients and the respective adjusted plant data are expressed in Figs 6, 7, 8, 9 and 10.

It must be understood that even though the main differences due to two indicator plants were eliminated by statistical pooling (adjustment of the average macronutrient content), other causes of uncontrolled variation, such as those due to different plant varieties, varying ages of plants at sampling, fertilization, etc., remained, and good correlations between the plant and soil macronutrient contents can hardly be expected. In fact, the macronutrient correlations based on the original plants are generally poorer than those of most micronutrients where many of the factors causing uncontrolled variation in micronutrient contents of plants were eliminated by replacing the two original plants and numerous varieties by pot-grown wheat of one variety, sampled at the same time (see Section 1.2). Furthermore, compared to fertilization with macronutrients, the effects of micronutrient fertilization were negligible because fertilizers containing micronutrients were very seldom used.

The general effects of recent N, P and K fertilization are illustrated by dividing the data into three fertilization classes (Figs 6, 7 and 8), but the data concerning farmyard manure application were too fragmentary to be taken into account. The influence of different wheat and maize varieties is discussed in Section 1.2.4.

The data in Figs 6 to 10 give only a general picture of macronutrient status in various countries against the international background and will be discussed in more detail at national levels in Part II.

Although the total **nitrogen** content of topsoil cannot always be considered a good indicator of the nitrogen status of a soil, the correlation between the N content of the plant and total soil N in the whole material is relatively good (Fig. 6, regression a). Nitrogen fertilization is a factor with a strong effect on plant N content but affects little the total soil N (Table 4) and consequently is likely to impair the correlation. Therefore, to obtain a general outline of the effect of applied nitrogen on the N content of the original plant samples, additional regressions were computed for two subgroups: one consisting of plants that had received more than 100 kg N/ha (Fig 6, regression b) and the other of plants that had received less than 50 kg N/ha (regression c). There is a considerable difference between the elevations of the two sub-regression lines (b and c). Most of the countries where high rates of nitrogen had been applied to the sampled crops (Hungary, Belgium, Korea, Italy, Zambia) stand clearly above the general regression line (a), while most countries of low nitrogen fertilizer use lie below that line.

In case of **phosphorus** the plant-soil correlation is better than that for other macronutrients. Although the effects of recent phosphorus fertilizer applications can be seen from Fig. 7, it is evident that the applied P not only increased the P content of the plant but also the P extractable from the soil by 0.5 M $NaHCO_3$. Since applied P is not leached as easily from soil as N, and plants remove much less P from soils than N or K, many countries with a traditionally high use of phosphates have built up the P reserves of their soils during the past decades. According to FAO statistics (FAO, 1980a) phosphate fertilizer consumption per hectare in 1961—78 has been considerably higher in New Zealand, Belgium, Korea, Finland, Hungary and Italy than in the other countries included in this study. In spite of relatively strong fixation of applied P in soils, the effects of recent phosphate applications on plant P contents are quite substantial as shown by the differences in the elevation of regression lines b and c representing high (> 30 kg P/ha) and low (< 10 kg P/ha) phosphate application levels, respectively.

Since a large proportion of analysed samples which had received high (> 30 kg P/ha) phosphate dressings came from the countries with a traditionally high phosphate consumption, the average $NaHCO_3$ extractable P content of these soil samples was double

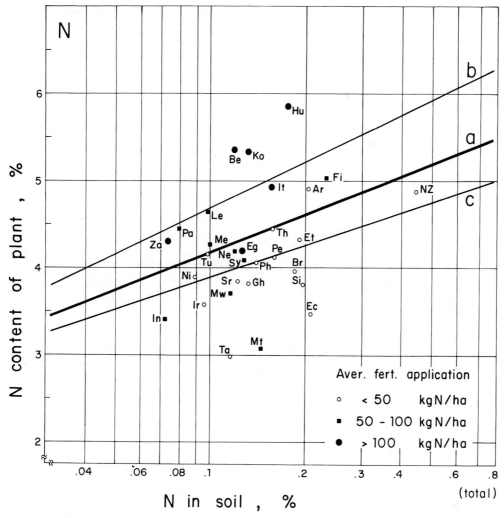

Fig. 6. Regressions of nitrogen content of original wheat and maize (pooled) on total nitrogen content of soil a) for the whole international material, b) for crops fertilized with >100 kg N/ha, and c) for crops fertilized with <50 kg N/ha. Regression equations are given in Table 4. The national mean values for plant and soil N contents of the 30 countries are plotted in the graph indicating also the general level of nitrogen fertilizer application to the sampled crops in each country:

Ar	= Argentina,	Be	= Belgium,	Br	= Brazil,	Ec	= Ecuador,
Eg	= Egypt,	Et	= Ethiopia,	Fi	= Finland,	Gh	= Ghana,
Hu	= Hungary,	In	= India,	Ir	= Iraq,	It	= Italy,
Ko	= Korea, Rep.,	Le	= Lebanon,	Mt	= Malta,	Mw	= Malawi,
Me	= Mexico,	Ne	= Nepal,	NZ	= New Zealand,	Ni	= Nigeria,
Pa	= Pakistan,	Pe	= Peru,	Ph	= Philippines,	Si	= Sierra Leone,
Sr	= Sri Lanka,	Sy	= Syria,	Ta	= Tanzania,	Th	= Thailand,
Tu	= Turkey,	Za	= Zambia.				

that of samples receiving only small (< 10 kg P/ha) phosphate dressings (Table 4). This difference accounts for the divergence of regression line *a* from lines *b* and *c*. According to the present data P deficiency is likely to occur more frequently in Pakistan, Iraq, Ghana and India than elsewhere, although in most countries soils with low P status are common.

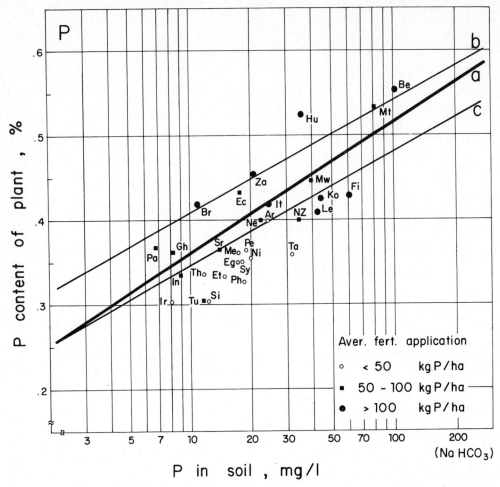

Fig. 7. Regressions of phosphorus content of original wheat and maize (pooled) on NaHCO$_3$ extractable phosphorus content of soil a) for the whole international material, b) for crops fertilized with >30 kg P/ha, and c) for crops fertilized with <10 kg P/ha. For regression equations see Table 4 and for abbreviations the text for Fig. 6.

The **potassium** contents of plants were almost tenfold those of phosphorus. Even an average crop therefore takes more K from the soil than is usually applied in potassium fertilizers. Consequently, soil potassium reserves even in countries of traditionally high potassium fertilizer consumption have not been increased appreciably, if at all. According to FAO statistics (FAO, 1980a), among the 30 countries represented in this study, those with the highest potassium consumption per hectare in 1961—78 were Belgium, Hungary, Finland, Korea and Italy. The crops sampled from these countries for this study were also fertilized with high rates of potassium. The locations of these countries in the "international K field" (Fig. 8), however, do not show such high soil K contents as could be expected from their traditionally high potassium fertilizer consumption. On the other hand, the mean K contents of plants from countries where K had been generously applied to the sampled crops are usually above the main regression line *(a)*. Without doubt, the greatly varying rates of recent K fertilizer application are one of the main reasons for the relatively poor plant K—soil K correlation in the whole material (Table 4).

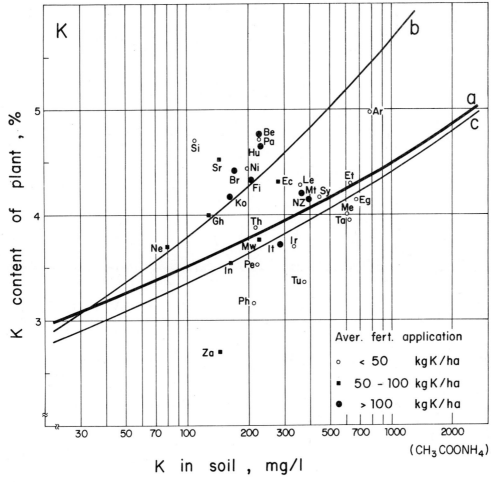

Fig. 8. Regressions of potassium content of original wheat and maize (pooled) on CH_3COONH_4 exchangeable potassium content of soil a) for the whole international material, b) for crops fertilized with >30 kg K/ha, and c) for crops fertilized with <10 kg K/ha. For regression equations see Table 4 and for abbreviations the text for Fig. 6.

Since the majority of crops sampled had received no or only a minimal (< 10 kg K/ha) dressing of K, the regression (line *c*) for low rates of applied K differs little from that for all samples (line *a*). Potassium fertilization at high rates considerably increased the K contents of plants as indicated by the regression line *b*.

In general, soils in countries of high mean exchangeable K contents (e.g. Argentina, Egypt, Ethiopia, Tanzania, Mexico and Syria) are fine to medium textured (average texture index 40—62), while soils in countries of low exchangeable K (Nepal, Sierra Leone, Ghana, Sri Lanka and Zambia) are relatively coarse (mean TI 30—39). See also Fig. 12. In the last mentioned countries potassium fertilization is likely to be one of the primary requirements for obtaining reasonably high yields.

Calcium is essential, not only to correct soil acidity but also as a nutrient element necessary for normal plant growth. Soil Ca plays an essential role in regulating soil pH. All countries shown in Fig. 9 where the average exchangeable Ca content of soils was less than 1000 mg/l had a mean $pH(CaCl_2)$ of 5.1 or lower, countries with a soil Ca

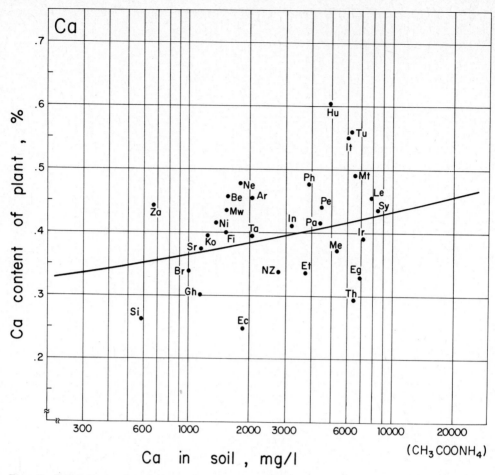

Fig. 9. Regression of calcium content of original wheat and maize (pooled) on CH_3COONH_4 exhangeable calcium content of soil for the whole international material. For regression equation see Table 4 and for abbreviations the text for Fig. 6.

content of 1000—1600 mg/l had an average pH of 5.2—5.8, and those with soil Ca over 6500 mg/l had a $pH(CaCl_2)$ of 7.4 or higher. The relationship between Ca content of plants and soil pH is similar although not as clear (Fig. 11).

Liming was most frequently practised in Korea and Finland where 9 out of every 10 sampled soils received lime in recent years preceding sampling. In Belgium, Brazil, New Zealand and Hungary liming frequencies (7, 6, 5 and 3 out of 10 sampled soils, respectively) were also high. Liming was uncommon in other countries.

In countries of very low soil Ca status the low pH may also be accompanied by toxic or excess contents of elements such as Mn and Al. Owing to the general abundance of exchangeable Ca in soils in relation to Ca requirements of plants, the absorption of Ca by plants may rarely be directly limited by low soil Ca contents, and further, the Ca contents of soils are poorly reflected in the Ca contents of plants. It is apparent that true Ca deficiency is relatively rare and of less importance than the indirect malfunctions due to shortage of soil Ca.

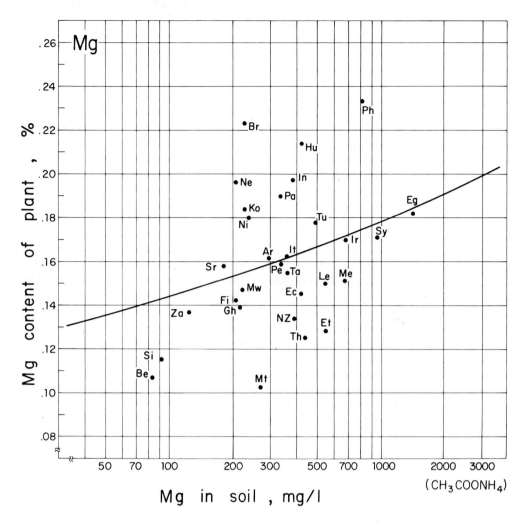

Fig. 10. Regression of magnesium content of original wheat and maize (pooled) on CH_3COONH_4 exchangeable magnesium content of soil for the whole international material. For regression equation see Table 4 and for abbreviations the text for Fig. 6.

The correlation between plant **magnesium** and exchangeable soil Mg was only slightly better than that of Ca. Even though in almost every country there are a few soils with low contents of exchangeable Mg, their relative frequency is very low in countries such as Egypt, Syria, Philippines, Iraq, Mexico, Lebanon, Turkey and Hungary, where fine textured soils (national mean texture indexes 46—60) with medium to high pH($CaCl_2$) (national mean 6.0—7.7) predominate. In typical low Mg countries such as Belgium, Sierra Leone and Zambia, soils with a low pH and coarse texture (national mean pH($CaCl_2$) 4.9—5.8, TI 27—33) are in a majority.

The internal variations within each country of plant and soil macronutrient contents are given in Part II.

Table 4. Regressions of pooled plant N, P, K, Ca and Mg contents (y) on respective soil macronutrients (x). For regression curves, see Figures 6—10.

	n	$\bar{y} \pm s$	$\bar{x} \pm s$	Regression	r	Fig.
N, whole material	3732	4.27 ± 1.16	0.134 ± 0.086	y = 5.61 + 1.43 log x	0.280***	6 a
N, fert. > 100 kg N/ha	1102	4.80 ± 1.11	0.131 ± 0.086	y = 6.47 + 1.78 log x	0.334***	6 b
N, fert. < 50 kg N/ha	1955	3.99 ± 1.05	0.138 ± 0.081	y = 5.11 + 1.22 log x	0.274***	6 c
P, whole material	3732	0.375 ± 0.121	21.4 ± 29.4	y = 0.205 + 0.155 log x	0.556***	7 a
P, fert. > 30 kg P/ha	763	0.457 ± 0.123	35.1 ± 32.7	y = 0.275 + 0.133 log x	0.450***	7 b
P, fert. < 10 kg P/ha	2089	0.347 ± 0.106	17.2 ± 28.4	y = 0.212 + 0.134 log x	0.523***	7 c
K, whole material	3732	4.03 ± 1.12	346 ± 324	log y = 0.329 + 0.108 log x	0.304***	8 a
K, fert. > 30 kg K/ha	709	4.31 ± 1.01	204 ± 139	log y = 0.229 + 0.175 log x	0.398***	8 b
K, fert. < 10 kg K/ha	2810	3.97 ± 1.12	396 ± 352	log y = 0.286 + 0.120 log x	0.338***	8 c
Ca, whole material	3732	0.428 ± 0.177	4027 ± 3004	log y = −0.663 + 0.075 log x	0.183***	9
Mg, whole material	3732	0.173 ± 0.072	466 ± 451	log y = −1.028 + 0.093 log x	0.219***	10

2.2.4 Macronutrient contents of plants and soils in relation to four soil characteristics

In order to obtain a broad assessment of factors influencing the macronutrient contents in this material, regressions of N, P, K, Ca and Mg contents of plants (pooled) and soils (total N, CH_3COONH_4 exchangeable Ca, Mg, and K, and $NaHCO_3$ extractable P) were computed on four important soil characteristics. The regression lines are given in Figures 11—14 and the respective equations in Table 5.

Table 5. Plant and soil macronutrients as functions of four soil characteristics in the whole international material (n = 3731). For regression curves, see Figs 11—14.

Variables y	x	Regression	r	Variables y	x	Regression	r	Fig.
Plant N	pH	log y = 0.718 − 0.130 log x	−0.074***	Soil N	pH	y = 0.281 − 0.0221 x	−0.289***	11
"	TI	log y = 0.485 + 0.0788 log x	0.099***	"	TI	log y = −1.76 + 0.510 log x	0.390***	12
"	Org.C	y = 4.24 + 0.932 log x	0.213***	"	Org.C	log y = −0.959 + 0.765 log x	0.890***	13
"	CEC	log y = 0.491 + 0.0881 log x	0.161***	"	CEC	log y = −1.60 + 0.484 log x	0.528***	14
Plant P	pH	log y = −0.371 − 0.0116 x	−0.090***	Soil P	pH	y = 1.59 − 0.0734 x	−0.190***	11
"	TI	y = 0.412 − 0.000835 x	−0.113***	"	TI	y = 28.4 − 0.158 x	−0.089***	12
"	Org.C	log y = −0.451 + 0.0824 log x	0.152***	"	Org.C	log y = 1.08 + 0.631 log x	0.384***	13
"	CEC	y = 0.392 − 0.000617 x	−0.074***	"	CEC	log y = 0.835 + 0.193 log x	0.110***	14
Plant K	pH	y = 4.58 − 0.677 log x	−0.047**	Soil K	pH	log y = 1.41 + 1.19 log x	0.256***	11
"	TI	log y = 0.509 + 0.0485 log x	0.064***	"	TI	log y = 0.906 + 0.916 log x	0.432***	12
"	Org.C	log y = 0.585 + 0.0797 log x	0.163***	"	Org.C	log y = 2.38 + 0.305 log x	0.220***	13
"	CEC	log y = 0.548 + 0.0287 log x	0.055***	"	CEC	log y = 1.37 + 0.739 log x	0.501***	14
Plant Ca	pH	y = 0.157 + 0.332 log x	0.146***	Soil Ca	pH	log y = 0.065 + 4.14 log x	0.773***	11
"	TI	y = 0.455 − 0.000604 x	−0.056***	"	TI	log y = 1.43 + 1.25 log x	0.513***	12
"	Org.C	y = 0.439 − 0.00841 x	−0.059***	"	Org.C	y = 4240 − 178 x	−0.073***	13
"	CEC	y = 0.438 − 0.000362 x	−0.030 n.s.	"	CEC	y = 609 + 124 x	0.603***	14
Plant Mg	pH	log y = −0.942 + 0.0221 x	0.149***	Soil Mg	pH	log y = 0.471 + 2.49 log x	0.495***	11
"	TI	y = 0.176 − 0.000066 x	−0.015 n.s.	"	TI	log y = 0.364 + 1.32 log x	0.581***	12
"	Org.C	log y = −0.792 − 0.0899 log x	−0.144***	"	Org.C	y = 483 − 15.1 x	−0.041*	13
"	CEC	log y = −0.771 − 0.000638 x	−0.056***	"	CEC	log y = 1.15 + 0.984 log x	0.620***	14

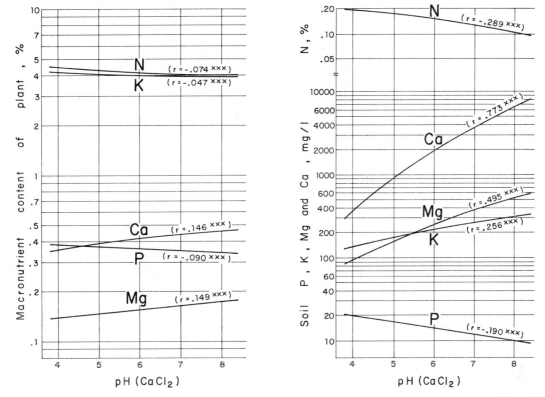

Fig. 11. Plant and soil macronutrients as functions of soil pH in the whole international material (n = 3731). Regression equations are given in Table 5.

In respect to soil pH (Fig. 11) the firmest relationship was obtained for exchangeable soil Ca, which increased from about 400 to 6000 mg/l when the pH increased from 4 to 8[1]. The respective increase in plant Ca content was from about 0.35 to 0.45 %. If these increases (400 to 6000 and 0.35 to 0.45) are compared to the regression given in Fig. 9, it can be seen that between soil Ca levels of 400 and 6000 mg/l the respective increase in the Ca content of plants is about from 0.34 to 0.42 %. The increase in plant Ca is therefore almost entirely due to an increase in exchangeable soil Ca, and pH seems to have practically no effect on absorption of Ca by the plants.

The increasing soil Mg and decreasing soil N and P contents as functions of pH are accompanied by similar relationships between the Mg, N, and P contents of plants and the pH. Comparing these to the respective plant-soil regressions given in Figs 6, 7 and 10 (as in the case of Ca) it can be seen that the changes in the Mg, N, and P contents of plants are almost entirely due to the respective changes in the soil Mg, N and P contents, and the effect of pH on the absorption of these elements by plants is minimal.

Potassium is the only macronutrient the availability of which to plants seems to be affected by soil pH. In spite of increasing exchangeable K contents of soils with increasing pH, the K contents of the plant decrease. While these relationships are statistically not as firm as those for the other macronutrients, they still point to pH as being one cause for the relatively poor correlation between the results of the plant and soil K analyses.

[1] Although soil Ca is given as a function of pH the cause and effect are likely to be the opposite.

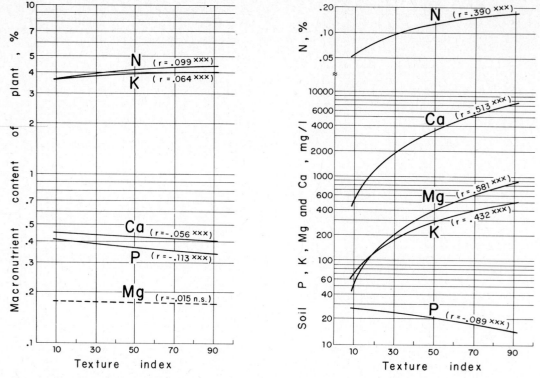

Fig. 12. Plant and soil macronutrients as functions of soil texture (TI) in the whole international material (n = 3731). Regression equations are given in Table 5.

The contents of soil total N and exchangeable Ca, Mg, and K increased significantly from coarse to fine textured soils but the $NaHCO_3$ extractable P contents decreased (Fig. 12). Similar changes in the plant contents of N, P and K took place, indicating that texture has a relatively small effect on the availability of these elements to plants.

In spite of the strong increases in the exchangeable soil Ca and Mg toward fine textured soils, there were no corresponding increases in the Ca and Mg contents of plants. Instead these showed a slight decrease, and indicate that plants can absorb these elements more easily from coarse than from fine textured soils. Clearly the varying relative availability of Ca and Mg in texturally different soils is one of the factors impairing the plant-soil correlations for these elements.

The plant and soil macronutrient data given as functions of the organic carbon content of soil in Fig. 13 are relatively consistent with each other. The N, K and P contents of both the soil and the plant increased and those of Ca and Mg decreased with increasing organic carbon content of soils. The magnitudes of the changes are also of the same order as those given in Figs 6 to 10. For example, the total N content of soil increases from about 0.03 to 0.45 % and the N content of plants from about 3.6 to 5.0 % when the organic carbon content increases from 0.2 to 6.4 % (Fig. 13). Likewise, with the total soil N content increasing from 0.03 to 0.45 % (Fig. 6, line *a*) the plant N content increases from about 3.4 to 5.1 %. Thus, the increase in the plant N content is due to the respective increase in the soil total N content, and the varying soil organic carbon contents do not interfere with the plant N—soil N correlation to any marked extent. The same is true for the relationships between the other macronutrients and soil organic carbon content.

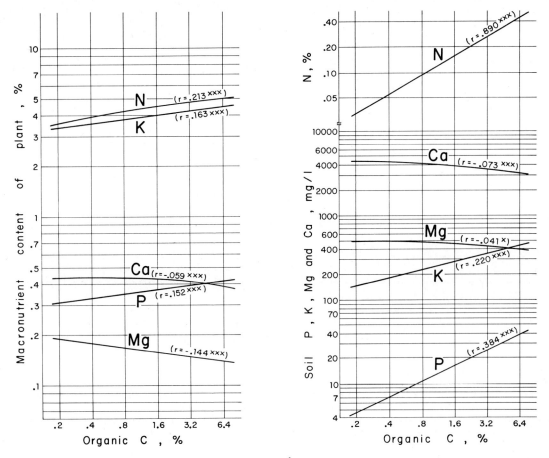

Fig. 13. Plant and soil macronutrients as functions of soil organic carbon content in the whole international material (n = 3731). Regression equations are given in Table 5.

The soil contents of all five macronutrients increased with increasing cation exchange capacity (Fig. 14). For N, Ca, Mg and K these relations are more firm than for P. Of the plant macronutrients, only the contents of N and K increased with increasing CEC but Ca, Mg and P decreased. The plant macronutrient—CEC regressions are generally less firm than the respective relations between soil macronutrients and CEC.

Because of the pronounced effects of texture and organic matter content on CEC (Figs 3 and 4), the effects of these factors on soil and plant macronutrients are indirectly reflected also in macronutrient—CEC relations.

The dissimilarity between the regressions of plant and soil Ca, Mg and P contents on CEC (Fig. 14) can be considered as one of the factors interfering with the plant-soil correlations for these elements.

The above regression data of the plant and soil macronutrient contents on four soil characteristics are summarized in Table 6.

With rising pH the N and P contents decreased but those of Ca and Mg increased. With increasing texture index the N and K contents increased but P contents decreased. With increasing soil organic matter content the N, P and K contents increased but those of Ca and Mg decreased. With increasing CEC the N and K contents increased.

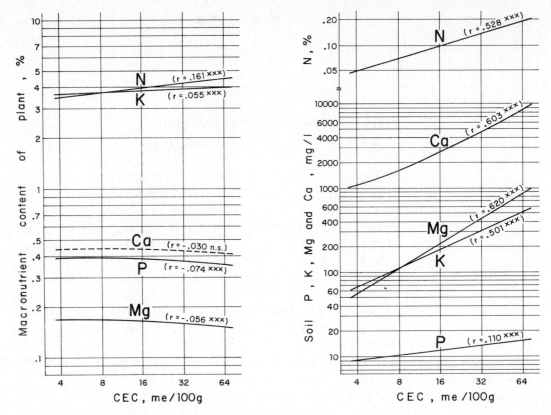

Fig. 14. Plant and soil macronutrients as functions of cation exchange capacity in the whole international material (n = 3731). Regression equations are given in Table 5.

These relationships are supported by both the plant and the soil macronutrient analyses, and the soil characteristics specified are unlikely to have any marked adverse effect on the plant-soil correlations for the particular elements. In the cases of pH—K, TI—Ca, TI—Mg, CEC—P, CEC—Ca and CEC—Mg the soil characteristics have opposite effects on the plant and soil nutrient contents, and are likely to interfere with the plant-soil correlation for those nutrients. Taking these effects into consideration may help with the interpretation of the data from plant and (or) soil analyses.

Table 6. Directions (+ or — symbols) of regressions of plant and soil macronutrient contents on four soil characteristics.

Soil characteristic	Macronutrient									
	N		P		K		Ca		Mg	
	Plant	Soil	Plant	Soil	Plant	Soil	Plant	Soil	Plant	Soil
pH(CaCl$_2$)	—	—	—	—	—	+	+	+	+	+
Texture index	+	+	—	—	+	+	—	+	—	+
Organic C	+	+	+	+	+	+	—	—	—	—
CEC	+	+	—	+	+	+	—	+	—	+

2.3 Micronutrients

2.3.1 Plant analysis versus soil analysis

Opinions of scientists differ as to the relative value of plant and soil analysis. In the present study both techniques were used on all soils and the evaluation of results is based on both (see Section 1.4.3).

The analysis of over 3500 original plants from the field and the respective soils, plus over 3500 pot plants grown on the same soils, gave an opportunity to compare the data and to draw certain conclusions at theoretical and practical levels.

To understand better the characteristics of these two widely used techniques, plant analysis and soil analysis, for diagnosing the micronutrient status of soils, one must realize that **these analyses are based on fundamentally different principles.**

The most essential difference between these analyses is that the micronutrient fractions to be analysed are obtained by different ways. Micronutrient absorption by a plant is a process taking place under laws of biochemistry and plant physiology while chemical soil extraction obeys the laws of chemistry. Consequently, the same micronutrient fractions do not appear in the analysis, nor is it always the aim that they should. Therefore, it is understandable that **many soil factors react differently during these two processes.**

Furthermore, the amounts (total) of micronutrients found in plants represent micronutrient fractions in the soil which have been available to the plant during its period of growth. Depending on the analytical method and on the micronutrient analysed, the results of soil analyses also include varying amounts of soil micronutrient reserves. The total contents of soil micronutrients, even though having an influence on the soluble or on plant available amounts, are in general, poor estimates of the available fractions. A variety of extraction methods has therefore been developed to obtain more reliable estimates of the fractions available to plants. Even the extractable fractions, although representing only a small portion of the total, exceed the requirements of plants by a considerable margin.

From a theoretical point of view, the amounts of micronutrients absorbed by plants from soils are to be considered as the **most reliable measure** of the fraction available to plants since the process of absorption is one taking place under the laws of nature. Soil analysis again is to be considered as an **attempt to imitate** plants. If there is contradiction between the results there should be no doubt who is wrong, the plant or the soil chemist.

If both analyses give similar estimates of the micronutrient status of soils, i.e. the mutual correlation between the results of plant and soil analyses is good, the quantitative differences between the micronutrient fractions analysed from plants and extracted from soils are of little importance.

When there is **contradiction** between the results obtained by the two techniques, it is important to find the main reason(s) for it, i.e. which soil factor impairs the correlation between the results of soil and plant analyses by affecting the results of plant analyses in one way and those of soil analyses in another way. Such a contradiction indicates that the soil extraction method used fails to take into account the effect of a soil factor regulating the availability of a micronutrient to plant. In order to improve the method of soil analysis, its results must be corrected so that they are in accord with the results of plant analyses with regard to the effects of the soil factor in question. In other words, a

correction coefficient is required for eliminating or minimizing the deviating effects of the particular soil factor(s).

The prerequisite for introducing a correction coefficient is that both the plant material and the soil material meet certain requirements:
— large numbers (preferably thousands) of plant-soil sample pairs
— widely varying soil characteristics[1]
— comparable plant material consisting of one indicator plant species, the same variety, same sampling time (physiological age), same parts of plant analysed, minimum contamination, and comparable growing conditions[2].

The present material (soils and pot-grown plants) can be considered to meet these requirements rather well.

Differences between the absorption of most microelements by a plant and the extraction of that microelement from the soil by chemical treatment are caused by more than one soil factor, but in practice all these cannot be taken into account. Therefore, in this study only the factors with the greatest effects (the key factors) are used as a basis for correcting the results of soil analyses.

Examples of the effects of two correction factors (pH and TI), however, are given in case of Mo (Section 2.3.2) and an example of correcting the results of two extraction methods (DTPA and AAAc-EDTA) for one soil factor (pH) is given in the case of Mn (Section 2.3.6).

Identification of the soil factor(s) responsible for impaired correlation between the results of plant and soil analyses was carried out by computing regressions for both the results of plant analyses and the results of soil analyses as functions of all six soil characteristics determined for all the international material. Comparing the regressions of plant and soil micronutrient contents as functions of various soil factors, it was usually possible to identify visually the factors with the most impairing effects on the plant-soil correlation. In verification, the actual impairing effects of all soil factors involved were computed, quantified and compared.[3] According to these data the soil factors impairing the plant-soil correlation most severely appeared to be as follows:

Micronutrient	Soil factor*	Soil extraction method
B	CEC	Hot water
Cu	Organic carbon content	AAAc-EDTA
Mn	pH	AAAc-EDTA and DTPA
Mo	pH	AO-OA
Zn	pH	AAAc-EDTA

* In the cases of AAAc-EDTA exractable soil Fe and DTPA extractable soil Zn, none of the correction coefficients calculated for the six soil factors improved the respective plant-soil correlation appreciably.

Identification of soil factors responsible for impaired correlation between the results of plant and soil analyses is the first step toward a more systematic interpretation of analytical results. These factors must be taken into account in one way or another when interpreting the results of soil analyses.

[1] For example, it would not be justifiable to make general conclusions about the effect of pH, if the range of variation in pH of the soil samples were one or two pH units only.
[2] For details, see Section 1.2.
[3] In the next Section (2.3.2) the data concerning Mo are presented in detail to give a fuller account of the principles and procedures for quantifying the impairing effects of various soil factors, for identifying the soil factor with the greatest effects and for eliminating these effects by determining and applying a correction coefficient.

For example, in case of Mo the soil pH is one such factor. The results of soil Mo analyses must be either:
(a) interpreted differently at different soil pH levels, or
(b) the interfering effects of pH must be eliminated by making the effect of pH on soil chemical extraction equal to its effect on Mo absorption by plants.

Should course (a) be adopted, a number of parallel stepwise interpretations would be needed. Thus, for example, extractable soil Mo contents obtained from a soil of pH 4 would have to be interpreted quite differently to the contents measured from soils with a pH of say 5, 6 or 7.

In order to simplify the interpretation of soil analytical data, the second course of action (b) was adopted and applied to the results of this study. Accordingly, the interference of soil pH was eliminated by introducing a "correction coefficient" for pH to the results of the soil Mo analyses so that the extractable soil Mo values as a function of pH became similar to plant Mo values as a function of pH. The introduction of this coefficient is not aimed to eliminate the influence of a given soil factor (in the case of Mo, the pH) but to liberate soil Mo values from an interfering factor (pH) and to minimize its impairing effects on the plant Mo—soil Mo correlation. With the application of a pH correction to soil Mo values, the need for different parallel interpretations at different soil pH levels vanishes.

As pointed out above, soil factors exercising the greatest interference with the plant-soil correlation vary from one micronutrient to another depending also on the method of extraction. In the next Sections (2.3.2 to 2.3.7) the effects of six soil factors on the six micronutrients are discussed. Departing from the otherwise alphabetic order of the elements, the data concerning Mo is given first and in more detail than those of other micronutrients. The underlying reasons for so doing were that the soil Mo analysis needs to be corrected more than those of other micronutrients and also because in the case of Mo two correction coefficients (for pH and for texture) are introduced.

2.3.2 Molybdenum

2.3.2.1 General aspects

The average Mo contents of the three types of indicator plant, the AO-OA extractable Mo contents of the respective soils, and the best fitting regressions of plant Mo and soil Mo for all the international material are shown below. The corresponding national averages are given in Appendixes 2, 3 and 4.

Indicator plant	n	Molybdenum content in plant DM mean ± s (ppm)	in resp. soils mean ± s (mg/l)	Regression of plant Mo (y) on soil Mo (x)	Correlation (r)
Original maize	1966	0.86 ± 1.35	0.212 ± 0.273	y = 0.715 + 0.661 x	0.134***
Original wheat	1766	0.94 ± 1.03	0.204 ± 0.229	y = 0.836 + 0.525 x	0.117***
Pot-grown wheat	3537	0.32 ± 0.36	0.210 ± 0.257	y = 0.529 + 0.246 x	0.245***

The differences in the average Mo contents of soils between the three groups are small. Mo differed from the other micronutrients in that the mean Mo content of maize was lower than that of the original wheat. This may not so much be due to the lesser ability of maize to absorb Mo from soils as to the generally lower pH of maize soils (mean pH 6.40) than of soils where the wheats were originally grown (mean pH 6.91). For much the same reason, the Mo contents of the original wheats exceeded those of the pot-grown wheats (mean soil pH 6.64) by a considerable margin. The effect of soil pH on extractable Mo and Mo content of plants will be discussed later in this chapter. The small amounts of soil in pots (see Section 1.2.6) may have further lowered the Mo contents of the pot-grown wheat as compared to the contents of the original wheat plants.

The regression of Mo contents of pot-grown wheat on soil Mo is shown in Fig. 15, where the national mean values of plant and soil Mo are also plotted. There was a substantial improvement in the correlation when the original plants were replaced by the pot-grown wheat plants. In spite of this, the correlation (r = 0.245***) was weaker than those for the other five micronutrients.[1] Nor was the respective correlation between the **Mo uptake** (pot-grown wheat) and soil Mo (y = 0.711 + 0.341 log x; r = 0.243***) any better than that in Fig. 15.

It is clear that soil Mo analysis (ammonium oxalate-oxalic acid, pH 3.3) **as such** does not give a sufficiently reliable index of Mo availability to plants.[2] For instance, the average extractable Mo contents of Pakistani and Brazilian soils were approximately equal, but the Mo contents of plants from Pakistan were about 20 times higher.

Studies concerning the reasons for the contradictory results of soil and plant analyses are presented in Section 2.3.2.2.

2.3.2.2 Soil factors affecting the molybdenum contents of plants and soils

To obtain first a general picture of how various soil factors are related to the amounts of Mo absorbed by plants (Mo contents of pot-grown wheat) and to AO-OA extractable soil Mo, both the plant Mo and soil Mo contents were computed as functions of the six soil characteristics (pH, texture, organic carbon content, CEC, electrical conductivity and $CaCO_3$ equivalent) determined from all the soil samples. These regressions are presented graphically in Fig. 16 (graphs on the left and in the centre), and the respective regression equations and correlation coefficients are given in Table 7. To give a better visual picture of these relations the soil data were classified into 11 to 18 classes (columns) with respect to each soil characteristic.[3] The number of samples falling into each class is given.

Visual comparisons of the regressions show that the effect of a soil characteristic on plant Mo usually differs considerably from its effect on AO-OA extractable soil Mo. Only the effects of CEC on plant Mo and soil Mo seem to be in conformity; the other soil factors affect the absorption of Mo by plants to an extent considerably different to their effect on the extractability of soil Mo by AO-OA extractant. These differences account for the conflicting results between the soil and plant analyses. AO-OA extraction is thus insensitive to the effects of those soil factors on the availability of Mo to plants and, therefore, gives a poor estimate of Mo availability to plants. In fact, three of the soil factors studied (electrical conductivity, pH and $CaCO_3$ equivalent) are better correlated

---

[1] Because of the very large number of samples (n = 3535 — 3538) used in this study, correlation coefficients of 0.033, 0.043 and 0.055 would have been significant at the 5 %, 1 % and 0.1 % level, respectively.

[2] At an early stage of this study other extraction methods for soil Mo were examined on limited materials, but none of these proved promising enough for acceptance.

[3] The reason for the low location of some regression curves (e.g. curve *a*) in relation to the columns is that the columns indicate arithmetic mean values and the regression is in a logarithmic form.

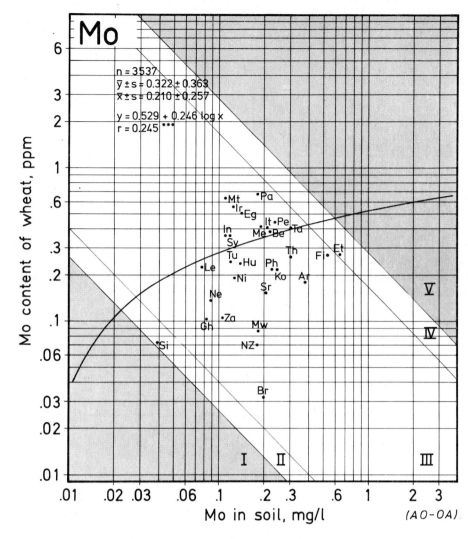

Fig. 15. Regression of Mo content of pot-grown wheat (y) on ammonium oxalate-oxalic acid extractable soil Mo (x) for the whole international material. Abbreviations for national mean values:

Ar = Argentina,   Be = Belgium,   Br = Brazil,   Eg = Egypt,
Et = Ethiopia,   Fi = Finland,   Gh = Ghana,   Hu = Hungary,
In = India,   Ir = Iraq,   It = Italy,   Ko = Korea, Rep.,
Le = Lebanon,   Mt = Malta,   Mw = Malawi,   Me = Mexico,
Ne = Nepal,   NZ = New Zealand,   Ni = Nigeria,   Pa = Pakistan,
Pe = Peru,   Ph = Philippines,   Si = Sierra Leone, Sr = Sri Lanka,
Sy = Syria,   Ta = Tanzania,   Th = Thailand,   Tu = Turkey,
Za = Zambia.

to Mo content of plants (r values 0.498***, 0.459*** and 0.410***, respectively, Table 7) than is the AO-OA extractable soil Mo (r = 0.245***, Fig. 15). Also the other three soil factors studied (texture, organic carbon content and CEC) seem to have significant effects on the Mo content of plants (Fig. 16, regressions f, i and l). Furthermore, all six soil factors studied affect significantly the AO-OA extractable contents of soil Mo (Fig. 16, regressions a, e, h, k, n and q).

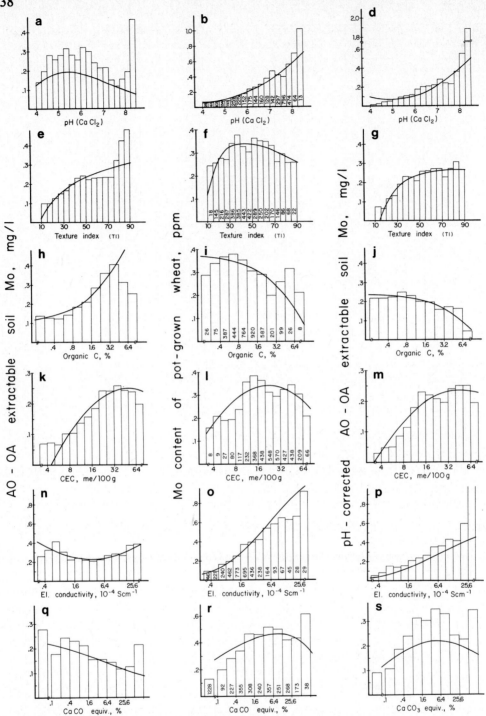

Fig. 16. Relationships of ammonium oxalate-oxalic acid extractable soil Mo and plant Mo to various soil characteristics in the whole international material.
Graphs on the left: AO-OA extractable soil Mo as a function of six soil factors.
Graphs in the middle: Plant Mo as a function of six soil factors.
Graphs on the right: pH-corrected AO-OA extractable soil Mo as a function of six soil factors.
In regard to the six soil characteristics the material is classified into 11—18 classes. The columns indicate arithmetic mean Mo values of each class. The number of samples (n) within each class is given. The equations and correlation coefficients for regressions are given in Table 7. See also Footnote 3 on p. 36.

Table 7. Equations and correlation coefficients for regressions of AO-OA extractable soil Mo (uncorrected and pH-corrected) and plant Mo on six soil characteristics. Regression curves are given in Figs 16 and 17.

Variables y	x	n	Regression	Correl. coeff.	Regr. curve
Soil Mo (uncorr.)	pH	3537	$\log y = -6.741 + 16.40 \log x - 11.15 (\log x)^2$	.306***	a
Soil Mo (uncorr.)	pH	18	$y = -8.558 + 4.429 x - 0.722 x^2 + 0.0383 x^3$	.796***	a_1
Plant Mo	pH	3537	$y = 0.574 - 0.251 x + 0.0312 x^2$	.459***	b
Plant Mo/soil Mo	pH	3537	$\log y = -2.300 + 0.360 x$	.770***	c
pH-corr. soil Mo	pH	3537	$y = 0.909 - 0.336 x + 0.0338 x^2$	.397***	d
Soil Mo (uncorr.)	TI	3535	$y = -0.274 + 0.299 \log x$	.199***	e
Plant Mo	TI	3535	$y = -1.559 + 2.344 \log x - 0.722 (\log x)^2$	.075***	f
pH-corr. soil Mo	TI	3535	$y = -1.328 + 1.642 \log x - 0.423 (\log x)^2$	.195***	g
Soil Mo (uncorr.)	Org. C	3537	$y = 0.102 + 0.0914 x - 0.00353 x^2$	.255***	h
Plant Mo	Org. C	3537	$y = 0.380 - 0.0483 x + 0.00157 x^2$	.098***	i
pH-corr. soil Mo	Org. C	3537	$y = 0.244 - 0.0280 x + 0.000753 x^2$	.078***	j
Soil Mo (uncorr.)	CEC	3537	$y = -0.390 + 0.738 \log x - 0.213 (\log x)^2$	.177***	k
Plant Mo	CEC	3537	$y = -0.202 + 0.809 \log x - 0.302 (\log x)^2$	.066***	l
pH-corr. soil Mo	CEC	3537	$y = -0.251 + 0.584 \log x - 0.175 (\log x)^2$	.115***	m
Soil Mo (uncorr.)	El. cond.	3537	$y = 0.219 - 0.135 \log x + 0.121 (\log x)^2$	.122***	n
Plant Mo	El. cond.	3537	$\log y = -0.856 + 0.784 \log x - 0.213 (\log x)^2$	.498***	o
pH-corr. soil Mo	El. cond.	3537	$\log y = -1.043 + 0.776 \log x - 0.217 (\log x)^2$	.498***	p
Soil Mo (uncorr.)	$CaCO_3$ eq.	3537	$\log y = -0.844 - 0.107 \log x - 0.0254 (\log x)^2$	.307***	q
Plant Mo	$CaCO_3$ eq.	3537	$y = 0.406 + 0.0910 \log x - 0.0243 (\log x)^2$	.410***	r
pH-corr. soil Mo	$CaCO_3$ eq.	3537	$\log y = -0.700 + 0.118 \log x - 0.0820 (\log x)^2$	.580***	s

In this connection it should be remembered that although all six soil characteristics studied seem to affect significantly both the plant Mo and AO-OA extractable soil Mo, some of these correlations are only "secondary" or "pseudocorrelations" due to the significant (r values from 0.093*** to 0.903***) mutual correlations between the six soil characteristics (see Section 2.1). Some of these are referred to later in connection with Mo as well as other micronutrients.

One of the principal differences between the effects of the above soil characteristics and extractable soil Mo on plant Mo seems to be that these soil characteristics act as regulators for the availability of Mo to plants, while extractable soil Mo values are more influenced by the Mo reserves of soil but fail to take into account the effects of the factors regulating the availability of Mo to plants. Therefore, a combination of one or more of the above soil factors and AO-OA extractable Mo would give a better estimate of the amounts of Mo available to plants than any of them alone.

So far the twelve regressions of AO-OA extractable soil Mo (Fig. 16 a, e, h, k, n, q) and plant Mo (b, f, i, l, o, r) on the six soil characteristics discussed above, indicate how soil Mo and plant Mo are correlated to these factors, but only broadly quantify the interference of these factors with the plant Mo—soil Mo correlation.

To quantify the interfering effect of a soil factor its effect on AO-OA extractable soil Mo must first be **equalized** to its effect on plant Mo, i.e. a **correction coefficient** for soil Mo values is required. Applying this coefficient to all the international material gives a measure of the interfering effect of the soil factor in question in the form of a new correlation coefficient. The more the plant—soil Mo correlation is improved the greater the interference of the factor in question has been. Furthermore, by comparing the effects

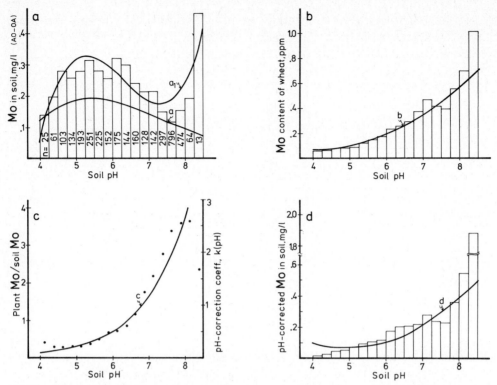

Fig. 17. Relationships of AO-OA extractable soil Mo and plant Mo to soil pH(CaCl$_2$).
Graph a: Soil Mo as a function of pH.
Regression curve a is calculated for the whole material (n = 3537) and curve a$_1$ for mean values of pH classes (see text).
Graph b: Plant Mo as a function of pH.
Graph c: Ratio of plant Mo to soil Mo as a function of pH.
The regression curve also indicates the coefficient, k(pH), for correcting soil Mo for pH (right-hand ordinate).
Graph d: pH-corrected soil Mo as a function of pH.
Columns and points indicate mean values of each pH class. n values of pH classes are given in Graph a.
Equations and correlation coefficients for the regression curves are given in Table 7.

of correction coefficients calculated for different soil factors, the factor which has most impaired the correlation can be **identified**. A concrete example of the procedures involved is given in the following:

The **effects of soil pH** on AO-OA extractable soil Mo are also given in Fig. 17 a. A large increase in soil Mo up to about pH 5—6 was first followed by a decrease, reaching a minimum at about pH 7.5—7.7, and then by another increase.[1]

The Mo contents of plants increased strongly toward alkaline soils (Fig. 17 b). This general tendency, however, was interrupted at about pH 7.5—7.7, i.e. at the very pH level where soil Mo values reached a minimum. In view of the large number of samples at this pH level, the connection between these minima is scarcely a coincidence but possibly due to soil chemical reactions affecting both the extractability of soil Mo and the availability of soil Mo to plants.

[1] Because of the abnormal distribution of samples into pH classes, the regression curve (a) calculated for the whole material ignores the latter increase. Therefore, another regression curve (a$_1$), based on mean values of pH classes, is also given.

The ratio of plant Mo/soil Mo plotted as a function of pH in Fig. 17 c, shows that there is a very close relationship between the plant Mo/soil Mo ratio and soil pH (r = 0.770***). It should be noted that this regression (curve c) was calculated from all the sample material, 3537 sample pairs. **The regression line (c) can be used to read off the correction of soil Mo values needed at different pH levels to eliminate the difference between pH effects on soil Mo and plant Mo analyses.**

However, if the regression equation of curve c (modified to y = $10^{-2.30 + 0.36\,pH}$) as such is used to correct AO-OA extractable soil Mo values, it would raise the soil Mo values from their original **average** level of 0.210 mg/l to an average of 0.297 mg/l. Therefore, to avoid this and also to restore the original average level for the corrected values, a **reversion coefficient** ($\frac{0.210}{0.297}$ = 0.707) has to be introduced into the equation. **The pH correction coefficient,** k(pH), is now given by:

$$k(pH) = 0.707 \times 10^{-2.30 + 0.36\,pH}$$

which in a simplified form is:

$$k(pH) = 10^{-2.45 + 0.36\,pH}$$

The numerical values for k(pH) can also be read off directly from the right hand ordinate of the regression curve c in Fig. 17. The k(pH) has been calculated for pH (CaCl$_2$). The relation between pH(H$_2$O) and pH(CaCl$_2$) in this material is given in Fig. 5 (Section 2.1).

When the soil Mo values in the whole international material are corrected with k(pH), the soil Mo—pH relation given in graph d is obtained. The latter is very similar to the plant Mo—pH relation given in Fig. 17, graph b, i.e. the effect of pH on soil Mo has not been eliminated but it has been made equal to its effect on plant Mo.

The plant Mo—soil Mo—pH relations are further illustrated in Fig. 18, where the data given in graphs a and b of Fig. 17 are combined. The respective relations after pH correction (combining data of graphs d and b) are given in Fig. 19. The S-shaped line depicting plant Mo and soil Mo in pH classes (Fig. 18) which transversely crosses the regression line for the whole material, illustrates how severely the effect of pH impairs the plant Mo—soil Mo correlation. After pH-correction (Fig. 19) the new line is in good agreement with the calculated regression line.

Application of k(pH) to the whole international material raised the correlation between plant Mo and soil Mo from r = 0.245*** (uncorrected, Fig. 15) to r = 0.696***. The latter regression is given in Fig. 20 with national mean values.

After correction for pH, correlations between plant Mo and soil Mo within different countries were considerably higher than without correction (Table 8). The only exception was Malta, where the variation of soil pH was only from 7.48 to 7.64. The largest improvements in correlations were obtained in countries where the variation of soil pH was relatively wide. In five countries, non-significant correlations became significant: in Ghana r = 0.134 n.s. to r = 0.643***, Malawi 0.062 n.s. to 0.487***, Sierra Leone 0.040 n.s. to 0.586***, Sri Lanka —0.161 n.s. to 0.650** and Zambia —0.004 n.s. to 0.656***.

The effects of the other five soil factors on soil Mo and plant Mo were investigated in a way analogous to that of pH. Correction coefficients for each of the five soil factors were

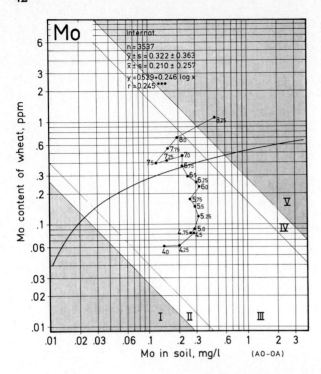

Fig. 18. Relationship of plant Mo to AO-OA extractable soil Mo in the whole material classified for pH. The points indicate mean plant Mo and soil Mo contents of various pH classes in Fig. 17 b and a. The lower limit of each pH class is given.

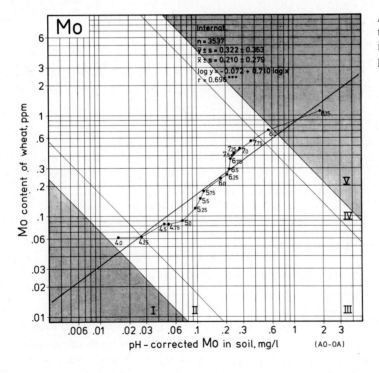

Fig. 19. Relationships of plant Mo to pH-corrected soil Mo (AO-OA) in the whole material classified for pH. See Fig. 17 b and d.

calculated and their effects were **tested and compared** by applying them to the whole international sample material. These comparisons show that application of the various correction coefficients improved the plant Mo—soil Mo correlation from an r value of 0.245*** (uncorrected) to the following r values:

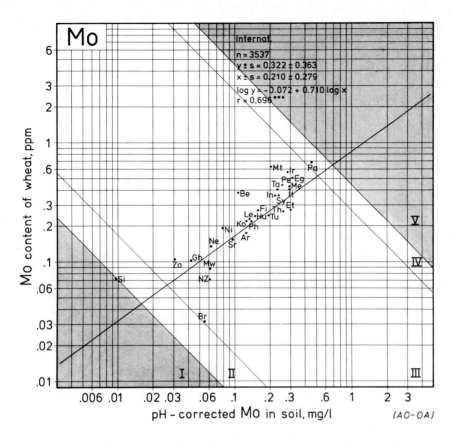

Fig. 20. Regression of Mo content of the pot-grown wheat (y) on the pH-corrected AO-OA extractable soil Mo (x) for the whole international material. National mean values are plotted in the graph. For abbreviations see Fig. 15.

Soil Mo corrected for	Value of r
pH	0.696***
Texture (TI)	0.277***
Organic carbon content	0.308***
CEC	0.260***
Electrical conductivity	0.460***
$CaCO_3$ equivalent	0.620***

Correction for pH proved to be more effective than correction for any other soil factor involved, although those for $CaCO_3$ and electrical conductivity also led to considerable improvements in the plant Mo—soil Mo correlation. These, however, are clearly due to high mutual correlations between pH and the two soil characteristics in question (see Section 2.1). This applies partly to the other soil factors as well. Therefore, the effects of pH correction on the relations between soil Mo and the five other soil factors were further investigated. These results, given in Fig. 16 (the right hand graphs), show the following:

Soil Texture. AO-OA extractable soil Mo contents increased quite substantially from coarse toward fine textured soils (Fig. 16 e). The ability of plants to absorb Mo reached a maximum in medium textured soils and a further increase in texture index (TI) resulted in lowered Mo contents of plants (Fig. 16 f). Evidently, Mo is fixed in heavy textured clay soils so firmly that its availability to plants is limited. This fixation, however, does not seem to affect the extractability of Mo by a chemical extractant. These correlations (Table 7, e and f) are less close than those between Mo and pH (a and b).

Correction of soil Mo values for pH brought the soil Mo—texture relation slightly closer in form to the respective relation between plant Mo and texture (curves g and f) but their essential differences still remain.

As the **organic carbon** content of soils increased, the Mo contents of plants decreased and there was a tendency for AO-OA extractable soil Mo to increase (Fig. 16, h and i). When soil Mo values were corrected for pH, they also became corrected for organic carbon, i.e. the effect of organic carbon on pH-corrected soil Mo (curve j) is very similar to its effect on plant Mo (curve i).

Cation exchange capacity. Both the AO-OA extractable soil Mo and Mo content of plants increased initially with increasing CEC (Fig. 16, k and l). With a further increment in CEC, there was a tendency for a decrease which in the case of plant Mo was very pronounced. Correction for pH also corrected slightly the soil Mo values for CEC.

Electrical conductivity and **$CaCO_3$ equivalent** of soils affected AO-OA extractable soil Mo values and Mo contents of plants quite differently (Fig. 16, n—o and q—r). After pH correction, however, these discrepancies were almost eliminated (o—p and r—s) because of strong mutual correlations between pH and electrical conductivity and pH and $CaCO_3$ equivalent (see Figs 3 and 4).

As shown above, due to mutual correlations between pH and the other five soil factors, the correction of AO-OA extractable soil Mo values for pH also reduced to varying degrees the discrepancies among the relationships of soil Mo and plant Mo to these soil factors.

To quantify the **remaining** interfering effects of various soil factors all the procedures explained above were repeated, this time with soil Mo values already corrected for pH. They showed that texture was the factor now causing most interference with the plant Mo—soil Mo correlation. The calculation of the **second** correction coefficient, that for texture k(TI), has been presented earlier in detail (Sillanpää, 1981). The formula for the texture correction coefficient is given by:

$$k(TI) = 10^{1.472 - 0.899 \log TI}$$

Application of the texture correction (in addition to the pH correction) to all the international material raised the correlation between plant Mo and soil Mo from an r value of 0.696*** (Fig. 20) to 0.739*** (Fig. 21).

A visual assessment of the efficacy of the pH and texture corrections can be made by comparing Figs 15, 20 and 21. Application of the texture correction at the national level gave the highest plant Mo—soil Mo correlations in 20 out of 29 countries (Table 8). Quantitatively, the improvement is not so marked as that of the pH correction, but substantial enough to be taken into account, especially in cases where soil materials with wide textural variation are to be analysed and accurate information on the Mo status of the soil is required.

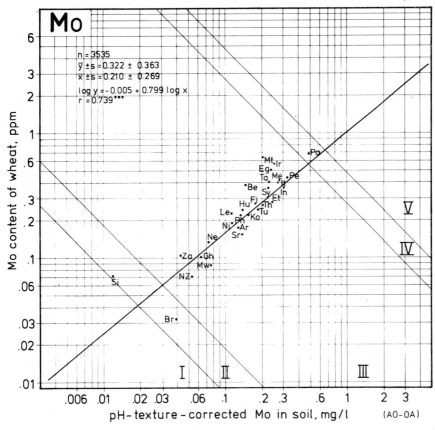

Fig. 21. Regression of Mo content of the pot-grown wheat (y) on the pH- and texture-corrected soil Mo (x) for the whole international material. National mean values are also given. For abbreviations see Fig. 15.

Table 8. Correlations (r values) between plant Mo and AO-OA extractable soil Mo in various countries. Soil Mo analyses are both uncorrected and corrected for pH and texture. Correction coefficients are: $k(pH) = 10^{-2.45 + 0.36\ pH}$ and $k(TI) = 10^{1.472 - 0.899\ \log TI}$. Correlations of the best fitting regressions are given. Regression models: (a) $y = a + bx$; (b) $y = a + b \log x$; (c) $\log y = a + bx$, and (d) $\log y = a + b \log x$.

Country	n	No correction regr. r	pH correction only regr. r	pH + TI correction regr. r	Country	n	No correction regr. r	pH correction only regr. r	pH + TI correction regr. r
Belgium	36	c −0.207 n.s.	c 0.036 n.s.	d 0.217 n.s.	Egypt	198	a 0.544***	a 0.614***	b 0.523***
Finland	90	d 0.341***	a 0.638***	a 0.743***	Iraq	150	b 0.778***	b 0.817***	d 0.800***
Hungary	201	b 0.172*	d 0.642***	d 0.664***	Lebanon	16	a 0.637**	a 0.779***	a 0.819***
Italy	170	a 0.418***	d 0.565***	d 0.539***	Syria	38	d 0.706***	d 0.742***	d 0.749***
Malta	25	c 0.914***	c 0.906***	c 0.827***	Turkey	298	a 0.664***	a 0.751***	a 0.727***
N. Zealand	35	a −0.121 n.s.	b 0.171 n.s.	b 0.240 n.s.	Ethiopia	125	a 0.480***	a 0.908***	a 0.949***
Argentina	208	a 0.583***	a 0.790***	a 0.709***	Ghana	93	a 0.134 n.s.	a 0.643***	a 0.601***
Brazil	58	c −0.078 n.s.	d 0.117 n.s.	c 0.287*	Malawi	97	c 0.062 n.s.	a 0.487***	a 0.597***
Mexico	242	b 0.252***	d 0.583***	d 0.632***	Nigeria	153	b 0.088 n.s.	d 0.645***	a 0.711***
Peru	68	a 0.484***	a 0.692***	d 0.651***	Sierra L.	48	b 0.040 n.s.	a 0.586***	a 0.668***
					Tanzania	163	d 0.545***	d 0.765***	d 0.846***
India	258	b 0.514***	b 0.613***	d 0.685***	Zambia	44	b −0.004 n.s.	a 0.656***	a 0.717***
Korea, Rep.	90	b 0.280**	d 0.623***	d 0.694***					
Nepal	35	a 0.335*	d 0.749***	b 0.794***					
Pakistan	237	d 0.598***	d 0.618***	d 0.587***					
Philippines	194	a 0.265***	d 0.662***	a 0.697***					
Sri Lanka	18	c −0.161 n.s.	a 0.650**	a 0.694**	Whole				
Thailand	149	b 0.388***	d 0.671***	d 0.677***	material	3537	b 0.245***	d 0.696***	d 0.739***

The mutual relations between plant Mo and AO-OA extractable soil Mo, and the effects of various soil factors on plant and soil Mo are summarized as follows:
— Owing to the heterogeneity of plant material, the original plants (maize and wheat) were unreliable indicators of the Mo status of soils.
— The heterogeneity was minimized by growing fresh plant samples (wheat) in pots on soils received from participating countries.
— Replacing the original plants by pot-grown wheat improved the plant Mo—AO-OA extractable soil Mo correlations from r values of 0.117—0.134*** to 0.245***.
— Soil Mo analysis (ammonium oxalate-oxalic acid, pH 3.3) as such does not give a reliable index of the Mo availability to plants as the data are much affected by soil pH.
— Both plant Mo and AO-OA extractable soil Mo were significantly affected by all six soil characteristics studied.
— Soil pH proved to be the factor which most disturbed the plant Mo and AO-OA extractable soil Mo correlation.
— Equalizing the effects of pH on soil Mo and plant Mo by correcting soil Mo values for pH raised the overall plant Mo—soil Mo correlation (r) from 0.245*** to 0.696*** and individually in 28 out of 29 countries.
— the pH correction also moderated and sometimes almost eliminated the differences between the effects of the other five soil factors on plant Mo and soil Mo.
— After correcting for pH, differences in soil texture had the greatest effect on plant Mo and soil Mo correlation.
— Texture correction (in addition to pH correction) improved the plant Mo—soil Mo correlation from an r value of 0.696*** to 0.739***, and in 20 out of 29 countries.
— When the results of this study are expressed by countries (Part II), the uncorrected as well as the pH-corrected AO-OA extractable soil Mo contents are presented.

2.3.3 Boron

2.3.3.1 General aspects

The national average B contents of original indicator plants, those of pot-grown wheat, and the average hot water extractable B contents of the respective soils are given in Appendixes 2, 3 and 4. Corresponding data for the whole of the international material as well as plant/soil regressions are tabulated below:

Indicator plant	n	Boron content in plant DM mean ± s (ppm)	in resp. soils mean ± s (mg/l)	Regression of plant B (y) on soil B (x)	Correlation (r)
Original maize	1966	9.24 ± 8.00	.65 ± .71	y = 5.25 + 6.15x	0.548***
Original wheat	1768	6.56 ± 7.80	.81 ± .75	y = 0.68 + 7.23x	0.694***
Pot-grown wheat	3538	6.09 ± 4.80	.73 ± .75	y = 2.63 + 4.75x	0.741***

On average, the B contents of maize were almost 50 per cent higher than those of wheat. The difference between the original and pot-grown wheat was small. This may be partly due to the small difference in soil B contents, and also to the small amount of soil in pots partly filled with quartz sand (see Section 1.2.6). The variation in B values was wide.

The standard deviations (s) of both plant and soil B are of the same magnitude as the respective mean values.

Because of the factors causing uncontrolled variation in analytical results from the original plants (see Section 1.2), the correlation between plant B and soil B was lower for the original plants than for pot-grown plants. The difference, however, was small compared with differences for other micronutrients. As in the case of other micronutrients, better quality information on the behaviour of B was obtained from pot-grown wheat.

The regression line of B content of pot-grown wheat on soil B for all the international material is given in Fig. 22. The correlation is highly significant (r = 0.741***). The

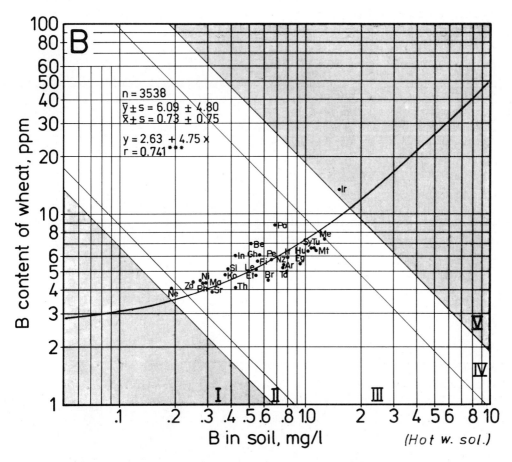

Fig. 22. Regression of B content of pot-grown wheat (y) on hot water extractable soil B (x) for all the international material. National mean values of plant and soil B are also given. Countries:

Ar = Argentina,	Be = Belgium,	Br = Brazil,	Eg = Egypt,
Et = Ethiopia,	Fi = Finland,	Gh = Ghana,	Hu = Hungary,
In = India,	Ir = Iraq,	It = Italy,	Ko = Korea, Rep.,
Le = Lebanon,	Mt = Malta,	Mw = Malawi,	Me = Mexico,
Ne = Nepal,	NZ = New Zealand,	Ni = Nigeria,	Pa = Pakistan,
Pe = Peru,	Ph = Philippines,	Si = Sierra Leone,	Sr = Sri Lanka,
Sy = Syria,	Ta = Tanzania,	Th = Thailand,	Tu = Turkey,
Za = Zambia.			

The boron content zones I—V are explained in Section 1.4.3.

national mean values of plant and soil B contents are also given in the graph. Because of a wide internal variation within most countries, the national mean values only loosely define the B status in different countries.

However, distinct differences between some of the countries can be seen. For example, B deficiency is likely to be more common in Nepal, Zambia, Nigeria and Philippines than in Iraq or Mexico. Detailed examination of the data by countries is given in Part II.

The regressions of micronutrient uptake as opposed to micronutrient content of pot-grown wheat, on micronutrients in soil were calculated for all six micronutrients. In general, the uptake correlations differed little from the content correlations. The greatest difference in favour of uptake was found in the case of boron, for which element the correlation (r) between the plant content and soil content was 0.741*** and the respective uptake—soil correlation 0.758***. The latter regression for the whole material is given in Fig. 23.

Comparison of the regressions in Figures 22 and 23 reveals no essential dissimilarities between the two methods of expressing plant B.

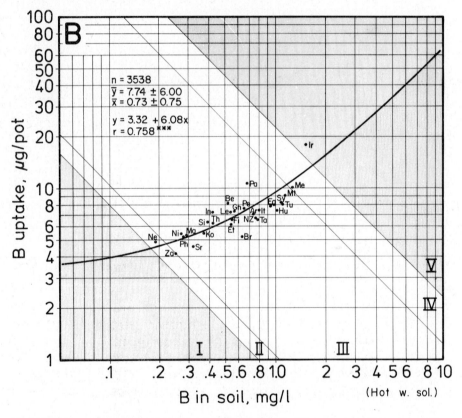

Fig. 23. Regression of B uptake by pot-grown wheat (y) on hot water extractable soil B (x) for all the international material. For abbreviations see Fig. 22.

2.3.3.2 Soil factors affecting the boron contents of plants and soils

The regressions of hot water extractable soil B and plant B on six soil factors are given in Figs 24 and 26. The regression equations and correlation coefficients are given in Table 10.

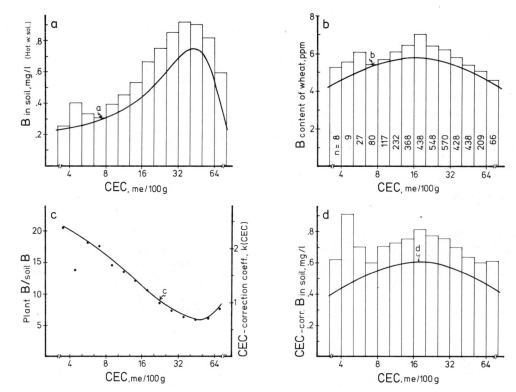

Fig. 24. Relationships of plant B and soil B to cation exchange capacity.
Graph a: Hot water soluble soil B as a function of CEC.
Columns indicate arithmetic mean soil B values of each CEC class.[1]
Graph b: Plant B as a function of CEC; n values within each CEC class are given in this graph.
Graph c: Ratio of plant B to soil B as a function of CEC. Points indicate the ratios within each CEC class. The regression curve calculated for the whole material (n = 3538) also indicates the coefficient, k(CEC), for correcting soil B for CEC (right-hand ordinate).
Graph d: Soil values as corrected with CEC correction coefficient.
Equations and correlation coefficients for the regression curves are given in Table 10.

[1] Because the columns indicate arithmetic mean values but the regression curves are logarithmic, the latter are often located at a lower level.

Both soil B and plant B were significantly correlated with all six soil factors studied. The highest correlations were found in the case of electrical conductivity and pH. The plant B and soil B contents, however, were similarly correlated to pH as well as to electrical conductivity. A closer statistical examination showed that cation exchange capacity had the most disturbing effects on plant B and soil B. Therefore, the CEC, obviously indirectly, was the most likely soil factor impairing the correlation between hot water soluble soil B and plant B.

Effects of CEC on soil B and plant B. Based on analytical data from 3538 sample pairs of soils and plants, the contents of hot water soluble soil B increased with increasing CEC reaching a maximum at CEC values of about 35—45 me/100 g and decreased with further increase of CEC (Fig. 24 a). The B content of plants was quantitatively somewhat less dependent on CEC and the maximum was already reached at CEC values around 20 me/100 g (Fig. 24 b).

The ratio of B content of plants to hot water soluble soil B is given as a function of CEC in Fig. 24 c. The regression (curve c) indicates the correction of soil B values needed at

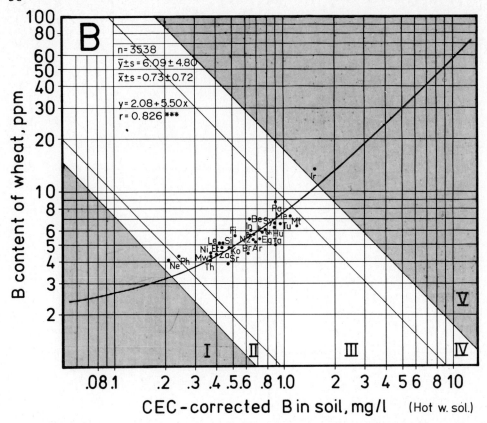

Fig. 25. Regression of B content of the pot-grown wheat (y) on CEC-corrected hot water soluble B (x) for the whole international material. National mean values of plant and soil B plotted in the graph. For abbreviations see Fig. 22.

various CEC levels to eliminate the difference in CEC effects between soil B and plant B analyses. The CEC correction coefficient for hot water soluble soil B is given by:

$$k(CEC) = 0.1164 \times 10^{1.4 - 0.026\,CEC + 0.000273\,CEC^2}$$

where the exponent is obtained from the regression equation in Table 10 c and the constant (0.1164) is a reversion coefficient for restoring the soil B values to their original level. The formula above can be simplified to:

$$k(CEC) = 10^{0.466 - 0.026\,CEC + 0.000273\,CEC^2}$$

Numerical values for k(CEC) can also be read off the right-hand ordinate of the regression curve in Fig. 24 c.

When soil B values in the whole material are corrected with k(CEC), the relationship between soil B and CEC given in Fig. 24 d is obtained. This relationship is very similar to the one between plant B and CEC given in Fig. 24 b. In other words, the effect of CEC on soil B has not been eliminated but it has been made equal to its effect on plant B.

Application of k(CEC) to all soil B values of the whole international material improved the correlation between plant B and soil B from an r value of 0.741*** (uncorrected) to 0.826*** (Figures 22 and 25). This improvement takes place in spite of the fact that B which normally is present in **anionic** form is corrected for **cation** exchange capacity. Obviously, the total contribution of all soil factors related to CEC are **indirectly** reflected by CEC.

Application of the CEC correction at the national level improved the plant B—soil B correlation in 24 out of 29 countries (Table 9). The improvement was most warranted in those countries (e.g. Thailand, Tanzania and Zambia) where the variation of soil CEC was wide.

Table 9. Correlations (r values) between plant B and hot water extractable soil B in various countries. Soil B analyses are both uncorrected and corrected for cation exchange capacity [k(CEC) = $10^{0.466 - 0.026 \text{ CEC} - 0.000273 \text{ CEC}^2}$]. Correlations of the best fitting regressions are given. Regression models: a) y = a + bx; b) y = a + b log x; c) log y = a + bx, and d) log y = a + b log x.

Country	n	Uncorrected Regr. model	r	CEC-corrected Regr. model	r	Country	n	Uncorrected Regr. model	r	CEC-corrected Regr. model	r
Belgium	36	a	0.521**	a	0.469**	Egypt	198	a	0.597***	a	0.761***
Finland	90	a	0.668***	a	0.605***	Iraq	150	a	0.945***	a	0.947***
Hungary	201	a	0.506***	a	0.579***	Lebanon	16	a	0.132 n.s.	b	0.255 n.s.
Italy	170	a	0.700***	a	0.740***	Syria	38	a	0.696***	d	0.715***
Malta	25	a	0.771***	c	0.784***	Turkey	298	a	0.851***	a	0.905***
New Zealand	35	a	0.179 n.s.	a	0.079 n.s.						
						Ethiopia	125	a	0.338***	a	0.418***
Argentina	208	a	0.524***	a	0.537***	Ghana	93	a	0.162 n.s.	c	0.208*
Brazil	58	a	0.501***	c	0.458***	Malawi	97	a	0.336***	a	0.363***
Mexico	242	a	0.793***	a	0.814***	Nigeria	153	a	0.381***	d	0.409***
Peru	68	a	0.560***	a	0.677***	Sierra Leone	48	a	0.357*	b	0.470***
						Tanzania	163	a	0.195*	d	0.346***
India	258	a	0.809***	a	0.828***	Zambia	44	a	0.236 n.s.	a	0.500***
Korea, Rep.	90	a	0.738***	a	0.771***						
Nepal	35	a	0.544***	a	0.577***						
Pakistan	237	a	0.910***	a	0.886***						
Philippines	194	a	0.371***	a	0.393***						
Sri Lanka	18	a	0.206 n.s.	a	0.236 n.s.						
Thailand	150	a	0.013 n.s.	d	0.223**	Whole material	3538	a	0.741***	a	0.826***

Effects of other soil factors on soil B and plant B. Soil pH seems to have a relatively small effect on hot water soluble soil B as also on the B content of the plant within the pH range of 4 to about 7.7. Nevertheless, a slight increase with rising pH can be observed, especially in soil B (Fig. 26, curves e and g). With a further rise in pH, however, there was a very marked increase in soluble soil B as well as in the contents of plant B.

Correction of soil B values for CEC did not appreciably affect the soil B—pH relationship, apparently because of the relatively low correlation between pH and CEC (Fig. 24 c and Fig. 4 a, Section 2.1).

Soil texture is a factor closely related to the cation exchange capacity of soils (R = 0.802***, Fig. 3 h). Consequently, the relations of soil B and plant B to soil texture (Fig. 26) are very similar to the respective B—CEC relations (Fig. 24). For the same reason, the correction of soil B values for CEC renders the relationship between soil B and texture index more like the relationship between plant B and texture (Fig. 26, curves i and j).

Organic carbon content of soils is another factor accounting for the cation exchange capacity of soils (R = 0.535***, Fig. 3 m). Since this relation is not so close as that between CEC and texture, the resemblance of B—organic carbon regressions (Fig. 26, curves k, l and m) to B—CEC regressions is not so good as in the case of texture. It is obvious, however, that the CEC correction decreases the difference between the regressions of soil B and of plant B on the organic carbon content of soil.

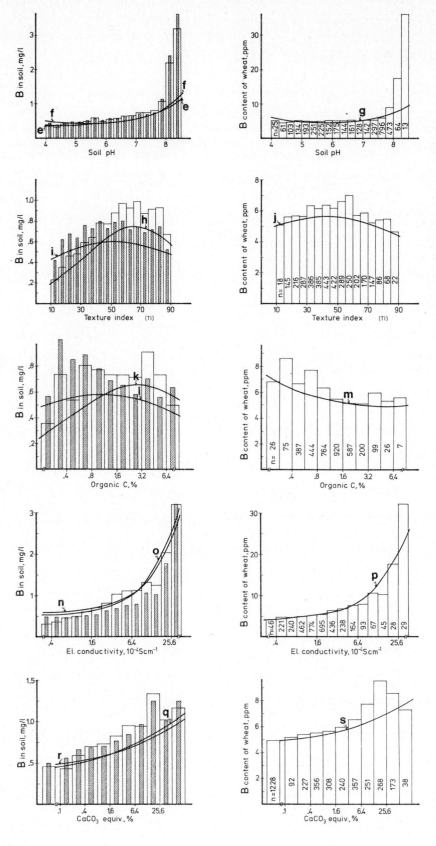

Fig. 26. Relationships of hot-water soluble soil B and plant B to various soil factors.
Graphs on the left: Uncorrected soil B (wide columns and regression curves e, h, k, n, and q) and CEC-corrected soil B (narrow columns and regression curves f, i, l, o, and r) as functions of various soil factors.
Graphs on the right: Plant B as a function of various soil factors. Number of samples (n) within each class is given. The equations and correlation coefficients for regression curves (e to s) are given in Table 10. See also footnote to Fig. 24.

Table 10. Equations and correlation coefficients for regressions of hot water soluble soil B (uncorrected and CEC-corrected) and plant B on various soil factors. Regression curves are given in Figures 24 and 26.

Variables y	x	n	Regression	Correl. coeff.	Regr. curve
Soil B (uncorr.)	CEC	3538	$\log y = -0.711 + 0.0274 x - 0.000320 x^2$	0.389***	a
Plant B	CEC	3538	$\log y = 0.375 + 0.637 \log x - 0.263 (\log x)^2$	0.153***	b
Plant B/soil B	CEC	3538	$\log y = 1.400 - 0.0260 x + 0.000273 x^2$	0.524***	c
CEC-corr. soil B	CEC	3538	$\log y = -0.760 + 0.868 \log x - 0.349 (\log x)^2$	0.105***	d
Soil B (uncorr.)	pH	3538	$\log y = 0.260 - 0.300 x + 0.0324 x^2$	0.405***	e
CEC-corr. soil B	pH	3538	$\log y = 0.972 - 0.508 x + 0.0475 x^2$	0.396***	f
Plant B	pH	3538	$\log y = 2.094 - 0.500 x + 0.0433 x^2$	0.427***	g
Soil B (uncorr.)	TI	3536	$\log y = -0.937 + 0.0246 x - 0.000187 x^2$	0.391***	h
CEC-corr. soil B	TI	3536	$\log y = -0.433 + 0.00794 x - 0.000075 x^2$	0.091***	i
Plant B	TI	3536	$\log y = 0.664 + 0.00403 x - 0.000047 x^2$	0.097***	j
Soil B (uncorr.)	Org. C	3538	$\log y = -0.249 + 0.284 \log x - 0.316 (\log x)^2$	0.234***	k
CEC-corr. soil B	Org. C	3538	$\log y = -0.238 + 0.00459 \log x - 0.146 (\log x)^2$	0.052**	l
Plant B	Org. C	3538	$\log y = 0.736 - 0.122 \log x + 0.084 (\log x)^2$	0.185***	m
Soil B (uncorr.)	El. cond.	3538	$y = 0.528 + 0.0729 x$	0.436***	n
CEC-corr. soil B	El. cond.	3538	$y = 0.511 + 0.0789 x$	0.489***	o
Plant B	El. cond.	3538	$y = 4.44 + 0.586 x + 0.00144 x^2$	0.594***	p
Soil B (uncorr.)	$CaCO_3$ eq.	3538	$\log y = -0.237 + 0.115 \log x + 0.0133 (\log x)^2$	0.413***	q
CEC-corr. soil B	$CaCO_3$ eq.	3538	$\log y = -0.234 + 0.0930 \log x + 0.0146 (\log x)^2$	0.348***	r
Plant B	$CaCO_3$ eq.	3538	$\log y = 0.735 + 0.0563 \log x + 0.0154 (\log x)^2$	0.338***	s

Electrical conductivity and **$CaCO_3$ equivalent** of soils have a high mutual correlation (Fig. 4, j and o). Furthermore, both of these characteristics are highly correlated with soil pH (Fig. 3, d and e; Fig. 4, f and k). Therefore, the relations of soil B and plant B to these factors resemble the respective B—pH relationships. A common feature of these relationships is the increase of both soil B and plant B with increasing electrical conductivity and with increasing $CaCO_3$ equivalent (Fig. 26). In short, both of these soil characteristics affect equally the soil B and plant B. Because of the relatively low correlation between CEC and the above soil factors (Fig. 4, d, e, i, n) the correction of soil B values for CEC does not materially affect the soil B—electrical conductivity or the soil B—$CaCO_3$ equivalent relationships.

The effects of the above six soil characteristics on soil and plant B can be summarized as follows:

— Determination of soil B by the hot water extraction method is clearly a useful way of estimating the B status of soils. The correlations between B content of plants and hot water soluble soil B (uncorrected and CEC-corrected) are higher than those for any other of the six micronutrients under study.

— Soil pH, electrical conductivity and $CaCO_3$ equivalent have similar effects on hot water soluble soil B as on the B content of plants. Therefore, these soil factors do not cause appreciable differences between the analytical results of soil B and plant B, and no corrections due to these factors were necessary. The correction of soil B for CEC does not materially affect the relationship of soil B to the above three soil characteristics.

— CEC, texture, and organic carbon have quite similar effects on hot water soluble B values but very different effects on the ability of plants to absorb B. This dissimilarity is most pronounced in the case of CEC, apparently because the nature of the other two characteristics (texture and organic carbon content) is partly reflected in CEC. Therefore, the correction of soil B values for CEC substantially corrects the soil B values for texture and organic carbon as well.

Correction coefficients were also calculated (analogously to the procedure for CEC) for the other five soil factors. These are not presented because none of them, as applied to the whole material, improved the soil B—plant correlation as fundamentally as does the CEC correction.

When the results of this study are expressed by countries (Part II), both uncorrected and CEC-corrected soil B values are given.

2.3.4 Copper

2.3.4.1 General aspects

The average Cu contents of the original and pot-grown plants the AAAc-EDTA extractable Cu contents of the respective soils in the whole material, and the best fitting plant/soil regressions are given below. The corresponding mean values by countries are given in Appendixes 2, 3 and 4.

Indicator plant	n	Copper content in plant DM mean ± s (ppm)	in resp. soils mean ± s (mg/l)	Regression of plant Cu (y) on soil Cu (x)	Correlation[1] (r)
Original maize	1966	11.6 ± 4.2	6.0 ± 7.9	log y = 0.97 + 0.114 log x	0.344***
Original wheat	1768	9.4 ± 5.7	6.1 ± 6.4	log y = 0.90 + 0.0062 x	0.254***
Pot-grown wheat	3537	7.0 ± 2.6	6.0 ± 7.0	log y = 0.62 + 0.309 log x	0.664***

[1] Respective correlations between plant Cu and DTPA extractable soil Cu were: 0.114***, 0.125*** and 0.518***. The mutual correlation between AAAc-EDTA extractable and DTPA extractable soil Cu was 0.829***. Because of their better correlation with plant Cu, the results given in this study are based on AAAc-EDTA extraction.

The average Cu content of maize exceeds that of the original wheat by 2.2 ppm and that of the pot-grown wheat by 4.6 ppm. There is no difference, though, in the mean values of extractable soil Cu between the three groups. The higher Cu contents of the original wheat as compared to the pot-grown wheat may partly be due to the small amount of soil in the pots (the pots were partly filled with quartz sand, see Section 1.2.6). The plants were not able to absorb from this small amount as much Cu as could plants grown on the same soils under field conditions. The coefficients of variation of the Cu contents of plants ranged from 36 to 61 percent in the three groups, and those of soils from 105 to 132 percent.

For reasons explained previously (Section 1.2), the correlation between the Cu contents of the original plants and the extractable Cu in soils was weak compared with that between the pot-grown wheat and soil. The regression for the latter is given in Fig. 27, where also the national mean values of plant and soil Cu contents have been inserted for a general comparison of the Cu status in the different countries. The correlation between **Cu uptake** and soil Cu (r = 0.674***) is about equal to that between the Cu content of plants and soil Cu. Also within individual countries, the differences between these two methods of expression are very small.

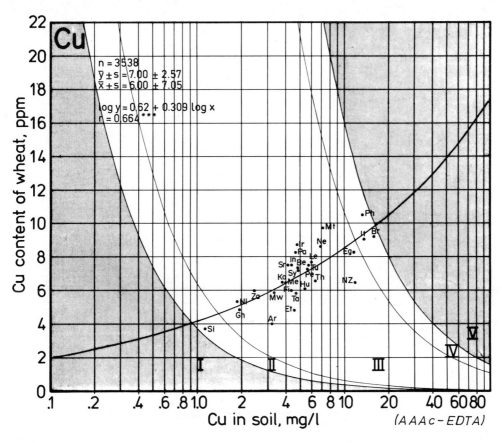

Fig. 27. Regression of Cu content of pot-grown wheat (y) on acid ammonium acetate-EDTA extractable soil Cu (x) for the whole international material. National mean values of plant and soil Cu are also given.
Countries:

Ar = Argentina,	Be = Belgium,	Br = Brazil,	Eg = Egypt,
Et = Ethiopia,	Fi = Finland,	Gh = Ghana,	Hu = Hungary,
In = India,	Ir = Iraq,	It = Italy,	Ko = Korea, Rep.,
Le = Lebanon,	Mt = Malta,	Mw = Malawi,	Me = Mexico,
Ne = Nepal,	NZ = New Zealand,	Ni = Nigeria,	Pa = Pakistan,
Pe = Peru,	Ph = Philippines,	Si = Sierra Leone,	Sr = Sri Lanka,
Sy = Syria,	Ta = Tanzania,	Th = Thailand,	Tu = Turkey,
Za = Zambia.			

2.3.4.2 Soil factors affecting the copper content of plants and soils

The Cu content of pot-grown wheat and AAAc-EDTA extractable soil Cu as functions of six soil characteristics are given in Figures 28 and 30. The regression equations and correlation coefficients are given in Table 12.

The AAAc-EDTA extractable soil Cu as well as the Cu content of wheat were significantly correlated with all six soil characteristics studied. In general, the effects of various soil factors on extractable soil Cu were in relatively good agreement with their respective effects on the Cu contents of plants. The most marked exception was the organic carbon content of the soil whose effect on the extractable soil Cu differed clearly from its effect on the Cu content of plants. Consequently, the different roles of soil organic carbon during the chemical extraction of soil Cu and during the Cu absorption by plants are likely to impair the plant Cu—soil Cu correlation.

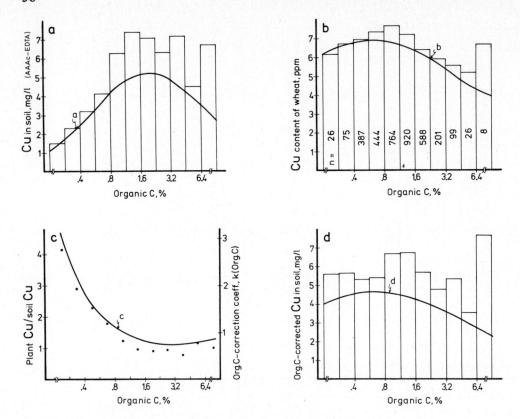

Fig. 28. Relationships of AAAc-EDTA extractable soil Cu and plant Cu to organic carbon content of soil.
Graph a: Soil Cu as a function of soil organic carbon content.
Graph b: Plant Cu as a function of soil organic carbon content.
Graph c: Ratio of plant Cu to soil Cu as a function of soil organic carbon content. The regression curve c also indicates the coefficient, k(org. C), for correcting soil Cu for organic carbon (right-hand ordinate).
Graph d: Soil Cu as corrected with k(org. C), as a function of organic carbon content. Columns and points indicate mean values of each organic carbon class; n values of classes are given in Graph b; equations and correlation coefficients for the regression curves are given in Table 12.

Effects of soil organic carbon content on soil Cu and plant Cu. AAAc-EDTA extractable soil Cu increased with increasing organic matter content of soils up to organic carbon contents of about 1—2 percent (Fig. 28 a). With further increases of organic carbon, the Cu contents of soil begin to decrease. On the basis of the results of this study, it is not clear whether this decrease is due to firmer Cu fixation by organic matter in soils of high organic matter content, or due rather to a progressive parallel decrease in the total Cu content toward organic soils. The latter possibility seems more likely. For instance, Sillanpää (1962 a, b) found a similar correlation between total Cu content of soils and soil organic matter, while the relative solubility of Cu increased slightly with increasing organic matter content of the soil. The Cu content of plants increases only slightly with increasing soil organic carbon in soils of very low organic matter content, while further increases in organic carbon cause a decrease in the Cu content of plants.

The ratio plant Cu/soil Cu (Fig. 28 c) indicates the difference between the effects of soil organic carbon on plant and soil Cu contents. The regression (curve c) also indicates the correction required to eliminate this difference. The correction coefficient for correcting soil Cu values for organic carbon is given by:

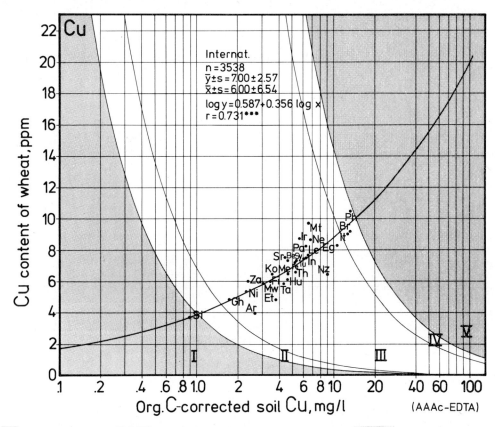

Fig. 29. Regression of Cu content of pot-grown wheat (y) on k(org. C)-corrected AAAc-EDTA extractable soil Cu (x) for the whole international material. National mean values of plant and soil Cu are plotted in the graph. For abbreviations see Fig. 27.

$$k(\text{org. C}) = 0.668 \times 10^{0.175 - 0.491 \log x + 0.470 (\log x)^2}$$

where the exponent is obtained from the regression equation in Table 12 (curve c where x = organic carbon content of soil) and the constant (0.668) is a reversion coefficient for restoring the soil Cu values to their original level. The formula above can be simplified to:

$$k(\text{org. C}) = 10^{-0.491 \log x + 0.470 (\log x)^2}$$

Numerical values for k(org. C) can be read off the right-hand ordinate of the regression curve in Fig. 28 c.

When soil Cu values in the whole material are corrected with k(org. C), the relationship between soil Cu and organic carbon given in Fig. 28 d is obtained. This relationship resembles basically the respective plant Cu—organic carbon relationship (Fig. 28 b), i.e. the effects of soil organic carbon on soil Cu and plant Cu have been equalized.

Application of k(org. C) to AAAc-EDTA extractable soil Cu values in the whole material improved the correlation between plant Cu and soil Cu from an r value of 0.664*** to 0.731*** (Figures 27 and 29). At the national level it improved the correlation in 21 out of 29 countries (Table 11). The improvement is not so pronounced as, for example, in the case of pH corrections applied to Mn and Mo. This may be partly due to the relatively narrow range of variation in organic carbon content in this material which included only a few peat soils.

Table 11. Correlations (r values) between plant Cu and AAAc-EDTA extractable soil Cu in various countries. Soil Cu analyses are both uncorrected and corrected for soil organic carbon contents [k(Org. C) = $10^{-0.491 \log x + 0.470 (\log x)^2}$, where x = Org. C, %]. Correlations of the best fitting regressions are given. Regression models: a) y = a + bx; b) y = a + b log x; c) log y = a + bx, and d) log y = a + b log x.

Country	n	Uncorrected Regr. model r		Org. C-corrected Regr. model r		Country	n	Uncorrected Regr. model r		Org. C-corrected Regr. model r	
Belgium	36	d	0.583***	d	0.686***	Egypt	198	d	0.573***	d	0.579***
Finland	90	d	0.633***	d	0.622***	Iraq	150	d	0.317***	d	0.400***
Hungary	201	b	0.289***	a	0.368***	Lebanon	16	b	0.149 n.s.	b	0.097 n.s.
Italy	170	b	0.804***	b	0.814***	Syria	38	d	0.644***	d	0.652***
Malta	25	d	0.760***	d	0.834***	Turkey	298	d	0.538***	b	0.523***
New Zealand	35	b	0.842***	b	0.817***						
						Ethiopia	125	d	0.830***	d	0.840***
Argentina	208	d	0.606***	d	0.693***	Ghana	93	b	0.726***	b	0.743***
Brazil	58	b	0.625***	d	0.649***	Malawi	97	d	0.503***	d	0.558***
Mexico	242	d	0.520***	d	0.587***	Nigeria	153	d	0.598***	d	0.651***
Peru	68	d	0.616***	d	0.568***	Sierra Leone	48	d	0.706***	a	0.751***
						Tanzania	163	d	0.727***	d	0.735***
India	258	b	0.396***	a	0.281***	Zambia	44	b	0.848***	b	0.870***
Korea, Rep.	90	b	0.456***	a	0.548***						
Nepal	35	d	0.410*	d	0.519**						
Pakistan	237	d	0.371***	d	0.282***						
Philippines	194	d	0.632***	d	0.687***						
Sri Lanka	18	d	0.723***	d	0.665***						
Thailand	150	d	0.742***	d	0.759***	Whole material	3538	d	0.664***	d	0.731***

Effects of other soil factors on soil and plant Cu.

Soil pH affects the AAAc-EDTA extractable soil Cu and Cu content of plants considerably less than the respective contents of Mn and Mo. However, a clear increasing trend of soil Cu and plant Cu toward alkaline soils can be noticed (Fig. 30 and Table 12, curves e, f, g). Correction with k(org. C) alters the soil Cu—pH relation slightly in the direction of the respective relationship between plant Cu and pH.

As **soil texture** becomes finer (texture index rises), the Cu contents of plants and especially the Cu contents of soils increase quite substantially (Fig. 30 and Table 12, curves h, i, j). The correction for organic carbon slightly moderates the effect of texture on soil Cu.

Cation exchange capacity, owing to its close correlation with texture (Figs 3 h and 4 b; see also Section 2.1), has very similar relations to soil and plant Cu as those of texture.

Electrical conductivity and **$CaCO_3$ equivalent** of soil are highly correlated with each other as well as with pH (Figures 3 and 4). Consequently, the relations between soil and plant Cu to these soil factors (Fig. 30, curves n—s) do not differ substantially from the Cu—pH relations. The copper contents of plants increased steadily with increasing electrical conductivity as well as with increasing $CaCO_3$ equivalent. Increases in the AAAc-EDTA extractable soil Cu at the highest levels of electrical conductivity and $CaCO_3$ equivalent levels tended to result in decreases in Cu contents of plants. The correction of soil Cu values for organic carbon had relatively small effects on the above relationships.

The results given in this Chapter can be summarized as follows:
— Soil Cu values determined from AAAc-EDTA extracts were better correlated with Cu contents of plants (r = 0.664***) than those determined from DTPA extracts (r = 0.518***). Therefore, the results presented in this study are based on the former extraction.

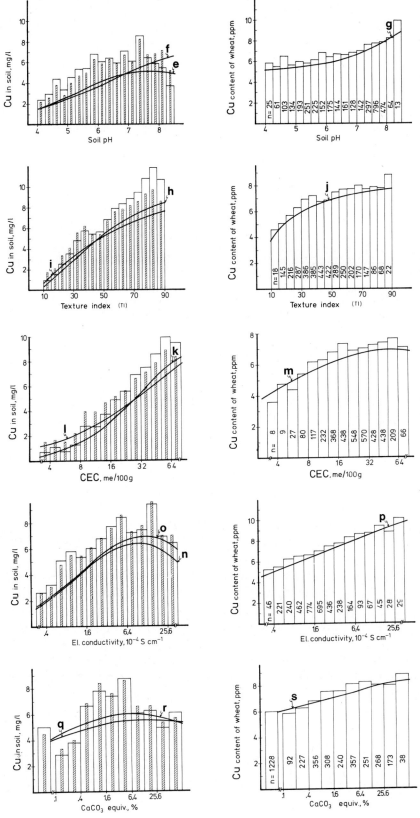

Fig. 30. Relationships of AAAc-EDTA extractable soil Cu and plant Cu to various soil factors.
Graphs on the left: Uncorrected soil Cu (wide columns and regression curves e, h, k, n, and q) and org. C-corrected soil Cu (narrow columns and regression curves f, i, l, o, and r) as functions of various soil factors.
Graphs on the right: Plant Cu as a function of various soil factors. Number of samples (n) within each class is given. The equations and correlation coefficients for regression curves (e to s) are given in Table 12. See also footnote to Fig. 24.

Table 12. Equations and correlation coefficients for regressions of AAAc-EDTA extractable soil Cu (uncorrected and Org. C-corrected) and plant Cu on various soil factors. Regression curves are given in Figs 28 and 30.

Variables y	x	n	Regression	Correl. coeff.	Regr. curve
Soil Cu (uncorr.)	Org. C	3538	$\log y = 0.655 + 0.406 \log x - 0.675 (\log x)^2$	0.315***	a
Plant Cu	Org. C	3538	$\log y = 0.830 - 0.0857 \log x - 0.205 (\log x)^2$	0.204***	b
Plant Cu/soil Cu	Org. C	3538	$\log y = 0.175 - 0.491 \log x + 0.470 (\log x)^2$	0.447***	c
Org. C-corr. soil Cu	Org. C	3538	$\log y = 0.657 - 0.0868 \log x - 0.239 (\log x)^2$	0.104***	d
Soil Cu. (uncorr.)	pH	3538	$\log y = -1.598 + 0.600 x - 0.0390 x^2$	0.330***	e
Org. C-corr. soil Cu	pH	3538	$\log y = -1.107 + 0.405 x - 0.0208 x^2$	0.447***	f
Plant Cu	pH	3538	$\log y = 0.840 - 0.0712 x + 0.00984 x^2$	0.357***	g
Soil Cu (uncorr.)	TI	3536	$\log y = -3.682 + 4200 \log x - 0.942 (\log x)^2$	0.574***	h
Org. C-corr. soil Cu	TI	3536	$\log y = -2.334 + 2.822 \log x - 0.602 (\log x)^2$	0.450***	i
Plant Cu	TI	3536	$\log y = -0.138 + 0.904 \log x - 0.193 (\log x)^2$	0.296***	j
Soil Cu (uncorr.)	CEC	3538	$\log y = -1.417 + 2.223 \log x - 0.524 (\log x)^2$	0.561***	k
Org. C-corr. soil Cu	CEC	3538	$\log y = -0.439 + 1.037 \log x - 0.179 (\log x)^2$	0.391***	l
Plant Cu	CEC	3538	$\log y = 0.378 + 0.536 \log x - 0.155 (\log x)^2$	0.191***	m
Soil Cu (uncorr.)	El. cond.	3538	$\log y = 0.548 + 0.543 \log x - 0.287 (\log x)^2$	0.368***	n
Org. C-corr. soil Cu	El. cond.	3538	$\log y = 0.562 + 0.532 \log x - 0.254 (\log x)^2$	0.389***	o
Plant Cu	El. cond.	3538	$\log y = 0.776 + 0.184 \log x - 0.0300 (\log x)^2$	0.349***	p
Soil Cu (uncorr.)	$CaCO_3$ eq.	3538	$\log y = 0.707 + 0.0666 \log x - 0.0311 (\log x)^2$	0.347***	q
Org. C-corr. soil Cu	$CaCO_3$ eq.	3538	$\log y = 0.749 + 0.0747 \log x - 0.0428 (\log x)^2$	0.441***	r
Plant Cu	$CaCO_3$ eq.	3538	$\log y = 0.836 + 0.0513 \log x$	0.383***	s

— Both the AAAc-EDTA extractable soil Cu and the Cu content of pot-grown wheat were highly correlated with all six soil characteristics (pH, texture, organic carbon content, CEC, electrical conductivity, and $CaCO_3$ equivalent) under study. In general, these soil factors affected the extractable soil Cu in much the same way as they affected the Cu content of plants. An exception, however, was the organic carbon content of soils.

— Correction of soil Cu values for organic carbon improved the plant Cu—soil Cu correlation for the whole material from an r value of 0.664*** to 0.731***.

— Similarly calculated correction coefficients for the other five soil characteristics did not improve the plant Cu—soil Cu correlations.

— When the results of this study are expressed by countries, both the uncorrected and the k(org. C)-corrected soil Cu values are given.

2.3.5 Iron

2.3.5.1 General aspects

Various factors caused uncontrolled variation in the analytical results of the original plants, as explained in Section 1.2. There was thus no correlation between the Fe contents of original plants and extractable soil Fe contents in the respective soils. The correlations (r values) between plant Fe and AAAc-EDTA extractable Fe for the original maize and wheat were —0.007 and —0.032, respectively. For DTPA extractable Fe the respective correlation coefficients were —0.028 and —0.030. The original plants were obviously contaminated, which affected the Fe analyses to a much greater extent than those of other micronutrients. Accordingly, the results of Fe analyses of the original plants were discarded.

The Fe contents of pot-grown wheat were more consistent with the contents of extractable soil Fe. The plant Fe—AAAc-EDTA extractable soil Fe and plant Fe—DTPA extractable soil Fe correlations (r values) in the whole material were 0.325*** and 0.263***, respectively.[1] The former relation is given in Fig. 31. The national mean values are shown in the graph.

The correlation between the Fe contents of pot-grown plants and AAAc-EDTA extractable soil Fe (r = 0.325***) is weak compared to correlations for other micronutrients in this study, but stronger than the correlation between Fe uptake and soil Fe ($\log y = 1.65 \times 0.106 \log x$; r = 0.266***).

2.3.5.2 Soil factors affecting the iron contents of plants and soils

The AAAc-EDTA extractable soil Fe and the Fe content of pot-grown wheat as functions of the six soil characteristics are given in Fig. 32. The respective regression equations and correlation coefficients are given in Table 13.

Contrary to the cases of the other five micronutrients, none of the six soil factors studied greatly affected the Fe contents of plants. The regression curves depart only slightly from the horizontal and consequently, the correlations are relatively weak.

Soil Fe seems to have been more affected by the various soil factors than plant Fe. Increases in **soil pH** and **$CaCO_3$ equivalent** values were accompanied by decreases in the extractable soil Fe contents. A similar, though milder relation between these soil characteristics and plant Fe can be noticed (Fig. 32 a, b, k, l).

Increasing **texture index, organic carbon content** and **CEC** were accompanied by increases in the contents of extractable soil Fe. The effects of these soil factors on Fe contents of plants are similar in direction to, but of lesser magnitude (Fig. 32 c, d, e, f, g, h). The effects of **electrical conductivity** on Fe (Fig. 32 i, j) are not marked.

None of the above six soil factors affected the plant Fe—soil Fe relations to such a degree that correction for these soil characteristics would have appreciably improved the correlation between plant Fe and soil Fe. This may be partly due to the insensitivity of plant Fe to the soil factors studied. Wheat is apparently a poorer indicator of the Fe status of soils than of the status of the other micronutrients studied. The oxidation-reduction conditions of soils are important in determining the behaviour of Fe in soils and its availability to plants. In this study the evaluation of these aspects was not possible.

---

[1] The mutual correlation (r) between AAAc-EDTA extractable and DTPA extractable soil Fe was 0.741***. Because of their better correlation with plant Fe, the AAAc-EDTA extractable soil Fe contents only will be presented in this study.

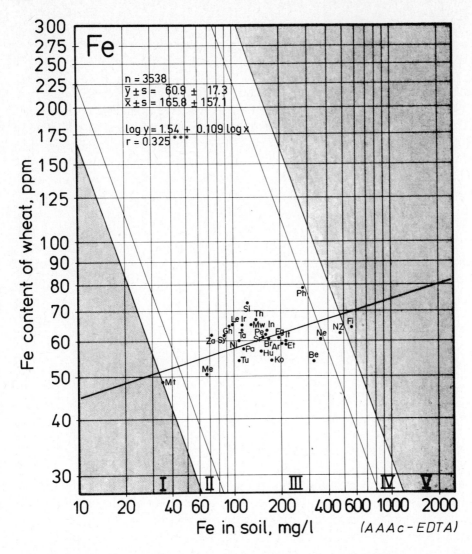

Fig. 31. Regression of Fe content of pot-grown wheat (y) on acid ammonium acetate-EDTA extractable soil Fe (x) for all the international material. National mean values of plant and soil Fe are also given. Countries:

Ar = Argentina,	Be = Belgium,	Br = Brazil,	Eg = Egypt,
Et = Ethiopia,	Fi = Finland,	Gh = Ghana,	Hu = Hungary,
In = India,	Ir = Iraq,	It = Italy,	Ko = Korea, Rep.,
Le = Lebanon,	Mt = Malta,	Mw = Malawi,	Me = Mexico,
Ne = Nepal,	NZ = New Zealand,	Ni = Nigeria,	Pa = Pakistan,
Pe = Peru,	Ph = Philippines,	Si = Sierra Leone,	Sr = Sri Lanka,
Sy = Syria,	Ta = Tanzania,	Th = Thailand,	Tu = Turkey,
Za = Zambia.			

The Fe content zones I—V are explained in Section 1.4.3.

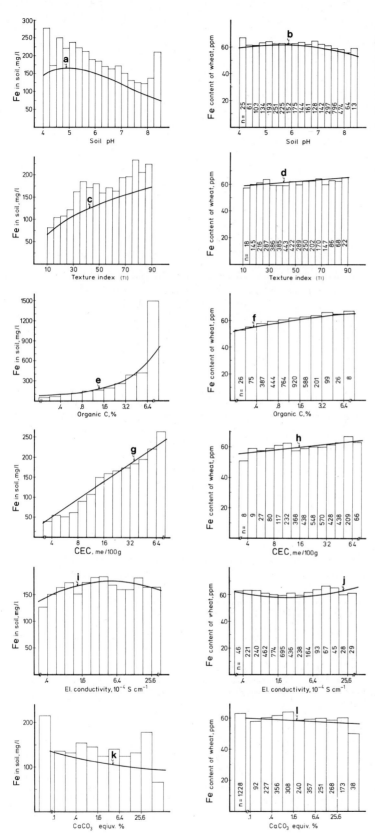

Fig. 32. AAAc-EDTA extractable soil Fe (graphs on the left) and plant Fe (graphs on the right) as a function of six soil characteristics. Number of samples (n) given within each class. The equations and correlation coefficients for regression curves (a—l) are given in Table 13.

Table 13. Equations and correlation coefficients for regressions of AAAc-EDTA extractable soil Fe and plant Fe on various soil factors. Regressions are given in Fig. 32.

Variables y	x	n	Regression	Correl. coeff.	Regr. curve
Soil Fe	pH	3538	$\log y = -0.965 + 9.127 \log x - 6.538 (\log x)^2$	0.300***	a
Plant Fe	pH	3538	$\log y = 1.580 + 0.0757 x - 0.00688 x^2$	0.136***	b
Soil Fe	TI	3536	$\log y = 1.374 + 0.442 \log x$	0.228***	c
Plant Fe	TI	3536	$y = 59.14 + 0.040 x$	0.038***	d
Soil Fe	Org. C	3538	$y = 62.2 + 80.2 x$	0.547***	e
Plant Fe	Org. C	3538	$\log y = 1.769 + 0.0629 \log x - 0.0176 (\log x)^2$	0.145***	f
Soil Fe	CEC	3538	$y = -37.62 + 147.7 \log x$	0.230***	g
Plant Fe	CEC	3538	$y = 52.74 + 5.949 \log x$	0.084***	h
Soil Fe	El. cond.	3538	$y = 162.4 + 33.05 \log x - 23.18 (\log x)^2$	0.049**	i
Plant Fe	El. cond.	3538	$\log y = 1.769 - 0.0274 \log x + 0.0368 (\log x)^2$	0.074***	j
Soil Fe	$CaCO_3$ eq.	3538	$\log y = 2.057 - 0.0586 \log x + 0.00395 (\log x)^2$	0.241***	k
Plant Fe	$CaCO_3$ eq.	3538	$\log y = 1.767 - 0.00868 \log x - 0.000533 (\log x)^2$	0.095***	l

2.3.6 Manganese

2.3.6.1 General aspects

The average Mn contents of the three types of indicator plant, the DTPA extractable Mn contents of the respective soils, and the best fitting plant Mn—soil Mn regressions for the whole material are given below. For the respective national averages, see Appendixes 2, 3 and 4.

Indicator plant	n	Manganese content in plant DM mean ± s (ppm)	in resp. soils mean ± s (mg/l)	Regression of plant Mn (y) on soil Mn (x)	Correlation[1] (r)
Original maize	1966	78 ± 48	43.2 ± 38.4	$y = 68.9 + 0.200 x$	0.161***
Original wheat	1768	74 ± 40	25.2 ± 33.1	$y = 70.0 + 0.143 x$	0.119***
Pot-grown wheat	3538	112 ± 155	34.7 ± 36.9	$\log y = 1.24 + 0.474 \log x$	0.552***

[1] Respective correlations between plant Mn and AAAc-EDTA extractable soil Mn were: —0.108***, 0.046* and 0.039*. The mutual correlation between DTPA extractable and AAAc-EDTA extractable soil Mn was 0.416***. Because of their better correlation with plant Mn, the final results given in this study are based on DTPA extraction.

The difference between the average Mn contents of the two original indicator plants is small and may be attributable more to differences in the respective soils than to the plant species. Contrary to the case with the other micronutrients, the average Mn content of the pot-grown wheat was about 50 per cent higher than that of the original wheat. Obviously, the difference is partly due to higher average extractable contents of Mn in soils in the pot experiment than in soils where the original wheats had grown. The effect of a higher Mn content of soils, however, may be nullified or even reversed by the small quantity of soil in the pots (Section 1.2.6). Another, more likely reason for the high Mn contents of the pot-grown wheats may be that the soils in the pots were in a less aerobic condition than they were in the field. The pot soils were kept near field moisture capacity during plant growth while in the wheat fields the soils must have been much drier. The availability of

Mn is strongly affected by oxidation-reduction reactions which largely depend on the soil moisture content. For example, Graven *et. al.* (1965) found a more than tenfold increase in Mn contents of plants due to flooding. See also Part II on the effect of irrigation on Mn in Iraq. The soil pH, which in the case of the pot soils was about 0.3 unit lower than in the original wheat soils (means 6.64 and 6.91, respectively), may also have been partly responsible for the increased Mn contents of the pot-grown wheats. The effect of pH will be discussed later in this Chapter. The variation in Mn contents of the pot-grown wheat is much wider than that for the other two plant types. In soils, the standard deviations (s) are of the same magnitude as the respective mean values.

The correlations between the Mn content of the original indicator plants and DTPA extractable soil Mn are highly significant but poor as compared to that between the Mn content of pot-grown wheat and soil Mn. The latter regression (with national mean plant Mn and soil Mn values) is given in Fig. 33.

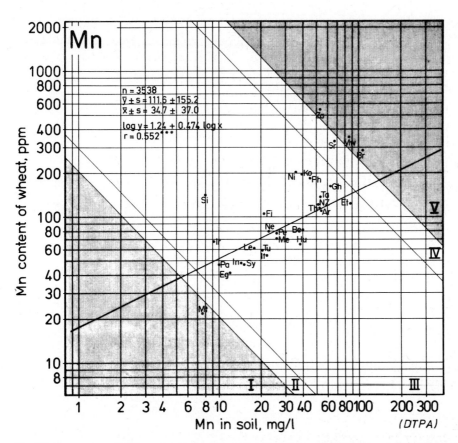

Fig. 33. Regression of Mn content of pot-grown wheat (y) on DTPA extractable soil Mn (x) for the whole international material. National mean values of plant and soil Mn are also given.

Countries:
Ar = Argentina,	Be = Belgium,	Br = Brazil,	Eg = Egypt,
Et = Ethiopia,	Fi = Finland,	Gh = Ghana,	Hu = Hungary,
In = India,	Ir = Iraq,	It = Italy,	Ko = Korea, Rep.,
Le = Lebanon,	Mt = Malta,	Mw = Malawi,	Me = Mexico,
Ne = Nepal,	NZ = New Zealand,	Ni = Nigeria,	Pa = Pakistan,
Pe = Peru,	Ph = Philippines,	Si = Sierra Leone,	Sr = Sri Lanka,
Sy = Syria,	Ta = Tanzania,	Th = Thailand,	Tu = Turkey,
Za = Zambia.			

The **Mn uptake**—soil Mn correlation in the whole international material (log y = 1.33 + 0.481 log x; r = 0.565***) differs only slightly from that given in Fig. 33. In national materials, with a few exceptions, the differences between the two methods of expression are also small.

2.3.6.2 Soil factors affecting the manganese content of plants and soils

DTPA extractable soil Mn and Mn content of pot-grown wheat as functions of six soil factors are given in Figs 34 and 38. The respective regression equations and correlation coefficients are given in Table 15.

The DTPA extractable soil Mn as well as the Mn content of the plant correlated significantly with all six soil factors studied. The effects of some of these soil factors on soil Mn agreed fairly well with their respective effects on the Mn content of plants. The most striking exception was the pH of soil which affected the soil Mn and the plant Mn very differently. This difference impaired the plant Mn—soil Mn correlation.

Effects of soil pH on soil Mn and plant Mn. The Mn contents of plants decreased steadily with rising pH (Fig. 34 b). This relationship is very close (R = 0.644***). The DTPA extractable soil Mn contents first increased with rising pH, reaching a maximum between pH 5 and 6, and thereafter decreased strongly toward alkaline soils (Fig. 34 a). This correlation is likewise close (R = 0.695***).

The ratio plant Mn/soil Mn (Fig. 34 c) illustrates the difference in pH effects at different pH levels. The equation for the regression curve (Fig. 34 c and Table 15 c) multiplied by a reversion coefficient (0.457) gives the correction coefficient, k(pH), required to eliminate the difference between the pH effects on plant Mn and soil Mn:

$$k(pH) = 10^{7.06} - 2.20 \, pH + 0.164 \, pH^2$$

k(pH) can also be obtained directly from the right-hand ordinate of the curve in Fig. 34 c.

When k(pH) is applied to DTPA extractable soil Mn values in the whole material (n = 3538) we obtain the relation between soil Mn and pH given in Graph d. Comparison of the relations given in Graphs b and d (Fig. 34) shows that the effect of pH on soil Mn has been equalized with its effect on Mn content of plants. In other words, the pH-corrected soil Mn values and plant Mn values are similarly correlated to soil pH.

It should be noted that the pH correction coefficient is calculated for pH($CaCl_2$). The relationship between pH(H_2O) and pH($CaCl_2$) in this material is given in Fig. 5 and Section 2.1.

The plant Mn—DTPA extractable soil Mn—pH relationships are also expressed in Fig. 35 where the data given in Graphs a and b (Fig. 34) are combined. The respective relationships after pH correction (combining data of Graphs d and b) are given in Fig. 36.

As can be seen from Fig. 35, the DTPA extractable soil Mn values e.g. at pH levels 4 to 5 and 6 to 7 are very similar in magnitude. The Mn contents of plants grown in acid (pH 4 to 5) soils, however, are several times higher than those grown in soils having pH 6 to 7. This clear conflict between the results of soil and plant analyses was eliminated by correcting the soil Mn values for pH (Fig. 36).

As numerous investigators have reported, the possibility of Mn deficiency or a response to Mn fertilization occurring in acid soils is very unlikely. Hence, the correction of soil Mn for pH will doubtless improve the reliability of soil analyses.

Application of a pH correction to DTPA extractable soil Mn values in the whole of the international material improved the correlation between plant Mn and soil Mn from an r value of 0.552*** to 0.713*** (Figures 33 and 37). At the national level the correlations

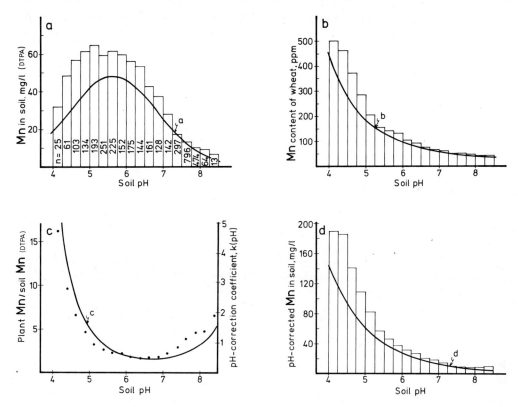

Fig. 34. Relationships of DTPA extractable soil Mn and plant Mn to soil pH(CaCl$_2$).
Graph a: Soil Mn as a function of pH.
Graph b: Plant Mn as a function of pH.
Graph c: Ratio of plant Mn to soil Mn as a function of pH.
The regression curve also indicates the coefficient, k(pH), for correcting soil Mn for pH (right-hand ordinate).
Graph d: pH-corrected soil Mn as a function of pH.
Columns and points indicate mean values within each pH class; n values of pH classes are given in Graph a; equations and correlation coefficients for the regression curves are given in Table 15.

were improved in 25 out of 29 countries (Table 14). These were most substantial in countries having generally acid soils with wide pH variation, e.g. Belgium, Finland, Brazil, Peru, Korea, Philippines and all the African countries. In countries where the pH correction was ineffective (India, Pakistan, Egypt and Iraq) alkaline soils with a relatively narrow pH range were predominant.

Effects of other soil factors on soil Mn and plant Mn. The total Mn contents of soils increased toward fine **textured** soils while the relative solubility of Mn (extractable Mn as a percentage of total Mn) increased toward coarse textured soils (Sillanpää, 1962 a, b). Clearly, the plant Mn—texture relations given in Fig. 38 (curves e, f, g) result from these two opposing functions. The pH correction of soil Mn values (curves e and f) brought the soil Mn—texture relation closer to that between plant Mn and texture (curve g). Generally, the Mn—texture correlations are much weaker than those between Mn and pH (Table 15).

As **organic carbon content** of soil increased, the Mn content of plants and the DTPA extractable soil Mn content tended to increase (Fig. 38, curves h, i, j). The columns indicate a tendency for soil Mn content to decrease at the highest organic carbon levels,

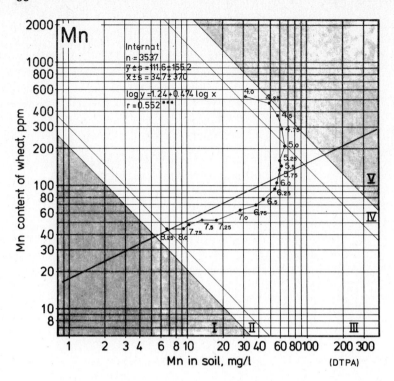

Fig. 35. Relationship of Mn content of pot-grown wheat to DTPA extractable soil Mn in the whole material classified according to pH. The points indicate mean plant Mn and soil Mn contents of various pH classes. The lower limit of each pH class is given.

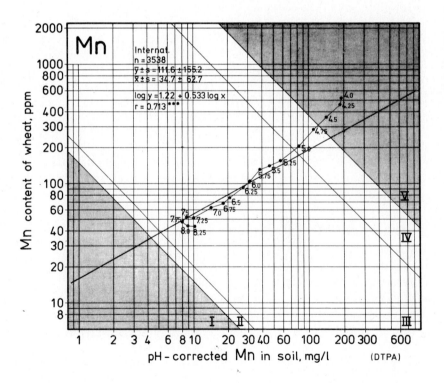

Fig. 36. Relationship of plant Mn to pH-corrected soil Mn (DTPA) in the whole material classified according to pH.

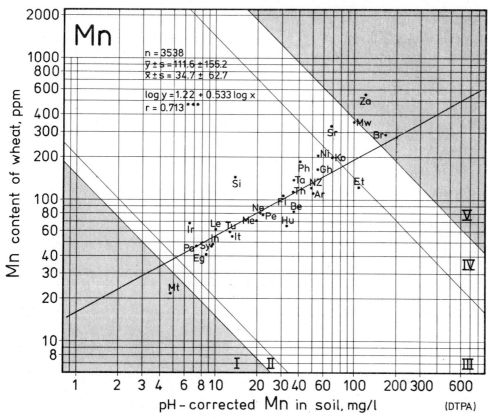

Fig. 37. Regression of Mn content of pot-grown wheat (y) on pH-corrected DTPA extractable soil Mn (x) for the whole international material. National mean values of plant and soil Mn are plotted in the graph. For abbreviations see Fig. 33.

Table 14. Correlations (r values) between plant Mn and DTPA extractable soil Mn in various countries. Soil Mn analyses are both uncorrected and corrected for pH(CaCl$_2$) [k(pH) = $10^{7.06 - 2.20\,pH + 0.164\,pH^2}$]. Correlations of the best fitting regressions are given. Regression models: a) y = a + bx; b) y = a + b log x; c) log y = a + bx; d) log y = a + b log x.

Country	n	Uncorrected Regr. model	r	pH-corrected Regr. model	r	Country	n	Uncorrected Regr. model	r	pH-corrected Regr. model	r
Belgium	36	c	0.347*	c	0.737***	Egypt	198	d	0.171*	d	0.139*
Finland	90	a	0.104 n.s.	a	0.639***	Iraq	150	d	0.057 n.s.	d	0.046 n.s.
Hungary	201	c	0.573***	c	0.742***	Lebanon	16	d	0.537*	d	0.642**
Italy	170	d	0.462***	d	0.531***	Syria	38	d	0.384*	d	0.455**
Malta	25	b	−0.128 n.s.	b	−0.090 n.s.	Turkey	298	a	0.195***	a	0.232***
New Zealand	35	c	0.403*	a	0.593***						
						Ethiopia	125	c	0.469***	d	0.665***
Argentina	208	a	0.447***	a	0.664***	Ghana	93	a	0.430***	a	0.739***
Brazil	58	c	0.433***	d	0.704***	Malawi	97	d	0.485***	d	0.769***
Mexico	242	d	0.568***	a	0.807***	Nigeria	153	c	0.395***	d	0.729***
Peru	68	c	0.508***	a	0.945***	Sierra Leone	48	b	0.269 n.s.	b	0.635***
						Tanzania	163	d	0.126 n.s.	d	0.387***
India	258	a	0.438***	a	0.356***	Zambia	44	d	0.266 n.s.	d	0.607***
Korea, Rep.	90	c	0.663***	a	0.935***						
Nepal	35	d	0.368*	b	0.606***						
Pakistan	237	a	0.287***	a	0.198**						
Philippines	194	d	0.517***	a	0.815***						
Sri Lanka	18	c	0.674***	d	0.871***						
Thailand	150	c	0.671***	a	0.827***	Whole material	3538	d	0.552***	d	0.713***

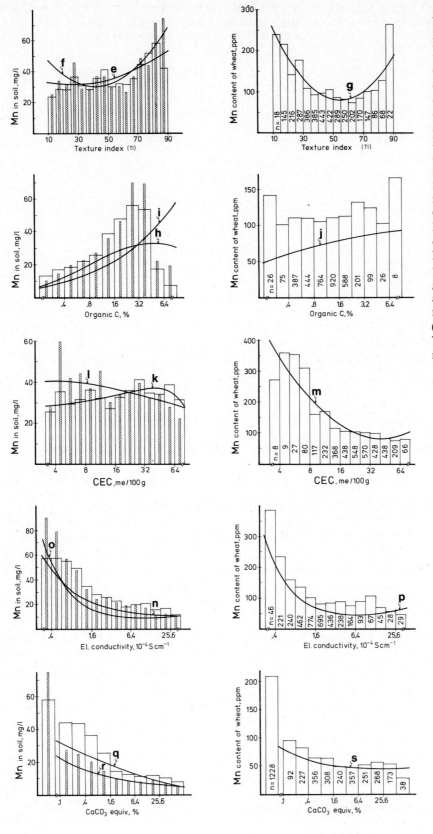

Fig. 38. Relationships of DTPA extractable soil Mn and plant Mn to various soil factors.
Graphs on the left: Uncorrected soil Mn (wide columns and regression curves e, h, k, n, and q) and pH-corrected soil Mn (narrow columns and regression curves f, i, l, o, and r) as a function of five soil factors.
Graphs on the right: Plant Mn as a function of five soil factors. Number of samples (n) within each class is given. The equations and correlation coefficients for regressions (e to r) are given in Table 15. See also footnote to Fig. 24.

Table 15. Equations and correlation coefficients for regressions of DTPA extractable soil Mn (uncorrected and pH-corrected) and plant Mn on various soil factors. Regression curves are given in Figs 34 and 38.

Variables y	x	n	Regression	Correl. coeff.	Regr. curve
Soil Mn (uncorr.)	pH	3538	log y = −3.075 + 1.684 x − 0.149 x^2	0.695***	a
Plant Mn	pH	3538	log y = 4.900 − 0.716 x + 0.0382 x^2	0.644***	b
Plant Mn/soil Mn	pH	3538	log y = 7.600 − 2.200 x + 0.1640 x^2	0.513***	c
pH-corr. soil Mn	pH	3538	log y = 3.892 − 0.486 x + 0.0126 x^2	0.781***	d
Soil Mn (uncorr.)	TI	3536	y = 35.2 − 0.291 x + 0.0055 x^2	0.114***	e
pH-corr. soil Mn	TI	3536	y = 59.1 − 1.384 x + 0.0165 x^2	0.098***	f
Plant Mn	TI	3536	y = 339 − 9.44 x + 0.0860 x^2	0.218***	g
Soil Mn (uncorr.)	Org. C	3538	log y = 1.358 + 0.492 log x − 0.372 (log x)^2	0.312***	h
pH-corr. soil Mn	Org. C	3538	log y = 1.218 + 0.595 log x − 0.0256 (log x)^2	0.329***	i
Plant Mn	Org. C	3538	log y = 1.865 + 0.187 log x − 0.0848 (log x)^2	0.148***	j
Soil Mn (uncorr.)	CEC	3538	y = 27.0 + 0.490 x − 0.00599 x^2	0.056***	k
pH-corr. soil Mn	CEC	3538	y = 39.9 + 4.712 log x − 5.973 (log x)^2	0.044**	l
Plant Mn	CEC	3538	y = 745 − 817.8 log x + 251.7 (log x)^2	0.267***	m
Soil Mn (uncorr.)	El. cond.	3538	log y = 1.441 − 0.549 log x + 0.182 (log x)^2	0.389***	n
pH-corr. soil Mn	El. cond.	3538	log y = 1.348 − 0.807 log x + 0.395 (log x)^2	0.446***	o
Plant Mn	El. cond.	3538	log y = 1.965 − 0.695 log x + 0.392 (log x)^2	0.491***	p
Soil Mn (uncorr.)	CaCO$_3$ eq.	3538	log y = 1.313 − 0.227 log x + 0.0356 (log x)^2	0.623***	q
pH-corr. soil Mn	CaCO$_3$ eq.	3538	log y = 1.088 − 0.240 log x + 0.0225 (log x)^2	0.720***	r
Plant Mn	CaCO$_3$ eq.	3538	log y = 1.724 − 0.123 log x + 0.054 (log x)^2	0.666***	s

but owing to the small number of samples at the highest organic carbon levels the regression curves do not reflect this tendency.

Cation exchange capacity has a relatively weak effect on DTPA extractable soil Mn (Fig. 38, curves k, l). Correction for pH slightly altered the direction of the regression curve, bringing it somewhat closer to the plant Mn—CEC regression curve (m).

As **electrical conductivity** and **CaCO₃ equivalent** of soils increased the Mn contents of plants as well as the extractable soil Mn decreased (Fig. 38, curves n, o, p, q, r, s). Correlations between Mn and these soil characteristics are strong, but somewhat lower than those between Mn and pH. Correction of DTPA extractable soil Mn for pH changed the relationship between soil Mn and both these soil factors slightly but distinctly toward the respective relations between plant Mn and these soil factors.

The above relations of plant Mn and DTPA extractable soil Mn to the six soil characteristics may be summarized as follows:
— As in the case of other micronutrients, the original plants (maize and wheat) due to their heterogeneity were poor indicators of the Mn status of soils.
— Determination of soil Mn by DTPA extraction gave a much higher correlation between plant Mn and soil Mn than did the AAAc-EDTA extractable Mn.
— Both plant Mn and DTPA extractable soil Mn were significantly correlated with all six soil characteristics studied.
— Correction of DTPA extractable soil Mn for any of the six soil factors would improve the correlation between plant Mn and soil Mn. (Correction coefficients for the other five soil factors were also calculated analogously to the procedure for correction for pH). The effects of these corrections, however, were small compared to that of pH correction. Therefore, these are not presented.

- Correction of DTPA extractable soil Mn for pH improved the plant Mn—soil Mn correlation substantially. It also moderated the differences between the effects of the other five soil factors on plant Mn and soil Mn.
- When the results of this study are classified by countries (Part II), both uncorrected and pH-corrected soil Mn values are presented.
- The extractable Mn contents of soils in the whole material of this study were analysed by two extraction methods (DTPA and AAAc-EDTA). The main differences between these methods as well as their usefulness are discussed in the next Section.

AAAc-EDTA extractable soil Mn as an indicator of Mn status of soils. The very poor correlation between the Mn contents of plants and AAAc-EDTA extractable soil Mn (r = 0.039*) is mainly due to the predominant role of pH in plant and soil Mn analyses. The Mn contents of plants decreased strongly with rising pH (Fig. 39 b). The AAAc-EDTA extractable soil Mn increased with rising pH to about pH 7 and decreased thereafter toward alkaline soils (Fig. 39 a). Both of the above relations are very firm (R values 0.644*** and 0.404***, respectively). Thus, in acid soils the plant Mn seems to be negatively correlated, and in alkaline soils positively correlated with the AAAc-EDTA extractable soil Mn.

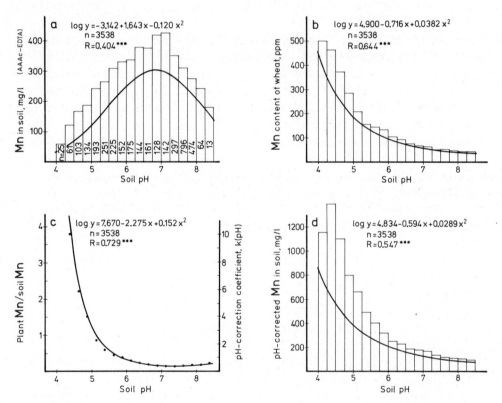

Fig. 39. Relationships of AAAc-EDTA extractable soil Mn and plant Mn to soil pH(CaCl$_2$).
Graph a: Soil Mn as a function of pH.
Graph b: Plant Mn as a function of pH.
Graph c: Ratio of plant Mn to soil Mn as a function of pH.
The regression curve also indicates the coefficient, k(pH), for correcting soil Mn for pH (right-hand ordinate).
Graph d: pH-corrected soil Mn as a function of pH.
Columns and points indicate mean values of each pH class; n values of pH classes are given in Graph a.

This AAAc-EDTA extractable Mn—pH relationship is basically of the same form as that between DTPA extractable Mn and pH. There is, however, an essential difference. The DTPA extractable Mn contents reached a maximum at pH 5—6 while the maximum for AAAc-EDTA Mn was reached first at about pH 7. This means that DTPA extractable soil Mn is positively correlated with plant Mn over a wider pH range than the AAAc-EDTA Mn. This also explains the poor correlation between plant Mn and AAAc-EDTA extractable soil Mn in this material consisting predominantly of soils approaching neutrality or alkalinity.

As in the case of DTPA extractable soil Mn, the correction coefficient for pH was calculated for AAAc-EDTA Mn. The ratio of plant Mn to AAAc-EDTA extractable soil Mn is given in Fig. 39 c. The regression equation in Graph c, multiplied by the reversion coefficient (2.642), gives the pH correction coefficient for AAAc-EDTA extractable soil Mn:

$$k(pH) = 10^{8.092 - 2.275\,pH + 0.152\,pH^2}$$

The pH correction coefficient is also given in Graph c (right-hand ordinate). The pH-corrected soil Mn as a function of pH is given in Graph d. This relation closely resembles the one between plant Mn and pH (Graph b) as well as that between pH-corrected DTPA Mn and pH (Fig. 34 d).

Application of the pH correction to AAAc-EDTA extractable soil Mn values in the whole material (n = 3538) improved the plant Mn—soil Mn correlation from an r value of 0.039* (uncorrected) to 0.588***. At the national level the respective correlations were (for uncorrected AAAc-EDTA Mn) non-significant or even negative in 22 out of 29 countries (Table 16). After pH correction, the correlations remained non-significant in two countries only (Malta and Iraq) and became highly significant in 20 countries.

Table 16. Correlations (r values) between plant Mn content and AAAc-EDTA extractable soil Mn in various countries. Soil Mn analyses are both uncorrected and corrected for pH [$k(pH) = 10^{8.092 - 2.275\,pH + 0.152\,pH^2}$]. Correlations of the best fitting regressions are given. Regression models: a) y = a + bx; b) y = a + b log x; c) log y = a + bx, and d) log y = a + b log x.

Country	n	Uncorrected Regr. model	r	pH-corrected Regr. model	r	Country	n	Uncorrected Regr. model	r	pH-corrected Regr. model	r
Belgium	36	b	—0.099 n.s.	c	0.645***	Egypt	198	d	0.668***	d	0.666***
Finland	90	b	—0.117 n.s.	a	0.603***	Iraq	150	a	0.024 n.s.	b	0.022 n.s.
Hungary	201	d	0.028 n.s.	d	0.720***	Lebanon	16	b	0.500*	b	0.502*
Italy	170	d	0.185*	d	0.462***	Syria	38	d	0.351*	d	0.352*
Malta	25	b	—0.376 n.s.	b	—0.375 n.s.	Turkey	298	d	0.200***	d	0.261***
New Zealand	35	d	0.245 n.s.	d	0.439***	Ethiopia	125	b	—0.123 n.s.	a	0.572***
Argentina	208	c	0.150*	d	0.565***	Ghana	93	a	—0.027 n.s.	a	0.722***
Brazil	58	d	0.211 n.s.	d	0.553***	Malawi	97	c	0.094 n.s.	d	0.718***
Mexico	242	d	0.066 n.s.	a	0.659***	Nigeria	153	a	—0.024 n.s.	d	0.648***
Peru	68	c	0.124 n.s.	a	0.867***	Sierra Leone	48	b	0.018 n.s.	b	0.575***
						Tanzania	163	a	—0.317***	c	0.204**
India	258	c	0.075 n.s.	c	0.151*	Zambia	44	d	—0.075 n.s.	d	0.556***
Korea, Rep.	90	d	0.150 n.s.	a	0.895***						
Nepal	35	c	—0.184 n.s.	b	0.496**						
Pakistan	237	b	0.161*	a	0.186**						
Philippines	194	a	0.076 n.s.	a	0.754***						
Sri Lanka	18	c	0.312 n.s.	d	0.773***						
Thailand	150	c	—0.194*	a	0.632***	Whole material	3538	c	0.039*	d	0.588***

Comparison of correlations between plant Mn and AAAc-EDTA Mn corrected for pH, given in Table 16, with respective correlations between plant Mn and DTPA Mn (uncorrected and pH-corrected, Table 14) shows that plant Mn correlated best with pH-corrected DTPA Mn. The r value for the whole material is 0.713*** and in 28 out of 29 countries the r values exceed those obtained by any other method. The second highest correlations were obtained with pH-corrected AAAc-EDTA and the third highest by uncorrected DTPA, r values 0.588*** and 0.552*** for the whole material, respectively. At the national level the former is better in 21 and the latter in 7 countries.

Comparison of relations between extractable Mn contents of soils obtained by different methods (Table 17) shows that the two methods that correlated best with the Mn contents of plants (pH-corrected DTPA and pH-corrected AAAc-EDTA) also have the highest mutual correlation.

Table 17. Mutual correlations (r values) between extractable Mn contents of soils in the whole material (n = 3538) as determined by four methods.

Extraction method	AAAc-EDTA (uncorrected)	pH-corrected AAAc-EDTA	pH-corrected DTPA
DTPA (uncorrected)	0.416***	0.672***	0.751***
pH-corrected DTPA	0.159***	0.840***	—
pH-corrected AAAc-EDTA	0.415***	—	—

It is clear that the essential difference between the two basic extraction methods, DTPA and AAAc-EDTA, lies in their divergent responses to soil pH. Therefore, the relationship of each of the two methods to pH mainly determines their usefulness for estimating the Mn status of soils. The difference between the methods, however, can be largely eliminated by correcting the analytical data for soil pH.

Soil analysis giving results that are in harmony with the results of plant analysis is naturally a more reliable index of the availability of a nutrient to a plant than a method giving contradictory results. Often a method which is well suited for one nutrient fails in the case of another. Nevertheless, the above results with Mn show that even a method which in itself gives poor results may be successfully used if its special features and its behaviour with respect to various soil characteristics are known and properly taken into account.

2.3.7 Zinc

2.3.7.1 General aspects

The average Zn contents of the three types of indicator plant in the whole international material, those of the respective soils determined by two extraction methods (AAAc-EDTA and DTPA) and the plant Zn—soil Zn regressions are given below.

Indicator plant	n	Zinc content in plant DM mean ± s (ppm)	in resp. soils mean ± s (mg/l)	Regression of plant Zn (y) on soil Zn (x)	Correlation[1] (r)
			AAAc-EDTA		
Original maize	1966	35.7 ± 47.2	4.13 ± 8.62	log y = 1.39 + 0.211 log x	0.391***
Original wheat	1768	27.4 ± 11.3	3.48 ± 7.72	y = 23.3 + 12.62 log x	0.405***
Pot-grown wheat	3537	18.3 ± 10.8	3.81 ± 8.26	y = 15.0 + 0.87 x	0.665***
			DTPA		
Original maize			2.14 ± 6.49	log y = 1.46 + 0.195 log x	0.381***
Original wheat		(as above)	1.77 ± 7.30	log y = 1.42 + 0.181 log x	0.473***
Pot-grown wheat			1.97 ± 7.05	log y = 1.23 + 0.321 log x	0.732***

[1] The mutual correlation between AAAc-EDTA extractable and DTPA extractable soil Zn was high (r = 0.819***), and the differences between their relations to plant Zn were small. In order to get more information on their usefulness, the relations of both methods to various soil factors were investigated.

The difference between the Zn contents of the original maize and original wheat may be partly due to maize being better able than wheat to absorb Zn from soils and partly to the higher contents of available Zn in maize soils. The average Zn content of the pot-grown wheat is one-third lower than that of the original wheat plants. Evidently the plants were unable to absorb from the small amounts of soil in pots (see Section 1.2.6) as much Zn as plants grown in field conditions. The standard deviation of Zn contents of maize (± 47 ppm) was much wider than for wheat (about 11 ppm). The variation of extractable soil Zn was considerably wider than in the cases of other micronutrients.

As for the other micronutrients, the correlations between the Zn content of pot-grown wheat and extractable soil Zn were much closer than those for the original maize and wheat. The former regressions (with national mean values) are shown in Figs 40 and 41.

The correlations of **Zn uptake** with AAAc-EDTA and with DTPA extractable soil Zn were very similar to those given in Figs 40 and 41, with r values of 0.669*** and 0.707***, respectively. At the national level, too, the differences were small.

2.3.7.2 Soil factors affecting the zinc contents of plants and soils

The regressions of plant and soil Zn on six soil characteristics are given in Figs 42 and 44 and in Table 19. The correlations were always highly significant with r varying from 0.055*** to 0.475***. In general, the organic carbon content of soils pH were better correlated with Zn than were the other soil factors studied. As in the case of the other micronutrients, correction coefficients were calculated for each of the six soil factors studied, as well as for the two extraction methods. The coefficients were tested over the whole soil and plant material. Possibly due to the relatively good agreement between the plant Zn and uncorrected (AAAc-EDTA and DTPA) soil Zn values, the corrections were less effective than in the cases of some other micronutrients, e.g. Mn and Mo.

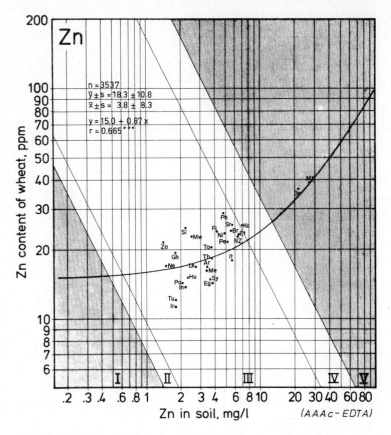

Fig. 40. Regression of Zn content of pot-grown wheat (y) on acid ammonium acetate-EDTA extractable soil Zn (x) for the whole international material. National mean values of plant and soil Zn are also given.
Countries:
Ar = Argentina,   Be = Belgium,   Br = Brazil,   Eg = Egypt,
Et = Ethiopia,   Fi = Finland,   Gh = Ghana,   Hu = Hungary,
In = India,   Ir = Iraq,   It = Italy,   Ko = Korea, Rep.,
Le = Lebanon,   Mt = Malta,   Mw = Malawi,   Me = Mexico,
Ne = Nepal,   NZ = New Zealand, Ni = Nigeria, Pa = Pakistan,
Pe = Peru,   Ph = Philippines, Si = Sierra Leone, Sr = Sri Lanka,
Sy = Syria,   Ta = Tanzania, Th = Thailand, Tu = Turkey,
Za = Zambia.
The Zn content zones I—V are explained in Section 1.4.3.

Compared to the other soil factors studied, soil pH caused the greatest difference between AAAc-EDTA extractable soil Zn and plant Zn. In the following paragraphs, the effects of pH and those of the other five soil characteristics are discussed.

Effects of soil pH on soil Zn and plant Zn. Comparison of the regression lines and columns in Graphs a, b and e (Fig. 42) shows that the correspondence between the effects of pH on plant Zn and DTPA Zn is greater than for the effects of pH on plant Zn and AAAc-EDTA Zn. The contents of plant Zn decreased with rising pH over almost the whole pH range. DTPA Zn decreased from about pH 5 upwards, while in the case of AAAc-EDTA, Zn values increased up to about pH 6 and only began to fall in the alkaline pH range. pH thus had a similar influence on plant Zn and DTPA Zn over a relatively wide pH range, whereas the effects of pH on plant Zn and AAAc-EDTA Zn resembled each other within only a rather narrow pH range on the alkaline side. There-

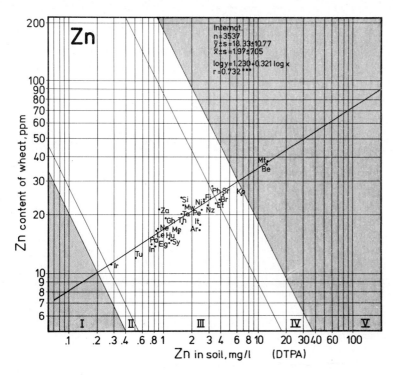

Fig. 41. Regression of Zn content of the pot-grown wheat (y) on DTPA extractable soil Zn (x) for the whole international material. For abbreviations see Fig. 40.

fore, the pH correction coefficient calculated for DTPA Zn was ineffective, but that calculated for AAAc-EDTA Zn improved the plant Zn—AAAc-EDTA Zn correlation.

The ratio of plant Zn to AAAc-EDTA Zn as a function of pH (Fig. 42c) indicates quantitatively the systematic difference between the pH effects on the two forms of Zn, and the regression curve (c) shows the correction required to eliminate this difference. The correction coefficient for pH is given by:

$$k(pH) = 0.1903 \times 10^{4.1 - 0.976\, pH + 0.068\, pH^2}$$

where the exponent is the regression equation for curve c (Fig. 42 and Table 19) and the constant (0.1903) is a reversion coefficient for restoring the AAAc-EDTA Zn values to their original level. The simplified formula is given by:

$$k(pH) = 10^{3.379 - 0.976\, pH + 0.068\, pH^2}$$

The numerical values for k(pH) can also be read from the right-hand ordinate of Fig. 42c and the regression line. The k(pH) applies to pH(CaCl$_2$). For the relation between pH(H$_2$O) and pH(CaCl$_2$) see Fig. 5 and Section 2.1.

When the AAAc-EDTA Zn values are corrected for pH in the whole material, we obtain the soil Zn—pH relation given in Fig. 42d. This curve shows that the effects of pH on plant Zn and on AAAc-EDTA Zn have been made almost equal.

Application of k(pH) to AAAc-EDTA Zn values improved the plant Zn—soil Zn correlation from an r value of 0.665*** (Fig. 40) to 0.706*** (Fig. 43) in the whole material and in 23 out of 29 countries (Table 18). The highest correlations, however, were

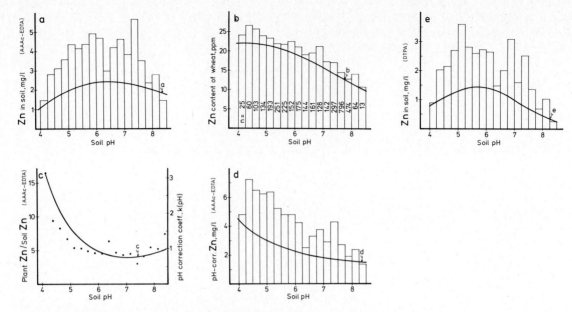

Fig. 42. Relationships of extractable soil Zn and plant Zn to soil pH(CaCl$_2$).
Graph a: AAAc-EDTA extractable soil Zn as a function of pH.
Graph b: Plant Zn as a function of pH.
Graph c: Ratio of plant Zn to AAAc-EDTA Zn as a function of pH.
The regression curve also indicates the coefficient, k(pH), for correcting AAAc-EDTA Zn for pH (right-hand ordinate).
Graph d: pH-corrected AAAc-EDTA Zn as a function of pH.
Graph e: DTPA extractable soil Zn as a function of pH.
Columns and points indicate mean values of each pH class. n values of pH classes are given in Graph b. Equations and correlation coefficients for regression curves are given in Table 19.

obtained between plant Zn and DTPA extraction (Fig. 41 and Table 18). The r value for the whole material is 0.732*** and in national materials the best correlation was obtained in 15 out of 29 countries. In general, the pH-corrected AAAc-EDTA extractable Zn was better correlated with plant Zn in countries where acid soils predominate and DTPA Zn in countries with alkaline soils.

In the case of Zn, the main difference between the DTPA and AAAc-EDTA extraction methods in favour of DTPA lies in the different reaction to pH. This difference can largely be eliminated by correcting AAAc-EDTA Zn for pH. In fact, correcting AAAc-EDTA Zn for pH raised the coefficient of correlation (r) between these two extraction methods from 0.819*** to 0.911***.

Effects of other soil factors in soil Zn and plant Zn. Figure 44 shows the relationships between uncorrected and pH-corrected AAAc-EDTA Zn (left-hand graphs), plant Zn (middle graphs), and DTPA Zn (right-hand graphs) and five soil characteristics.

Soil texture. The patterns of behaviour of Zn as a function of texture are quite similar irrespective of the method by which the Zn values have been determined (curves f, g, h, i, Fig. 44 and Table 19). With increasing texture index, the Zn values first decrease, reaching a minimum among medium textured soils and showing thereafter a tendency to increase towards fine textured clay soils. The effect of texture seems to be quantitatively more pronounced on extractable soil Zn values (f, g, i) than on plant Zn (h). Apparently

Table 18. Correlations (r values) in various countries between plant Zn and soil Zn determined by three different methods. Correlations of the best fitting regressions are given. Regression models: a) y = a + bx; b) y = a + b log x; c) log y = a + bx; d) log y = a + b log x.

Country	n	AAAc–EDTA-extractable Uncorrected Regr. model r	AAAc–EDTA-extractable pH-corrected[1] Regr. model r	DTPA-extr. Uncorrected Regr. model r	Country	n	AAAc–EDTA-extractable Uncorrected Regr. model r	AAAc–EDTA-extractable pH-corrected[1] Regr. model r	DTPA-extr. Uncorrected Regr. model r
Belgium	36	d 0.803***	a 0.895***	d 0.829***	Egypt	198	d 0.739***	b 0.730***	d 0.784***
Finland	90	d 0.595***	a 0.768***	d 0.645***	Iraq	150	d 0.672***	a 0.660***	a 0.705***
Hungary	201	a 0.394***	b 0.540***	b 0.581***	Lebanon	16	d 0.715**	d 0.709**	d 0.792***
Italy	170	d 0.751***	b 0.795***	d 0.814***	Syria	38	a 0.843***	a 0.840***	b 0.881***
Malta	25	d 0.916***	b 0.941***	b 0.896***	Turkey	298	a 0.507***	a 0.536***	b 0.712***
New Zealand	35	d 0.678***	b 0.684***	b 0.712***	Ethiopia	125	d 0.773***	a 0.846***	d 0.788***
Argentina	208	a 0.641***	a 0.676***	a 0.635***	Ghana	93	a 0.795***	b 0.827***	b 0.805***
Brazil	58	a 0.778***	a 0.825***	a 0.698***	Malawi	97	d 0.542***	d 0.607***	d 0.625***
Mexico	242	d 0.615***	b 0.651***	b 0.658***	Nigeria	153	d 0.758***	a 0.873***	a 0.869***
Peru	68	a 0.699***	a 0.760***	b 0.752***	Sierra Leone	48	a 0.900***	a 0.893***	a 0.936***
					Tanzania	163	d 0.604***	b 0.630***	b 0.669***
India	258	a 0.556***	a 0.560***	a 0.584***	Zambia	43	d 0.693***	b 0.787***	b 0.752***
Korea, Rep.	90	a 0.758***	d 0.815***	d 0.761***					
Nepal	35	a 0.626***	b 0.591***	b 0.666***					
Pakistan	237	d 0.682***	b 0.706***	b 0.734***					
Philippines	194	d 0.702***	a 0.827***	a 0.812***					
Sri Lanka	18	a 0.943***	a 0.948***	a 0.944***	Whole material	3537	a 0.665***	a 0.706***	d 0.732***
Thailand	150	d 0.660***	d 0.761***	d 0.842***					

[1] $k(pH) = 10^{3.379 - 0.976\,pH + 0.068\,pH^2}$

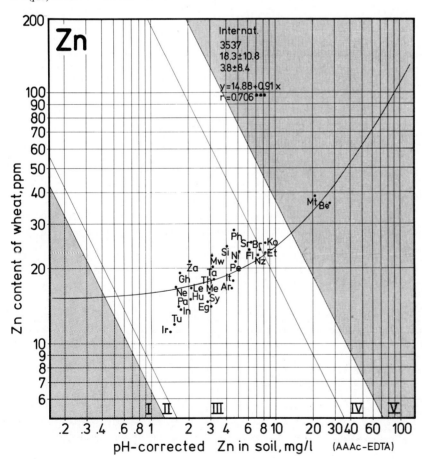

Fig. 43. Regression of Zn content of the pot-grown wheat (y) on pH-corrected AAAc-EDTA extractable soil Zn (x) for the whole material. For abbreviations see Fig. 40.

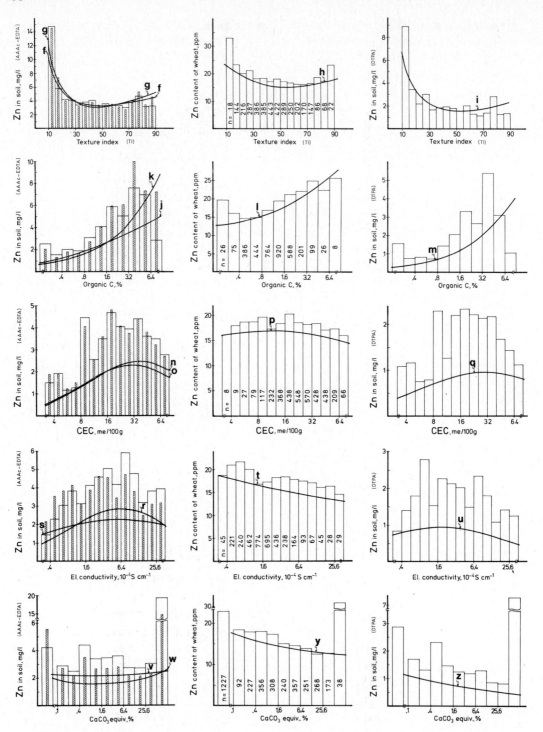

Fig. 44. Relationships of plant Zn and soil Zn to various soil factors. Graphs on the left: Uncorrected AAAc-EDTA extractable Zn (wide columns and curves f, j, n, r, v) and pH-corrected AAAc-EDTA Zn (narrow columns and curves g, k, o, s, w) as a function of five soil characteristics.
Graphs in the centre: Plant Zn as a function of five soil characteristics.
Graphs on the right: DTPA extractable soil Zn as a function of five soil factors. The number of samples (n) within each class is given in the centre graphs. The equations and correlation coefficients for regressions are given in Table 19. See also footnote to Fig. 24.

Table 19. Equations and correlation coefficients for regressions of AAAc-EDTA extractable soil Zn (uncorrected and pH-corrected), plant Zn, and DTPA extractable soil Zn on various soil factors. Regression curves are given in Figs 42 and 44.

Variables y	x	n	Regression	Correl. coeff.	Regr. curve
Soil Zn (AAAc-EDTA, uncorr.)	pH	3537	$\log y = -5.238 + 14.00 \log x - 8.713 (\log x)^2$	0.128***	a
Plant Zn	pH	3537	$\log y = 0.960 + 0.177 x - 0.0204 x^2$	0.475***	b
Plant Zn/soil Zn (AAAc-EDTA)	pH	3537	$\log y = 4.100 - 0.976 x + 0.0680 x^2$	0.402***	c
Soil Zn (AAAc-EDTA, pH-corr.)	pH	3537	$\log y = 1.660 - 0.327 x + 0.0181 x^2$	0.267***	d
Soil Zn (DTPA)	pH	3537	$\log y = -2.971 + 1.101 x - 0.0969 x^2$	0.378***	e
Soil Zn (AAAc-EDTA, uncorr.)	TI	3535	$y = 46.50 - 51.63 \log x + 15.43 (\log x)^2$	0.089***	f
Soil Zn (AAAc-EDTA, pH-corr.)	TI	3535	$y = 66.02 - 76.14 \log x + 23.04 (\log x)^2$	0.118***	g
Plant Zn	TI	3535	$\log y = 1.439 - 0.00883 x + 0.000075 x^2$	0.184***	h
Soil Zn (DTPA)	TI	3535	$y = 31.07 - 34.33 \log x + 9.990 (\log x)^2$	0.082***	i
Soil Zn (AAAc-EDTA, uncorr.)	Org. C	3537	$\log y = 0.324 + 0.529 \log x - 0.141 (\log x)^2$	0.343***	j
Soil Zn (AAAc-EDTA, pH-corr.)	Org. C	3537	$\log y = 0.288 + 0.673 \log x + 0.0449 (\log x)^2$	0.434***	k
Plant Zn	Org. C	3537	$\log y = 1.202 + 0.210 \log x + 0.0725 (\log x)^2$	0.286***	l
Soil Zn (DTPA)	Org. C	3537	$\log y = -0.0731 + 0.735 \log x$	0.423***	m
Soil Zn (AAAc-EDTA, uncorr.)	CEC	3537	$\log y = -1.153 + 1.970 \log x - 0.629 (\log x)^2$	0.224***	n
Soil Zn (AAAc-EDTA, pH-corr.)	CEC	3537	$\log y = -1.069 + 1.916 \log x - 0.644 (\log x)^2$	0.176***	o
Plant Zn	CEC	3537	$\log y = 1.128 + 0.193 \log x - 0.0925 (\log x)^2$	0.075***	p
Soil Zn (DTPA)	CEC	3537	$\log y = 0.929 + 1.229 \log x - 0.417 (\log x)^2$	0.098***	q
Soil Zn (AAAc-EDTA, uncorr.)	El. cond.	3537	$\log y = 0.280 + 0.429 \log x - 0.263 (\log x)^2$	0.261***	r
Soil Zn (AAAc-EDTA, pH-corr.)	El. cond.	3537	$\log y = 0.292 + 0.170 \log x - 0.110 (\log x)^2$	0.099***	s
Plant Zn	El. cond.	3537	$\log y = 1.230 - 0.0723 \log x$	0.135***	t
Soil Zn (DTPA)	El. cond.	3537	$\log y = -0.0407 + 0.0924 \log x - 0.170 (\log x)^2$	0.089***	u
Soil Zn (AAAc-EDTA, uncorr.)	$CaCO_3$ eq.	3537	$\log y = 0.338 + 0.0169 (\log x)^2$	0.055***	v
Soil Zn (AAAc-EDTA, pH-corr.)	$CaCO_3$ eq.	3537	$\log y = 0.206 - 0.00801 \log x + 0.0555 (\log x)^2$	0.252***	w
Plant Zn	$CaCO_3$ eq.	3537	$\log y = 1.174 - 0.0599 \log x + 0.00717 (\log x)^2$	0.433***	y
Soil Zn (DTPA)	$CaCO_3$ eq.	3537	$\log y = -0.0896 + 0.129 \log x - 0.00995 (\log x)^2$	0.340***	z

this is due to the wider range of variation in soil Zn than in plant Zn contents. The shape of the curves may be the result of two opposing factors, the increasing total contents of Zn and decreasing relative solubility of Zn towards finer textured soils (Sillanpää 1962 a, b). Correction of AAAc-EDTA Zn for pH had a relatively small effect on the Zn-texture relationships (f, g).

As the **organic carbon content** of the soil increased, there was a simultaneous increase in all four regression curves (Fig. 44, j, k, l, m). All four correlations are relatively close (Table 19). There seems to be a tendency for extractable soil Zn to decrease at an organic C content of about 0.4 percent (see columns). This, however, is not confirmed by the regressions, probably because of the small number of samples in the highest organic C classes and the general lack of organic soils in this material. Correction of AAAc-EDTA Zn for pH steepened the slope of the regression curve (j, k) but did not otherwise change the character of the relationship.

Cation exchange capacity had only moderate effects on plant and soil Zn (Fig. 44 n, o, p, q and Table 19). With increasing CEC the Zn values first increased but the curves turn slightly downwards at high CEC levels. The maximum for plant Zn was reached at a somewhat lower CEC level than in the case of extractable soil Zn. Correction of AAAc-EDTA Zn for pH affected only slightly the Zn—CEC relationship.

As the **electrical conductivity** of soil increased, the Zn contents of plants decreased (curve t), while extractable soil Zn values show at first a tendency to increase and begin to decrease only at higher levels of electrical conductivity. This tendency is more pronounced in the case of AAAc-EDTA Zn (curve r), but was moderated by pH correction (curve s).

With the increasing **CaCO$_3$ equivalent** of the soil, there was a general tendency for a slight decrease in the values of plant Zn and DTPA Zn (curves y and z) while the other two curves (v and w) are more independent of CaCO$_3$. The highest CaCO$_3$ equivalent class with its extremely high Zn values was an exception to this tendency. This class, however, was very small (n = 38) and two-thirds of its samples came from a very restricted area (Malta).

The mutual relations between plant Zn and extractable soil Zn, as well as the influence of various soil factors on Zn, are summarized as follows:

— Replacing the original plants (maize and wheat) by pot-grown wheat improved the plant Zn—soil Zn correlations.

— Both the AAAc-EDTA and DTPA extraction methods for determining soil Zn gave good indexes of Zn availability to plants. The highest correlations between plant Zn and extractable soil Zn were obtained when DTPA extraction was used.

— Both plant Zn and extractable soil Zn contents were significantly affected by all six soil characteristics studied.

— Most of the difference between the AAAc-EDTA and DTPA extraction methods in favour of DTPA is attributable to their different reactions to soil pH.

— Correction of the AAAc-EDTA Zn values for pH largely eliminated the above difference and improved the plant Zn—AAAc-EDTA Zn correlations.

— The plant Zn—DTPA Zn and the plant Zn—pH-corrected AAAc-EDTA Zn correlations (r) for the whole international material were 0.732*** and 0.706***, respectively. The former were higher in 15 countries where alkaline soils predominate, and the latter in 14 countries with predominantly acidic soils.

— When the results of this study are classified by countries (Part II), both pH-corrected AAAc-EDTA and DTPA extractable soil Zn values are presented.

2.4 Mutual relations between micronutrients

The contents of various micronutrients in plants vary with plant species and varieties and depend on the total micronutrient supply in soils and on factors controlling their availability to plants. Although there might be some differences between plant species in their reactions to soil factors controlling micronutrient availability, it is reasonable to assume that soil factors limiting the availability of micronutrients to one plant species react similarly in cases of other species. The mutual relations between the plant contents of six micronutrients presented (Figs 45 and 46) were obtained from a pot trial in which wheat, cv. 'Apu', was used as an indicator plant. Respective relations between the six micronutrients in soils (r values) are given in Table 20.

As shown earlier, the micronutrient contents of plants depend both on the micronutrient contents of soils and on soil factors regulating their availability. The effects of these factors vary considerably from one micronutrient to another as well as in their relative degree of efficacy. In general, when there is good correlation between the plant contents of two micronutrients, the availability of these micronutrients is largely controlled by the same soil factor or factors. Because of the large number of soil factors involved, these relationships are complicated but in some cases the reasons for correlations can be explained with a relatively high certainty. The best example of these is the highly significant negative correlation ($r = -0.515***$) between Mo and Mn. The regressions are given in Fig. 46. The availability of both these micronutrients is so strongly affected by soil pH that the other factors are overshadowed. While the Mn contents of plants decrease greatly with rising pH ($R = 0.644***$, Fig. 34 b), the Mo contents increase ($R = 0.459***$, Fig. 16 b) and deficiencies of both Mn and Mo can therefore hardly exist in same soil. A deficiency of Mn is often combined with an excess of Mo and *vice versa,* as will be seen in Part II. Extractable soil Mn and Mo are highly significantly negatively correlated ($r = -0.390***$) if both are corrected for pH (Table 20). Without pH corrections this relationship would be equally significant ($r = 0.391***$) but positive, i.e. in strong contradiction to the plant Mn—Mo correlation.

The second highest correlation found between various plant micronutrients is that of Mn and Zn ($r = 0.420***$, Fig. 46). In this case also the soil pH exercises the leading effect on the availability of these micronutrients to plants. The plant contents of both these micronutrients decrease strongly with rising pH (Fig. 34 b and 42 b). Comparing the results given in Fig. 38 (graphs on the right) and Fig. 44 (graphs in the centre) it can be seen that there are also other soil factors e.g. texture and organic carbon content, which correlate with plant Mn much as they do with plant Zn. Therefore, in addition to good correlations (r values $0.346***$ and $0.299***$; Table 20) between extractable soil Mn and Zn, these factors may contribute to the good plant Mn—Zn correlation.

For the three next highest correlations between the plant contents of Fe and Zn ($r = 0.350***$), Fe and Cu ($r = 0.344***$), and Fe and Mn ($r = 0.329***$) the common soil factors are not as clearly defined, since the Fe contents of plants were less strongly affected by any of the soil factors studied than were the other micronutrients (see Fig. 32 and related text). In all three cases, Fe—Zn, Fe—Cu, and Fe—Mn, the extractable soil contents are correlated to a degree similar to the plant contents (r values $0.380***$, $0.270***$ and $0.222***$, respectively; Table 20).

Plant B was best correlated to plant Mo and Mn, in the former case positively ($r = 0.329***$) and in the latter negatively ($r = -0.244***$). Respective correlations

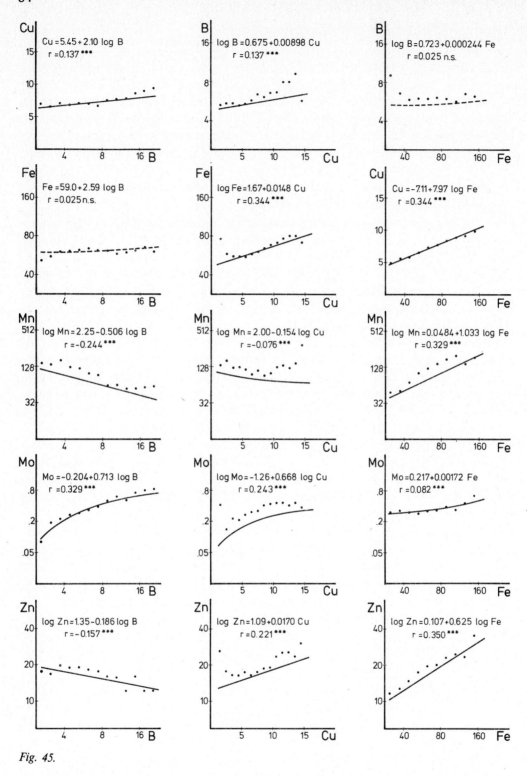

Fig. 45.

Figs 45 and 46. Mutual correlations between B, Cu, Fe, Mn, Mo, and Zn contents of pot-grown wheat (ppm) in the whole international material (n = 3536). Points indicate arithmetic mean contents of the micronutrients given on the y axis in content classes of the micronutrient given in the x axis.

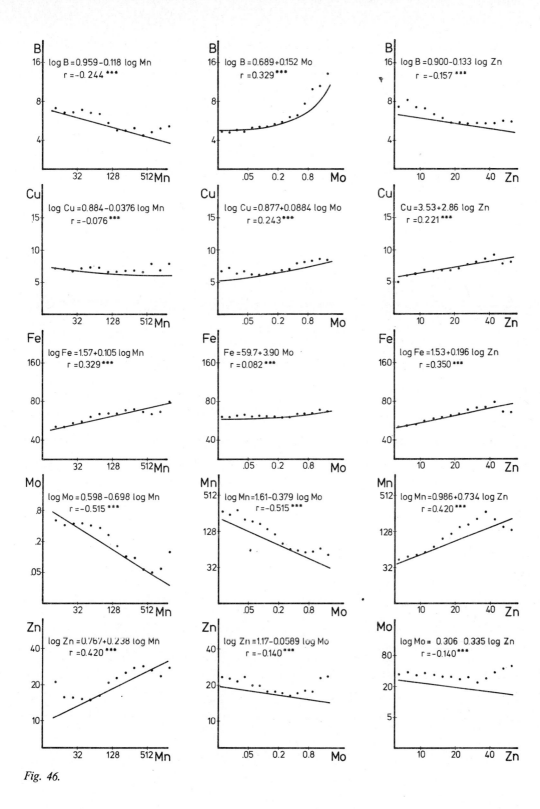

Fig. 46.

Table 20. Mutual correlations (r values) between extractable[1] soil micronutrients in the whole material (n = 3536). Correlations of the best fitting regressions are given. Regression models are indicated by letters: a) $y = a + bx$; b) $y = a + b \log x$; c) $\log y = a + bx$; and d) $\log y = a + b \log x$.

y \ x		B		Cu		Fe		Mn		Mo
Cu	c	0.069***		—						
Fe	d	—0.178***	d	0.270***		—				
Mn	d	—0.201***	d	—0.251***	d	0.222***		—		
Mo	d	0.365***	d	0.469***	d	0.037*	d	—0.390***		—
Zn_1	c	—0.089***	c	0.150***	d	0.347***	d	0.346***	d	0.015 n.s.
Zn_2	d	0.038*	c	0.186***	d	0.380***	d	0.299***	d	0.064***

[1] Extraction methods used:
B: hot water extr. + CEC correction
Cu: AAAc-EDTA + Org. C correction
Fe: AAAc-EDTA
Mn: DTPA + pH correction
Mo: AO-OA + pH correction
Zn_1: DTPA
Zn_2: AAAc-EDTA + pH correction[2]

[2] The mutual correlation (r) between the two extraction methods for Zn is 0.911***.

(r values 0.365*** and —0.201***) existed between the extractable soil contents of these micronutrients (Table 20).

In addition to the Cu—Fe correlation mentioned above, plant Cu correlated well with plant Mo (r = 0.243***) and with plant Zn (r = 0.221***). Especially good correlation was found between soil Cu and Mo (r = 0.469***); those between soil Cu and Zn (r values 0.150*** and 0.186***) were also relatively high. The other correlations between the plant contents of the six micronutrients were highly significant with only one exception (B—Fe), but compared to those specifically cited above these relations are less firm.

In general, taking into account the very small concentrations of micronutrients in soils, the possibility that one micronutrient directly affects the availability of another seems less likely than that their mutual relationships are indirectly determined by other soil factors affecting their behaviour. However, on the basis of the present data the possibility of direct chemical effects between two or more micronutrients cannot be entirely ruled out.

2.5 Micronutrient contents of plants and soils in relation to yields with special reference to the "concentration-dilution" phenomenon

Factors determining the yield of a plant are so numerous that the effects of a single micronutrient are almost inseparable from those of other factors simultaneously affecting the growth of the plants used in this study. However, there are certain matters which must be taken into consideration when dealing with yields of pot-grown wheat and their relationships to micronutrients. Samples of the plants acting as indicators had to be harvested at an early stage of growth when they had only developed vegetatively (see Section 1.2). At this time, the early dry matter (DM) yield may not always be a reliable measure of the final grain yield since the shortage of some micronutrients, if not severe, may appear first at relatively late stages of growth, thus limiting the grain more than the vegetative yield.

Taking into account the above comment on micronutrient-yield relations, in this section the yields are not presented as functions of micronutrients but *vice versa*. The micronutrient contents of plants and soils as functions of yield are of special interest in connection with the interpretation of the results of plant analyses. For example Cottenie (1980) pointed out that enrichment of a nutrient in a plant may be due to a high nutrient level in the soil or to reduced growth, and Tölgyesi and Mikó (1977) found that with increasing maize yields the uptake of nutrients increased but their concentration in plant DM decreased. The contents of the six micronutrients in plants and soils as functions of yield are presented in Fig. 47. To obtain a better quantitative illustration of the effects of yields on the micronutrient contents of plants and soils in relation to their

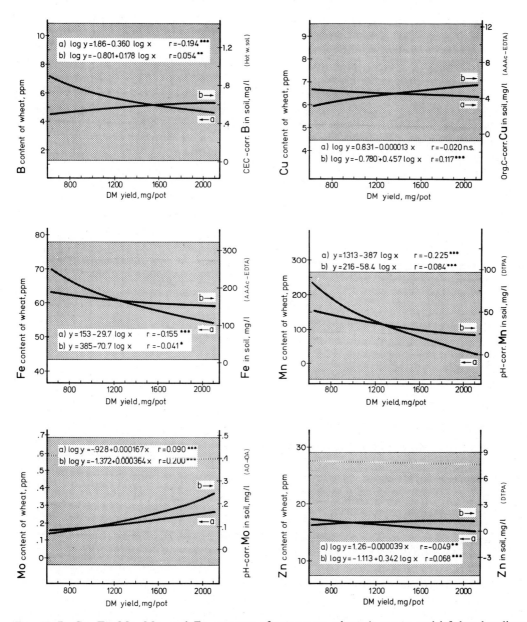

Fig. 47. B, Cu, Fe, Mn, Mo, and Zn contents of pot-grown wheat (a-curves and left-hand ordinate) and respective extractable soil micronutrients (b-curves and right-hand ordinate) as a function of yield (n = 3537). The hatched areas indicate the ranges of the standard deviations ($\bar{y} \pm s$) for both the plant and the soil micronutrient contents.

variations, the scales for the graphs have been adjusted so that the ranges of the standard deviations ($\bar{y} \pm s$) of the plant and the soil contents are equally wide and are indicated by the hatched areas. (For example, plant B content 6.09 ± 4.80 ppm and soil B 0.73 ± 0.72 mg/l, as in Fig. 25).

The behaviour of B, Cu and Zn is similar in that their soil concentrations increase with increasing yield but the concentrations in plants decrease. In the cases of Fe and Mn both the plant and soil contents decrease and the contents of Mo increase with increasing yield. The general direction of the regression curves, however, is of little importance because apart from the single micronutrient in question other factors are involved.

The most important information to be drawn from Fig. 47 concerns the mutual relationship between curves a and b. For all six micronutrients, the slopes of the *a* curves (regressions of plant contents on yield) show a relatively greater rate of decrease or a smaller rate of increase than the respective *b* curves (regressions of soil contents on yield). In other words, at low yield levels plants absorb more micronutrients per unit of produced plant mass in relation to available soil micronutrients than they do at high levels of yield. The principles of these relations, called here as **concentration-dilution** phenomenon, can be explained as follows:

Restricted plant growth (low yield) in most cases is often due to some factor other than the micronutrient in question. In such cases micronutrients which are available in soil in high or reasonable quantities tend to **concentrate** in the slowly produced plant mass, so raising the contents of these micronutrients to a higher level than when plant growth is normal or rapid. Therefore, as in this material, most of the very high plant micronutrient concentrations are to be found in those of restricted growth or in plants grown on soils where the micronutrient in question was abundantly available or, most often, where the effects of both these factors were combined.

When the plant growth is rapid (high yields), the plants are apparently unable to absorb micronutrients from soil in quantities related to the mass of DM produced. Consequently, portions of the micronutrients already absorbed are **diluted** in the increased DM mass and the micronutrient contents of the plants in relation to available micronutrients in soil decrease with increasing yields.

It is clear that the concentration-dilution phenomenon is in main principle the same as that prevailing during the growth of any plant in that variations in the micronutrient concentrations of a plant are related to the speed at which that plant produces new DM (see Section 1.2.2).

Because the extractable micronutrient contents of soils are independent of plant yield, the differences between the results of plant and soil analyses due to the concentration-dilution phenomenon are to be considered as a source of error affecting the results of plant analyses only. To compare the relative importance of the effects of yield (concentration-dilution) and the extractable soil micronutrient contents on the contents of micronutrients in plants, stepwise multiple regressions including these two variables were computed. The percentages of variation explained by the two characteristics were:

Characteristic	Percentage of variation explained					
	B	Cu	Fe	Mn	Mo	Zn
Soil micronutrient content alone	68.2	53.4	10.5	50.8	48.4	53.6
Soil micronutrient content + yield	69.4	54.6	10.6	51.0	48.5	54.6

The above comparisons show the relatively unimportant role of yield in explaining the **total variations** in plant micronutrient contents in this material.

For further estimation of the quantitative role of the concentration-dilution phenomenon, regressions of plant micronutrient contents on respective soil contents were computed separately for low yields (<1100 mg/pot) and high yields (>1500 mg/pot), each group consisting of about one-fifth of the whole material. Since in case of B the effect of yield was more pronounced than for other micronutrients, these regressions for B are given in Fig. 48 as example. The regression line calculated for high yields (curve c) is located at a level about 25 percent lower than that calculated for low yields (curve b). The locations of curves b and c on both sides of curve a (whole material) give, however, a rough visual picture of one source of variation in the plant B—soil B correlation, i.e. that of the concentration-dilution phenomenon. On the average, this difference in location of the curves is quantitatively relatively unimportant in comparison with the total variation of plant B contents. However, in cases of some single samples it may be quite substantial as mentioned before.

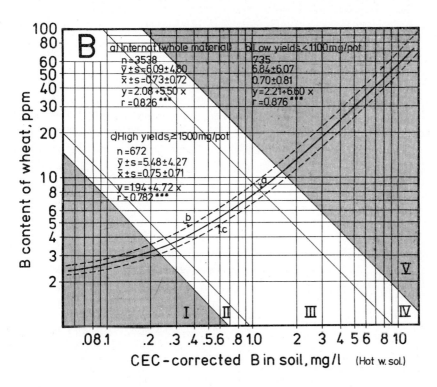

Fig. 48. Regressions of B content of pot-grown wheat on hot water soluble soil B corrected for CEC, (a) all samples, (b) low yielding samples, (c) high yielding samples.

2.6 Suitability of plant and soil analyses for large scale operations

A combination of plant and soil analyses offers a better means of estimating the micronutrient status of soils than either alone, but the improvement in quality is gained at the cost of quantity and *vice versa*. A decision to choose one or other of the two techniques must therefore often be made.

Plant analysis holds a **theoretical** advantage over soil analysis in that the micronutrient fractions found in plants have indeed been in the soil in forms available to plants. This advantage is offset by **practical difficulties** arising with work on a large scale. One such is the need for standardization of the plant material which must meet all requirements (see Sections 1.2 and 2.3.1) if comparable analytical data are to be obtained. Uniform standards are almost impossible to meet in practice. Even in small countries more than a dozen important crops are grown; comprehensive interpretation systems have been developed for only relatively few plant species, very little is known about differences between plant varieties, and no general agreement has been reached as to which parts of different plants should be analysed. One of the greatest drawbacks to using plant analysis to diagnose the micronutrient status of soils may be that plant samples, in order to obtain comparable results, must be taken when the plants are at the same physiological age. The sampling time is thus limited to a very short period. Plants grown in the field are more vulnerable to contamination than soil samples.

The limitations to the applicability of plant analysis on a large scale do not apply to soil analysis. Soil samples can be taken irrespective of the crop or variety growing on the land and the sampling time is not limited to any special period. Soil analysis is therefore well suited for use on a large scale but its value depends on how closely the data indicate the availability of the nutrients to plants. The basic differences between the two techniques for diagnosing the micronutrient status of soils are explained in Section 2.3 and the principles and quantitative measures needed to harmonize their results are presented in Sections 2.3.2 to 2.3.7. The aim has been to develop soil analyses in such a way that the findings would be consistent with those from the analysis of plants that meet all requirements and so combine in one method the advantages of both techniques. Whatever technique(s) may be adopted, the findings can only be considered indicative. There is a need for calibration by field trials with different crops to improve interpretation and establish critical deficiency and toxicity limits.

2.7 Estimation of critical micronutrient limits

As explained in Section 1.4.3 the zone limits presented in plant-soil micronutrient graphs are statistically defined, dividing the present international micronutrient data into five zones which do not however reveal critical deficiency or toxicity limits. Depending on micronutrient and crop, these zone limits may be too high or too low. Furthermore, the zone limits in this study are based on both the plant analyses and the soil analyses. In practical micronutrient studies both these data are seldom available and estimates of critical limits are usually based on one or other diagnosis alone.

In order to obtain a general understanding of the relative differences between the micronutrient contents of various plant species, an experiment was established at nine sites in different parts of Finland where 17 crops were grown side by side at each site. The micronutrient contents of the plants were analysed with the same methods as used in the present study, except that most of the plants were analysed at maturity. The results of the experiment will be published later (Yläranta and Sillanpää) but some preliminary data are given in Table 21.

Table 21. Two-year averages of micronutrient contents (ppm in DM) of 17 crops grown side by side at nine sites (some of the crops were successfully grown at only 7 or 8 sites).

Crop and part	No. of sites	B $\bar{x} \pm s$	Cu $\bar{x} \pm s$	Fe $\bar{x} \pm s$	Mn $\bar{x} \pm s$	Mo $\bar{x} \pm s$	Zn $\bar{x} \pm s$
Spring wheat, grain	9	1.7 0.5	6.7 2.2	60 15	86 37	0.20 0.11	56 21
" ", straw	9	2.2 0.8	3.4 1.2	69 21	87 43	0.45 0.43	31 26
Winter wheat, grain	9	1.6 0.2	6.2 2.4	66 40	76 27	0.26 0.12	48 11
" ", straw	9	2.1 0.8	3.8 1.8	86 47	106 56	0.43 0.27	33 20
Oats, grain	9	1.4 0.4	3.8 0.9	60 13	70 23	0.40 0.19	42 11
", straw	9	2.6 0.5	3.5 1.2	79 37	98 52	0.45 0.27	21 7
Barley, grain	9	1.4 0.2	7.1 2.2	52 10	24 6	0.33 0.16	37 14
", straw	9	3.4 0.6	5.0 2.3	73 15	61 37	0.37 0.31	28 20
Rye, grain	9	2.2 0.2	7.4 1.1	53 10	39 8	0.40 0.17	50 9
", straw	9	3.1 0.8	4.3 1.4	70 23	55 22	0.61 0.17	35 22
Timothy, silage, 1. cut	9	8.7 2.2	6.0 1.6	74 18	69 28	0.47 0.39	29 7
" , " 2. cut	8	8.0 2.2	5.9 1.4	81 22	113 31	0.91 0.57	30 8
" , dry hay	9	6.6 1.3	4.9 1.3	64 18	71 35	0.41 0.36	27 10
" , fresh growth	9	6.7 2.1	5.4 1.2	109 39	128 41	1.13 0.65	32 12
Rye grass, silage, 1. cut	8	8.5 2.4	7.1 2.4	164 50	92 31	0.81 0.56	22 5
" " , 2. cut	7	6.7 1.7	6.8 1.1	190 92	149 59	1.34 0.96	22 2
Red clover, dry hay	7	26.4 6.4	10.0 3.0	75 11	87 53	0.61 0.42	42 14
" , fresh growth	7	24.1 8.0	13.1 3.9	105 19	119 86	0.69 0.62	48 17
Pea, seed	8	7.5 1.0	9.3 1.9	86 22	21 8	1.84 1.43	46 6
" , stalk	9	29.4 9.2	8.7 2.4	153 34	133 100	1.24 1.20	81 47
Turnip rape, seed	7	13.2 2.2	4.9 0.6	73 8	36 13	0.27 0.10	42 7
" " , stalk	7	17.6 6.2	3.3 0.5	47 12	45 42	0.59 0.45	24 12
Turnip, root	9	20.8 4.9	4.5 0.8	44 12	19 13	0.19 0.12	29 14
Swede, root	9	18.6 4.1	3.1 0.5	45 18	19 10	0.14 0.06	19 8
" , tops	9	25.1 6.4	4.9 0.8	150 81	94 62	0.74 0.39	34 11
Sugar beet, root	7	13.5 0.8	4.4 1.2	37 15	92 54	0.07 0.03	29 18
" " , tops	7	26.3 7.8	8.6 1.6	162 43	277 178	0.39 0.22	109 69
Red beet, root	8	17.9 2.0	7.7 1.2	71 26	113 48	0.09 0.05	52 37
Carrot, root	9	19.3 2.8	4.6 0.7	54 29	28 11	0.04 0.01	20 5
Potato, tuber	9	6.2 0.6	5.6 1.6	41 10	8 2	0.15 0.08	16 4
Onion, bulb	9	12.1 2.0	4.7 0.7	29 7	24 13	0.16 0.25	25 11

These data show that different plant species can absorb widely different amounts of micronutrients from the same soils. Critical deficiency or toxicity limits established for one plant species cannot therefore be generally applied to others, and it would be necessary to develop different analytical interpretation systems for almost every plant species. These should include clearly specified stages of growth at which the plant is to be sampled and which part is to be analysed as well as details of analytical methods. For some crops and micronutrients systems of this kind have been fairly well established but are lacking or vague for the great majority of crops.

An extensive review of literature on the interpretation of results of plant analyses, including the six micronutrients dealt with in this study, was presented by Bergmann and Neubert (1976). A similar review concerning boron was recently published by Gupta (1979) who also stated that in practice there can be no single value or even a very narrow range of values to describe "critical" levels in crops. A value considered critical by workers in certain areas may not be critical under conditions in other areas, and for certain elements the margin between deficiency and sufficiency is so narrow as to cause an overlapping of values.

Most of the differences between the values considered critical for a certain crop by various investigators are clearly due to differences between the analytical procedures and methods adopted and not only to different environmental conditions. Reference to the above two reviews often shows differences of more than tenfold between the values considered critical (deficient or toxic). For example, were the lowest critical (deficiency) B level given for wheat (0.3 ppm) to be comparable to plant B values obtained in the present study, there would be no B deficiency in the 30 countries studied. Again, if the highest deficiency limit referred to (5 ppm) were to be correct, about half of these soils studied would be deficient in B. The toxicity B levels in wheat referred to by the above reviewers vary from 16 to 100 ppm.

Typical B contents of healthy and B-deficient plants were also reviewed for a number of crops by Brandenburg and Koronowski (1969). For example, the B contents of healthy sugarbeets on a DM basis were 25—40 ppm in leaves and 15 ppm in roots; while those for B-deficient plants were 13—20 and 13 ppm, respectively. Healthy mangel-wurzel contained 20—46 ppm in leaves and 17 in roots (B-deficient 7—18 and 16 ppm, respectively); turnips 30—40 ppm in leaves and 18—22 in roots (B-deficient 9—20 and 8—15 ppm, respectively); potato leaves 14—30 ppm (B-deficient 4 ppm); lucerne 20—29 ppm (B-deficient 7—19 ppm); apple 20—25 ppm in leaves and 16—26 in fruit (B-deficient 12—16 and 2—6 ppm, respectively); tobacco leaves 16—50 ppm (B-deficient 4 ppm); celery leaves 26—38 and tubers 19—29 ppm (B-deficient 15 and 13 ppm, respectively).

Since the situation with regard to other micronutrients studied is very similar to that of B, the range of plant micronutrient values obtained from the literature is undoubtedly too wide to help in establishing precise critical ranges for values obtained in this study.

Comparison of the results of soil analysis in the present study with those of others is also limited because of the wide variety of extraction methods used by different investigators. Boron is an exception in this respect; the hot water extraction method has been widely used and estimates of critical soil B values have been presented. For example, Reisenauer et. al. (1973), referring to several sources, and Kurki (1979) considered hot water extractable soil B values < 1 ppm not high enough for optimum plant growth, and Smilde (1976) has presented a value of 0.3 ppm. Severe B deficiency (more than 100 percent yield response to B) has been reported in soil with a hot water extractable soil B content of 0.15 ppm (Hong 1972). According to several authors (e.g. Bould and Hewitt

1963, Jackson 1964, Mitchell 1974, and Park and Park 1966) the deficiency limit may be in the range of 0.5 ppm depending on soil factors and plant species.[1]

Many investigators consider B deficiency to be more widespread than that of any other micronutrient. For example, in the United States the most commonly found micronutrient deficiency was that of B. It occurred in one or more crops in 41 states, and in Wisconsin alone nearly two million acres (about 800 000 ha) of alfalfa showed B deficiency at one time or another during the growing season (Berger 1962). Against this background, the lower five percent zone limit and even the lower 10 percent zone limit seem for many plants to be too low to be considered as deficiency limits or values below which B deficiency can be suspected. Tentatively in this study hot water extractable B values < 0.3—0.5 mg/l are considered suspect for B deficiency.

The margins between the amounts of B required for normal growth and those producing symptoms of excess or toxicity vary with plant species and are considered rather narrow for B in comparison to most other micronutrients. Critical, excessive or toxic, hot water extractable soil B values varying often from 2.5 to 5 ppm have been recorded (Anderson and Boswell 1968, Gupta and Munro 1969, Kurki 1979, Mortvedt and Osborn 1965). In general, data on B toxicity seem to be more rare than those on deficiency. Many of the plant—soil B values falling in Zone IV or the lower part of Zone V may therefore still not be considered high enough to indicate conditions of B excess, especially for crops with high tolerance to B.

Very little appears to be known about deficiency or toxicity limits for Mn based on DTPA soil extraction since the method has been in use for a relatively short period of time. Lindsay and Norvell (1978) when characterizing the possible usefulness of the DTPA soil test for Mn, suggested that the critical level of DTPA extractable Mn could be tentatively set at 1 ppm until further information is available. Compared to the DTPA data of the present study, this value seems very low since the minimum value recorded was 0.9 mg/l and in only a very few soils out of more than 3500 soils from 30 countries was Mn as low as 1 ppm. Nevertheless, Mn deficiency has been reported from numerous countries and symptoms of Mn deficiency have been described for more than 20 crops including oats, rye, wheat, rice, maize, peas, soya beans, potatoes, tomatoes, cotton, tobacco, sugarbeet, tea, sugar-cane, pineapples, pecan, peaches, spinach, citrus and a number of forest trees (e.g. Bergmann and Neubert 1976, Koronowski 1969, and Spraque 1964).

Against the above background, instead of DTPA values of < 1 mg/l, values of 2 or 3 seem a more appropriate deficiency limit for Mn and values of up to 4—5 mg/l may still be considered to indicate susceptibility. For pH-corrected DTPA the respective figures would be somewhat lower. The above figures are still tentative and subject to the soil material in this study being representative enough and the soil variation range wide enough for such estimates. The widely varying Mn requirements of different plant species should also be kept in mind.

DTPA values indicating excess or toxicity of Mn are not well established. Therefore, it cannot be stated whether the upper 10 or 5 percent zone lines of this study are too high or too low for predicting Mn excess. Tentatively, until further experience is gained, DTPA values exceeding 140—200 mg/l may be considered suspiciously high.

DTPA has been quite widely used to extract Zn from soils and, therefore, information on the interpretation of DTPA Zn values is more abundant than for other micronutrients.

---

[1] Taking into account that only about 0.2 percent of the soil material of this study were peat soils with a very low volume weight and also the effect of extraction ratio, the numerical ppm values can be considered about equal to those of mg/l.

Since Zn toxicity is generally not considered as serious a problem as Zn deficiency, the interpretation data given in literature concern mainly the estimation of critical deficiency level. Depending on the crops and other factors, various authors have proposed somewhat different DTPA values as critical deficiency limits, but in general, this limit seems to be relatively well established compared to most other micronutrients. Singh *et al.* (1977) considered 1.4 ppm DTPA extractable Zn as the critical limit for maize in pot trials. Randhava and Takkar (1975) proposed different deficiency limits for different Indian soils and crops (wheat, maize and rice) varying from 0.5 ppm (Deep and shallow black) to 1.0 ppm (Sierozem). Sedberry *et al.* (1980) found a response of rice to Zn on soils containing less than 1 ppm DTPA extractable Zn. Lindsay and Norvell (1978) considered 0.6 ppm sufficient for sorghum but a somewhat higher value (0.8 ppm) for maize. Viets and Lindsay (1973) estimated that less than 0.5 ppm was critical and 0.5—1.0 marginal for sensitive crops. Several other authors (Brown *et al.*, 1971, Rathore *et al.* 1978, Sakal *et al.*, 1979, Whitney *et al.*, 1973, Whitney, 1980) have come to the conclusion that DTPA extractable Zn values of about 0.5 ppm represent the critical deficiency limit for several crops. A Zn value of 0.5 ppm ($\approx$0.5 mg/l) of DTPA extractable Zn in this study would correspond to a Zn content of about 14 ppm in the pot-grown wheat and somewhat higher contents in original wheat and maize (see Zn graphs and Section 2.3.7). These figures are in relatively good agreement also with many recent data concerning deficiency limits based on plant Zn content. For example, Franck and Finck (1980), Rathore *et al.* (1978), Takkar *et al.* (1974) and Weir and Milham (1978) gave plant Zn contents varying from 12 to 20 ppm in wheat and maize as deficiency limits for Zn.

If a DTPA value of 0.5 mg/l is considered to be the deficiency limit for Zn and applied to the DTPA data of this study, it would mean that even the lower 10 percent zone line (corresponding to about 0.25 mg/l DTPA extractable Zn) given in the Zn graphs is located far too low to indicate the critical deficiency limit, and instead of 10 percent about one-third of the soils in the 30 countries could be suspected to be deficient in Zn. This would not be surprising since during recent years an increasing amount of information has emerged on Zn deficiency, especially from developing countries.

Zn toxicity may be relatively rare and little information on toxic DTPA Zn levels is available. According to Takkar and Mann (1978), DTPA values of 7 and 11 ppm and plant Zn contents of 60 and 81 ppm may be considered to be levels at which wheat and maize, respectively, are susceptible.

Since in this study the results given for soil Cu and Fe are based on AAAc-EDTA extraction which is a relatively new method, there are very few reference points in the literature concerning the deficiency or toxicity limits for this extractant. Therefore, direct comparisons to earlier published data are not possible. However, these elements were also extracted with DTPA from the whole of the international soil material of this study (see Sections 2.3.4 and 2.3.5) which makes some indirect approximations of deficiency limits possible.

For DTPA extractable soil Cu, 0.2 ppm has been considered as a limit below which plants are likely to suffer from Cu deficiency (Viets and Lindsay, 1973, Lindsay and Norvell, 1978). Of the DTPA Cu data of this study, varying from 0.03 to 100 mg/l (mean 1.8 mg/l), about 5 percent of soils show values lower than 0.2 mg/l indicating that in this material approximately 5 percent of soils are suspect for Cu deficiency. Thus, the lower 5 percent line given in the Cu graphs would give a rough estimate of the deficiency limit for Cu. When converted to AAAc-EDTA and organic C-corrected AAAc-EDTA Cu values, the deficiency limits would be in the range of 0.8—1.0 mg/l which according to the regression lines in the Cu graphs would correspond to wheat Cu contents of 4 ppm or slightly less. The latter figure is somewhat higher than is considered to be a critical deficiency level for wheat e.g. by Mitchell (1974) and Panin *et al.* (1971) who estimated Cu contents of 2.5—3.0 ppm as critical limits. The latter authors considered 11.3 ppm in wheat as a critical level of Cu toxicity but both lower and higher figures have been presented.

The DTPA Fe values measured from the soil material of this study varied from 1 to 586 mg/l, averaging 37 mg/l. Lindsay and Norvell (1978) expected soils with >4.5 ppm Fe not to show Fe deficiency. The present material includes about 5 percent of soils with DTPA Fe values lower than that. Since the correlation between DTPA Fe and AAAc-EDTA Fe is relatively good (r = 0.741***) the lower 5 percent line in Fe graphs could be tentatively considered as a line to separate cases of possible Fe deficiency and sufficiency. Converted to AAAc-EDTA extractable Fe values this would correspond to some 30—35 mg/l.

Estimation of available Mo status of soils based on AO-OA extraction is rather difficult because the results obtained with this method as such do not correlate well with the results of plant analyses of Mo (see Section 2.3.2). By correcting the AO-OA extractable soil Mo values for pH the plant-soil correlation is considerably improved, but there are no previous data for comparison or for assessment of any deficiency or toxicity limits for the pH-corrected soil Mo values. Furthermore, individual crops differ markedly in their Mo requirement, the *Cruciferae* and legumes usually considered to be the most demanding and monocots the least. Critical Mo deficiency or toxicity levels in different plant species are not well defined. For example, in the literature review by Bergmann and Neubert (1976) Mo contents of wheat considered as low varied from 0.09 to 0.7 and those considered as high from 0.18 to 1.5 ppm.

On the basis of the above it seems impossible to identify any soil or plant Mo values as critical limits. Therefore, until further information is available, plant—soil Mo values falling in the lowest and the highest Mo zones (I and V) may be considered indicative of possible Mo deficiency or excess, after taking into account the plant species concerned. The respective limits as applied to pH-corrected AO-OA extractable Mo data would be in the range of 0.01—0.02 mg/l for deficiency and 0.5—1.0 mg/l for excess.

The above discussion concerning the estimation of deficiency and toxicity limits for micronutrient data of this study can be briefly summarized as follows:

The statistically defined lower 5 and even the lower 10 percent limits given in B and Zn graphs seem to be located at too low a level to be considered as critical deficiency limits for B and Zn, nor may B and Zn values falling in Zone IV and lower parts of Zone V necessarily indicate an excess of these micronutrients.

For the other four micronutrients (Cu, Fe, Mn and Mo) the data obtained from literature either indicate that the lower 5 percent lines locate fairly well within the ranges below which deficiencies of these micronutrients are likely to exist, or the data are too heterogeneous to give clear indications of these lines being located at too low or too high levels. The same concerns the upper statistical 5 percent lines for Cu, Mn, and Mo and possible excess of these elements.

Converted into soil micronutrient data obtained by various extraction methods used in this study, the critical levels (mg/l) are estimated as shown in Table 22.

Table 22. Tentative critical levels of micronutrients determined by various methods.

Micronutrient and method	Range of deficiency	excess
B, hot water extraction	< 0.3—0.5	> 3—5
B, hot w. extr. + CEC correction	< 0.3—0.5	> 3—5
Cu, AAAc-EDTA	< 0.8—1.0	> 17—25
Cu, AAAc-EDTA + org. C correction	< 0.8—1.0	> 17—25
Fe, AAAc-EDTA	< 30—35	—
Mn, DTPA	< 2—5	> 140—200
Mn, DTPA + pH correction	< 2—4	> 150—220
Mo, AO-OA + pH correction	< 0.01—0.02	> 0.5—1.0
Zn, DTPA	< 0.4—0.6	> 10—20
Zn, AAAc-EDTA + pH correction	< 1.0—1.5	> 20—30

The values are tentative only and are subject to revision when more information is available. It is important when interpreting the results to take into account the varying micronutrient requirements of different crop species.

2.8 Plant and soil micronutrients and other soil data in relation to FAO/Unesco soil units

Classification of soils of the present study by FAO/Unesco soil "Orders" (Dudal, 1968) was carried out in the field by the sample collectors in the various participating countries. The data are incomplete because all sample collectors did not consider themselves competent enough to classify the soils. Out of 3744 soils 2265 were classified (Appendix 6). Further, there is bound to be a certain degree of heterogeneity in the classification of the soil material because more than 100 sample collectors participated in classifying the soils, all of them not equally highly qualified in this particular field. In some cases when the soils were classified by sample collectors according to the national classification system only, experts at FAO Headquarters were able to convert these into FAO/Unesco soil units. Since full profile descriptions were not available, coordination of the classified soil data and correction of possible errors were limited only to a few orders (e.g. Chernozems, Phaeozems and Rendzinas) having diagnostic features within the surface soil layers.

Because relatively little information on relations of micronutrients to FAO/Unesco soil units is available, it was considered justifiable to publish these results. However, taking into account the foregoing reservations, the data can only be considered as indicative, and therefore only the average micronutrient contents of soils and respective plants for various soil units are presented in Appendix 7. It should be noted that the background data in the graphs (regression curves, mean, standard deviations (s), zones) are calculated from the whole of the pot experiment data (n = 3538) but the soil unit data plotted in the graphs represent the number of soils classified (n = 2265) only. The average data concerning general soil characteristics and macronutrients are given in Appendix 6.

3. Summary for Part I

The original sample material collected from 30 countries totalled 7488 samples, half of which were soils and half plant samples growing on those soils. The plant (wheat and maize) samples, however, proved too heterogeneous to reflect reliably the micronutrient (B, Cu, Fe, Mn, Mg, Zn) status of soils where the plants had grown. Among the factors causing uncontrolled variation in analytical results were: differences due to the two plant species and cultivars (over 200 maize and over 200 wheat cultivars); variation due to differences in the physiological age of plants when sampled; differences in main nutrient fertilization, yield levels, irrigation, use of herbicides possibly containing micronutrients; contamination during sampling, sample pre-treatment and transportation. Therefore, to minimize the uncontrolled variation, fresh indicator plants (wheat, cv. 'Apu') were grown in pots on the original soils under uniform and controlled conditions. Owing to the superiority of this plant material, the discussion concerning micronutrients is based on the analytical results of the new indicator plants. After preliminary methodological studies from limited materials, soil micronutrients were extracted from all the soil samples as follows: B with the hot water extraction method, Mo with ammonium oxalate-oxalic acid (AO-OA), and Cu, Fe, Mn and Zn with both the acid ammonium acetate-EDTA (AAAc-EDTA) and the DTPA extraction methods.

The correlations between the results of plant and soil analyses for all six micronutrients were considerably improved when data from the new pot-grown plants were used instead of data from the original indicator plants (Table 23).

Table 23. Correlations between micronutrient contents of the three types of indicator plants (original maize and wheat and pot-grown wheat) and respective soils.

Micronutrient and extr. method		Original maize (n = 1966)	Original wheat (1768)	Pot-grown wheat (3537—3538)
B	hot water extr.	0.548***	0.694***	0.741***
Cu	AAAc-EDTA	0.344***	0.254***	0.664***
Cu	DTPA	0.114***	0.125***	0.518***
Fe	AAAc-EDTA	–0.007 n.s.	–0.032 n.s.	0.325***
Fe	DTPA	–0.028 n.s.	–0.030 n.s.	0.263***
Mn	AAAc-EDTA	–0.108***	0.046*	0.039*
Mn	DTPA	0.161***	0.119***	0.552***
Mo	AO-OA	0.134***	0.117***	0.245***
Zn	AAAc-EDTA	0.391***	0.405***	0.665***
Zn	DTPA	0.381***	0.473***	0.732***

In spite of the improvements in correlations, satisfactory conformity between the results of plant and soil analyses was not achieved.

To understand the contradictions still prevailing between the results of plant analyses and soil analyses, it must be realized that these two techniques are based on fundamentally different principles. Plant analyses give measures of micronutrients which have been available to and have indeed been absorbed by plants. Absorption of micronutrients by plants is a process that takes place under the laws of biochemistry and plant physiology while chemical soil extraction mainly follows the laws the chemistry. Accordingly the results of plant analyses (if based on reliable and comparable plant material) are to be considered as the most reliable measure of the soil micronutrient fractions available to plants. Soil analysis in more of a "shortcut" method and an attempt to imitate plants. Therefore, if the results of plant analyses obtained from reliable and comparable plant material contradict those of soil analyses, and if the responsible factor (which affects plant and soil analyses in different ways) can be identified and its effects quantified, it is the soil analysis which must be corrected accordingly.

In large scale micronutrient studies, results of both plant and soil analyses are seldom available and one or other technique must be chosen. The theoretical advantages of plant analyses are diminished by the practical difficulties of obtaining comparable plant material or of interpreting results from the analysis of heterogeneous materials. For example, comprehensive interpretation systems have only been developed for relatively few plant species, very little is known about differences between plant varieties and no general agreements have been reached as to which parts of different plants are to be analysed. Furthermore, to obtain comparable results plants must be sampled at the same physiological age which severely limits the sampling time. In contrast, soil samples can be taken irrespective of the crop or variety growing on it and the sampling time is not limited to any special period. Soil analyses are therefore well suited for large scale use, but their applicability depends on how closely their results agree with those from plants. The major aim in this study has therefore been to devise means of calibrating soil analyses so as to eliminate discrepancies between plant and soil analyses.

To identify the factors responsible for impairing the plant-soil micronutrient correlations, and to quantify and eliminate their contradictory effects, both the plant micronutrient data and the soil (extractable) micronutrient data were studied in relation to six soil characteristics (pH, texture, organic carbon content, CEC, electrical conductivity and $CaCO_3$ equivalent). These studies covering all the international soil samples, showed that in the case of hot water soluble B such a soil factor was the cation exchange capacity and for AAAc-EDTA extractable Cu the organic carbon content of the soil. The DTPA extractable Mn, AO-OA extractable Mo and AAAc-EDTA extractable Zn required correction for soil pH. The principles and procedures for such corrections are presented in detail. These adjustments should be considered as an essential part of soil analysis.

The mathematical equations of correction coefficients for the above micronutrient extraction methods are:

Boron: $\quad k(CEC) = 10^{0.466 - 0.026 x + 0.000273 x^2}$
where x = cation exchange capacity (me/100 g)

Copper: $\quad k(Org. C) = 10^{-0.491 \log x + 0.470 (\log x)^2}$
where x = soil organic carbon content (%),

Manganese: $\quad k(pH) = 10^{7.06 - 2.20 \, pH + 0.164 \, pH^2}$

Molybdenum: $\quad k(pH) = 10^{-2.45 + 0.36 \, pH}$

Zinc: $\quad k(pH) = 10^{3.38 - 0.976 \, pH + 0.068 \, pH^2}$

These five correction coefficients can also be obtained directly from graphs in Figs 49 and 50.

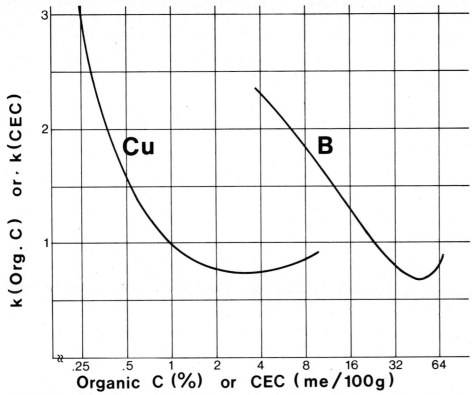

Fig. 49. Coefficients for correcting AAAc-EDTA extractable soil Cu for soil organic carbon content, k(org. C); and hot water soluble B for cation exchange capacity, k(CEC).

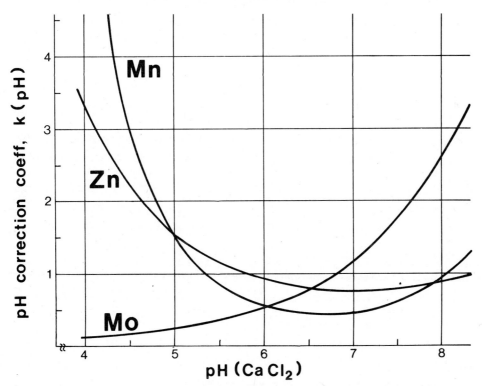

Fig. 50. Coefficients for correcting DTPA extractable soil Mn, AO-OA extractable Mo and AAAc-EDTA extractable Zn for soil pH.

Application of these correction coefficients to the whole of the international material (n = 3537 – 3538) improved the correlations (r values) between the results of plant analyses (pot-grown plant material) and soil analyses as follows:

Micronutrient and extr. method		r value without correction	Correction coefficient	r value with correction
B	hot water extr.	0.741***	k(CEC)	0.826***
Cu	AAAc-EDTA	0.664***	k(Org. C)	0.731***
Mn	DTPA	0.552***	k(pH)	0.713***
Mo	AO-OA	0.245***	k(pH)	0.696***
Zn	AAAc-EDTA	0.665***	k(pH)	0.706***

In addition to the uncorrected data the above correction coefficients are applied when presenting the micronutrient data by countries in Part II. Correction coefficients calculated for AAAc-EDTA extractable Fe and DTPA extractable Zn did not appreciably improve the respective plant-soil correlations and were not adopted. The most effective correction coefficient was that calculated for AAAc-EDTA extractable Mn for pH which improved the plant Mn—soil Mn correlation (r) from 0.039* to 0.588***. This also shows that even an extraction method which in itself gives poor results may be successfully used if its special features and its behaviour with respect to various soil factors (in this case the pH) are known and properly taken into account.

As the DTPA extractable Mn generally correlated better with plant Mn the results of AAAc-EDTA Mn are not presented in Part II.

The mutual relations between the contents of the six micronutrients in plants as well as those extractable from soils are presented. In general, if there is a good correlation between the contents of two micronutrients in plants or those extracted from soil, their availability to plants or their extractability from soils is largely controlled by the same soil factor or factors.

The micronutrient contents of plants and soils in relation to yield are presented with a special reference to the effect of yield on the micronutrient contents of plants. In general, plants with restricted growth (low yields) contain more micronutrients per unit plant mass produced in relation to available soil micronutrients than do the plants growing more rapidly (high yields). Rough quantitative estimates of this phenomenon called here the "concentration-dilution phenomenon" are given with special reference to its effects on the results of plant analyses.

Estimations of critical levels for various micronutrients are presented. These are to be considered tentative and are subject to revision when more information is available. Although the main emphasis of this study is put on micronutrients, the macronutrients (N, P, K, Ca, Mg) of soils and plants (original wheat and maize) were also analysed and data on their plant-soil relations as well as on their relations to various soil characteristics are presented, though in less detail.

PART II

PART II

NUTRIENT STATUS BY COUNTRIES

4. Introduction

Although the main emphasis of this study is on micronutrients, a number of additional analyses on soils and plants were carried out in order to obtain further information on factors affecting the behaviour of macronutrients in soils and plants. When expressing the results of the whole study by countries, these additional data are included, though in a more general way than data concerning micronutrients.

To facilitate comparison of the results from one country with those from all combined, the national results from each country are presented graphically and jointly with the respective data for all international samples. Accordingly, when presenting data on general soil properties and macronutrients, the frequency distributions of national results are expressed by columns and the respective international data superimposed on the same Figure as a slightly generalized frequency curve. Numerical data (number of samples, mean, standard deviation, minimum and maximum) for both the national and international material are also given in each graph.

The data on micronutrients are presented in greater detail. The micronutrient contents of each plant-soil sample pair are shown as points in each Figure; also given are the best fitting national regression line, regression formula, correlation coefficient, mean and standard deviation. Summarized data for all the international samples are given in each graph in order to make it easier to evaluate the national results against the background of the international data. Since it is not possible to plot the values of single sample pairs of the whole international data in one graph, five zones (I—V) are given to show, in broader terms, the distribution of the micronutrient in question in all samples. Of the whole international material, the lowest 5 percent of plant x soil content values fall in Zone I, the next lowest 5 percent in Zone II, the "normal" values (80 percent) in Zone III, and the highest 10 percent in Zones IV and V (5 percent in each). For more details concerning the five zones, see Section 1.4.3.

It must be noted that when data concerning the **macronutrient** contents of plants are presented, only the contents of the two **original indicator plants** are given, i.e. wheat and maize grown in various participating countries. Since no distinct differences in the macronutrient contents between spring wheat and winter wheat were found, the results for these crops were combined. In case of **micronutrients,** the study is based on the contents of

pot-grown wheat only. The reasons for this departure from uniformity of method are simple: because of the heterogeneity of the original plant samples (differences due to plant varieties, age at sampling, fertilization, contamination etc., see Sections 1.2.1 to 1.2.4) too much uncontrolled variation in the micronutrient contents of plants was involved to give a reliable picture of the actual micronutrient status in various countries. It is clear that this concerns the macronutrients as well, even if to a lesser extent. The analytical data on micronutrients presented in Part II therefore only refer to micronutrient contents of pot-grown wheat samples, although the original plants were also analysed. Unfortunately, the plant samples obtained from the pot experiment were too small to allow analyses of macronutrients in addition to micronutrient analyses.

To locate approximately the origin of the samples studied, national maps indicating the sampling sites in each participating country are usually included. The data for the maps were received from the cooperators in each country who also have more detailed geographical information on the sampling sites (town, village, farmer, field) in their possession for possible future use.

The national results presented in Part II are expressed in graphical forms which are self-explanatory. Therefore, these are furnished with rather short comments only.

5. Europe and Oceania

5.1 Belgium

5.1.1 General

The sites for sampling plants and soils in Belgium are shown in Fig. 51 and the frequency distributions of four important soil characteristics, texture (expressed as texture index), pH (CaCl$_2$), organic carbon content, and cation exchange capacity, are given in Fig. 52. The mean, standard deviation, minimum and maximum values for other soil properties (particle size distribution, pH(H$_2$O), electrical conductivity, CaCO$_3$ equivalent and volume weight) are given in Appendixes 2, 3 and 4, separately for soils from wheat fields, maize fields and soils used in the pot experiment.

Fig. 51. Sampling sites in Belgium (points = wheat fields, triangles = maize fields). Identification numbers of sample pairs are given.

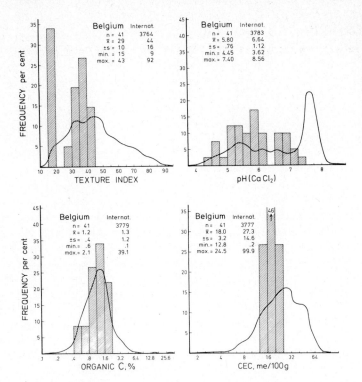

Fig. 52. Frequency distribution of texture, pH, organic carbon content and cation exchange capacity in Belgian soils (columns). Curves show the international frequency of the same characters. The number of samples (n), mean ($\bar{x}$), standard deviation (s) and minimum and maximum values are given for both national and international data.

Compared with all the international soil samples studied, the Belgian soils are relatively coarse textured and acid. The average texture index is 29 and pH 5.80 (international averages 44 and 6.67, respectively). The soil samples from the northeastern part of the country (samples 43321-34), classified as Arenosols and Podzols, are especially sandy and acid (clay content $\leq 5\%$, TI 15—19 and pH 4.45—5.90), so differing from soils in other parts of the country (mainly Luvisols and Regosols, TI 29—43 and pH 5.00—7.05). These differences are often reflected in other properties of soils and plants as will be discussed later.

The organic carbon contents of studied Belgian soils are at the average international level with a relatively narrow range of variation. The highest organic C contents were found in samples from the northeastern region.

The average cation exchange capacity (18.0 ± 3.2 me/100 g) is low compared to the international mean (27.3 ± 14.6) and varies only slightly from one sample to another. Electrical conductivity of Belgian soils is somewhat lower and $CaCO_3$ equivalent substantially lower than those in most other countries (Appendixes 2, 3 and 4). Low sodium contents are typical of most Belgian soils.

5.1.2 Macronutrients

The macronutrient contents of plants were analysed from the original wheat and maize samples only. The results of nitrogen, phosphorus, potassium, calcium and magnesium analyses of Belgian soils and plants are given as frequency distribution graphs in Figures 53—57, separately for wheat and maize and the respective soils. It should be noted that the differences in plant varieties, fertilization, climate and other local factors (see Section 1.2)

Figs 53–57. Frequency distributions of nitrogen, phosphorus, potassium, calcium and magnesium in original wheat and maize samples and respective soils (columns) of Belgium. Curves show the international frequency of the same characters.

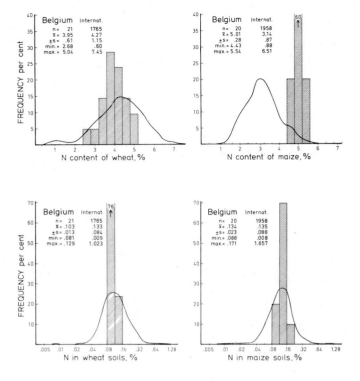

Fig. 53. Nitrogen, Belgium.

may have caused variation in the macronutrient contents of plants in addition to those due to soils.

The mean **nitrogen** contents of the original Belgian wheat and soils are somewhat below the international averages (Fig. 53). The nitrogen contents of Belgian maize (mean 5.01 %) are the highest in this study (Appendix 3) and those of the maize soils (0.134 %) about average internationally. Nitrogen was applied to Belgian crops (wheat 98 ± 21 and maize 118 ± 66 kg N/ha) at rates that were high compared to most other countries represented in this study. See also Fig. 6.

The average **phosphorus** content of Belgian maize samples (0.535 %) is the highest among the countries studied, exceeding the international mean content by a factor of 1.6. In the case of wheat, the Belgian average (0.501 %) is exceeded only by that of Malta (Appendixes 2 and 3). The average 0.5 N NaHCO$_3$ extractable P contents of Belgian maize soils (119.8 mg/l) and wheat soils (83.7 mg/l) are also higher than those of any other country. They are more than ten times the national mean P contents of soils in many developing countries and exceed the international mean contents by factors of 5.3 and 4.0, respectively.

According to information on phosphorus fertilization received from sample collectors, the Belgian wheat and maize crops had received 21—39 kg more P per hectare than the average for the crops studied in other countries (Appendix 5).

To what extent the high P contents of Belgian maize and wheat samples are attributable to the exceptionally high P contents of Belgian soils and how much to the most recent phosphorus application cannot be stated. Clearly, the exceptionally high P contents of Belgian soils are at least partly due to the liberal use of phosphates for many decades. See also Fig. 7 and related text. The higher extractable P contents of the coarse textured northeastern soils may be due to lower fixation of fertilizer P applied in past years.

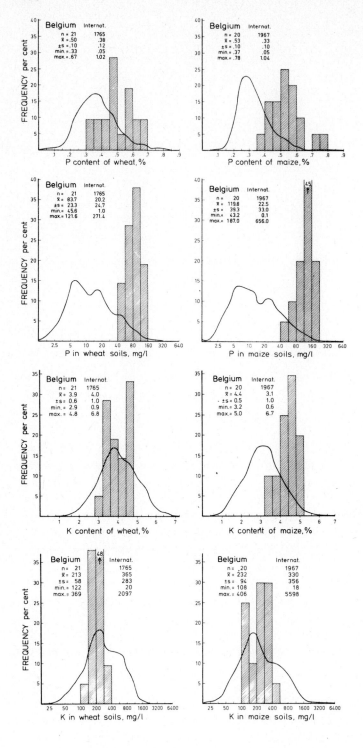

Fig. 54. Phosphorus, Belgium.

Fig. 55. Potassium, Belgium.

The mean **potassium** content of the original Belgian wheat samples is at the international average in spite of the relatively low 1 N CH_3COONH_4 exchangeable K content of Belgian wheat soils (Fig. 55). The K contents of Belgian maize samples are considerably higher than the international average of this material, notwithstanding that the exchangeable K

Fig. 56. Calcium, Belgium.

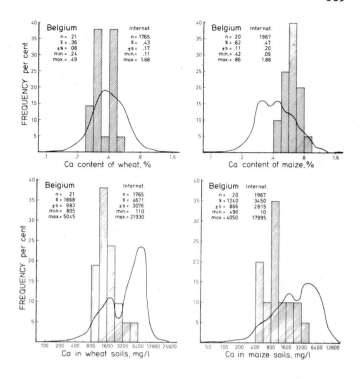

content of Belgian maize soils is below the international average. The high rates of K applied to Belgian wheat and maize crops (Belgian means 130 ± 29 and 122 ± 90 kg K/ha; international means 21 ± 44 and 7 ± 18 kg K/ha, respectively) may explain the discrepancy between the high plant and low soil K contents, but the relatively low pH of Belgian soils may also have played a part (see Section 2.2.4). For a general assessment of K status in Belgium against the international background, see Fig. 8 and related text (Section 2.2.3).

The **calcium** status of Belgian plants and soils is similar to that of K. The average Ca content of wheat is somewhat below, and the Ca contents of maize above the international averages, but the exchangeable Ca contents of soils are well below the international mean (Fig. 56). It is evident that there is always enough Ca in soils to meet the requirements of plants and that factors other than the Ca content of soil largely determine the uptake of Ca from soils. As mentioned in Section 2.2.4 (Fig. 12 and related text), Ca is relatively more available to plants in coarse than in fine textured soils. The Belgian maize crops sampled were grown on much coarser textured soils than wheat (mean TI values 23 and 36, respectively, Appendixes 2 and 3). This may at least partly explain the relatively high Ca contents of maize compared to wheat. Most of the sampled Belgian soils had recently been limed. See also Fig. 9.

The exchangeable **magnesium** contents of Belgian soils, on average, are lower than those of any other country included in this study (Appendixes 2, 3 and 4). The lowest soil Mg values come from the northeastern part of the country (samples 43321-34) but even the highest soil Mg value from Belgium (sample 43315, 146 mg Mg/l) is only about one-third of the international average. The low Mg contents of soils are reflected in the Mg contents of the plants. The mean Mg content from Belgian wheats (0.088 %) is the lowest national average in this study, and in the case of maize only three countries (New Zealand, Thailand and Sierra Leone) have slightly lower mean Mg contents than Belgium. Although

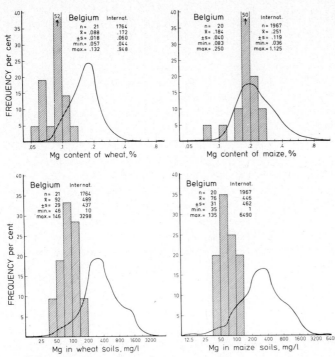

Fig. 57. Magnesium, Belgium.

the amount of Mg removed annually from soils by crops is relatively small (e.g. when compared to K), it is apparent that in Belgium where decade after decade the national crop yields have been among the highest in the world, considerably more Mg has been removed from soils than in most other countries. This may be one of the reasons contributing toward the present low Mg status of Belgian soils. No information is available on any possible Mg dressing or the Mg content of lime applied to the soils sampled for this study. To obtain a broad view of the Belgian soil Mg status against the international background, see Fig. 10 (Section 2.2.3).

5.1.3 Micronutrients

Boron. The B contents of individual pot-grown wheat samples and respective soils are given in Figures 58 and 59. The Belgian B points lie somewhat above the international regression line (Fig. 58) but come closer to it when the soil B values are corrected for CEC (Fig. 59). Because of the narrow range of variation of CEC in Belgian soils, the CEC correction only slightly affects the mutual relations between the points that indicate plant and soil B values of individual sample pairs. In view of the limited number of Belgian samples, the B situation in Belgium seems to be about average, internationally. Almost all the B values fall in the normal range (Zone III) with no extremely low or high B values. See also Figs 22 and 25 (Section 2.3.3).

Fig. 58. Regression of B content of pot-grown wheat (y) on hot water soluble soil B (x), Belgium. The points on the graph and the short regression line indicate the position of Belgian plant-soil B data relative to the respective international data (long regression line and Zones I—V; for details see Chapter 4).

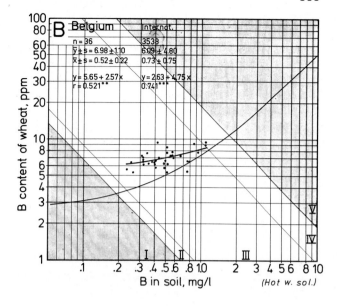

Fig. 59. Regression of B content of pot-grown wheat (y) on CEC-corrected soil B (x), Belgium.

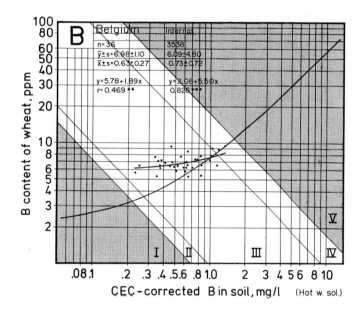

Copper. The Cu contents of Belgian plant and soil samples and their mutual regressions are given in Figures 60 and 61 against the background of the international data. The great majority of the Cu values fall within the "normal" international range (Zone III). The correction of soil Cu values for soil organic carbon improves the plant Cu—soil Cu correlation but does not otherwise change the position of Belgium in the "international Cu field" (Figs 27 and 29, Section 2.3.4). Only one sample pair falls in the highest Cu Zone (V) and one in Zone IV. The extremely high plant Cu content (20.6 ppm) of the latter (sample pair 43301) is obviously due to Cu contamination, because both the Cu content of that soil as well as that of the original plant (6.6 ppm) are "normal".

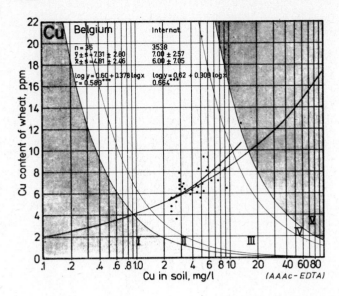

Fig. 60. Regression of Cu content of pot-grown wheat (y) on acid ammonium acetate-EDTA extractable soil Cu (x), Belgium.

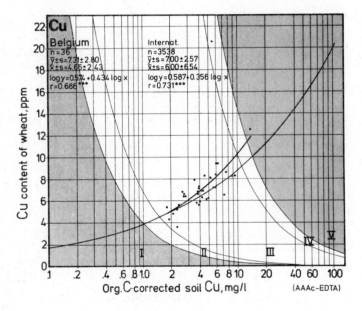

Fig. 61. Regression of Cu content of pot-grown wheat (y) on soil Cu corrected for organic carbon (x), Belgium.

Iron. The average Fe content of Belgian plants is slightly below the international mean, but the mean extractable Fe content of the soils is twice that of the international mean (Figs 31 and 62). Two-thirds of the Belgian Fe values fall in the normal range (Zone III) but one-third lies within the two highest Zones (IV and V).

Manganese. In spite of the limited material, the variations in Mn contents of Belgian plants and soils are relatively wide (Figs 63 and 64). Most of the values, however, fall within the normal range (Zone III), and the Belgian national means of plant and soil Mn (Figs 33 and 37) deviate less from the international means than those of most other countries. Because of the quite wide pH variation in Belgian soils (from 4.45 to 7.40), pH correction of DTPA extractable soils Mn values improved the correlation between plant Mn and soil Mn substantially (from $r = 0.347*$ to $0.737***$). The Mn contents of only one

Fig. 62. Regression of Fe content of pot-grown wheat (y) on acid ammonium acetate-EDTA extractable soil Fe (x), Belgium.

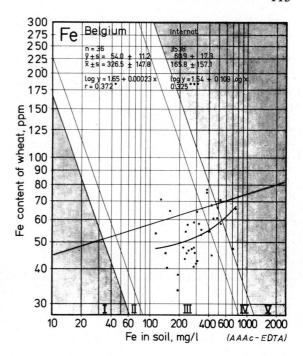

sample pair (43327) fall in the lowest Mn Zone (I) and those of four pairs (43321-22, 43328-29) in Zone IV. All these Belgian extreme values originate from the northeastern part of the country.

Molybdenum. The Mo contents of individual pot-grown wheat samples and respective soils are given in Figs 65 and 66. The ranges of variation in Mo contents of soils and plants are rather small in comparison to the whole international material. The highest Mo content found in Belgian wheat samples exceeds the lowest by a factor of 32 and in the whole international material by a factor of almost 1 000. Corresponding factors for the AO-OA extractable Mo contents of Belgian soils are 6 (uncorrected) and 17 (pH-corrected), and for the international material over 500 and 1 500, respectively. The correlation between plant and soil Mo contents in Belgian material is not significant. Correction of soil Mo values for pH changed the direction of the regression from negative to positive, but Belgium still remains one of the three countries where pH-correction is unable to improve the correlation to a statistically significant level (Table 8, Section 2.3.2.2).

The mean Mo content of plants is slightly above the international mean and that of soils (Fig. 66) is lower than the international average. All Belgian Mo values fall within the normal range (Zone III), and there are no extreme Mo contents that might indicate possible Mo deficiency or toxicity. See also Figures 15 and 20.

Zinc. The relationships between Zn contents of Belgian plants and the respective soils determined by two different extraction methods are presented in Figures 67 and 68. In both cases the correlations are very high and the mutual locations of individual points are very similar, irrespective of the extraction method used.

Almost all the Zn contents of plants and soils are above the international means. The Belgian national mean for plant Zn content is twice as high as the international mean and those for soil Zn (AAAc-EDTA and DTPA extractable) are over sixfold the respective

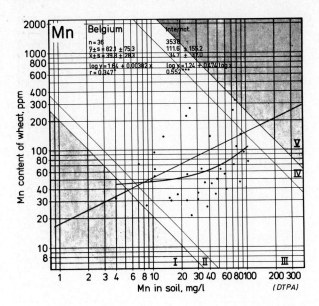

Fig. 63. Regression of Mn content of pot-grown wheat (y) on DTPA extractable soil Mn (x), Belgium.

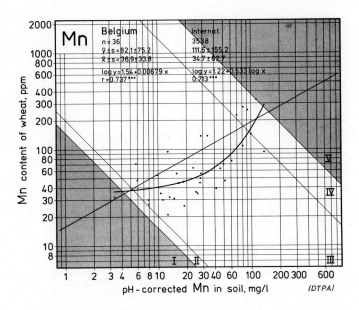

Fig. 64. Regression of Mn content of pot-grown wheat (y) on pH-corrected DTPA extractable soil Mn (x), Belgium.

international averages. Malta is the only other country where the plant and soil Zn contents are at the same high level as those of Belgium (Figs 41 and 43).

Half of the points in the Figures fall in the highest Zn range (Zone V) and the other half are in Zone IV or in the higher part of Zone III. The highest Zn values were recorded in plants and soils from the northeastern part of Belgium. It appears therefore that the possible Zn problems in Belgium are not those of deficiency but of excess. Industrial pollution is likely to be at least partly responsible for the high Zn values of Belgian soils and crops. For example, data from the UK indicate that as much as 370 to 4340 grammes of Zn per hectare/year may be deposited in non-urban sites (Cawse 1980, Wadsworth and Webber 1980).

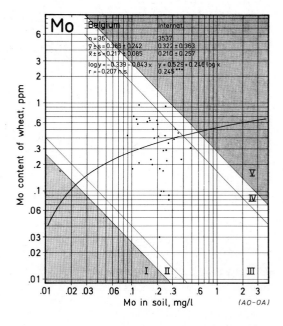

Fig. 65. Relationship between Mo content of pot-grown wheat (y) and ammonium oxalate-oxalic acid extractable soil Mo (x), Belgium.

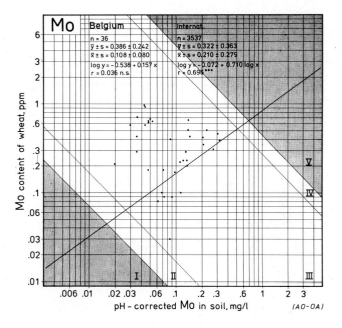

Fig. 66. Relationships between Mo content of pot-grown wheat (y) and pH-corrected AO-OA extractable soil Mo (x), Belgium.

5.1.4 Summary

The Belgian soils are generally acid, mostly coarse in texture, have a medium organic matter content and a low cation exchange capacity. The P contents of plants and soils are the highest recorded in this study. K, Ca and N are usually at the normal level but the N contents of maize are very high. The Mg contents of plants and soils are among the lowest of the 30 countries in this study. The B, Cu, Mn and Mo contents correspond more or less to the normal international level, Fe is somewhat high and the Zn content of most sampled Belgian soils and plants is exceptionally high.

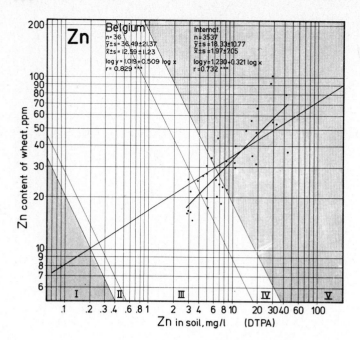

Fig. 67. Regression of Zn content of pot-grown wheat (y) on DTPA extractable soil Zn (x), Belgium.

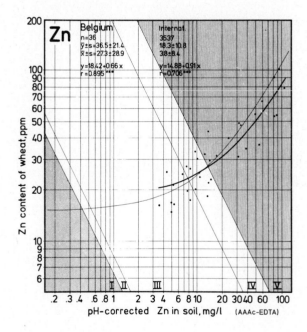

Fig. 68. Regression of Zn content of pot-grown wheat (y) on pH-corrected AAAc-EDTA extractable soil Zn (x), Belgium.

5.2 Finland

5.2.1 General

The wheat and soil sampling sites in Finland are shown in Fig. 69. The sampled area consists only of the southern part of the country since wheat is not successfully grown in the northern parts. Maize is not normally grown in Finland.

The frequency distributions of texture (TI), pH(CaCl$_2$), organic carbon content, and cation exchange capacity of Finnish soils are given in Fig. 70 and data on other soil properties in Appendixes 2 and 4.

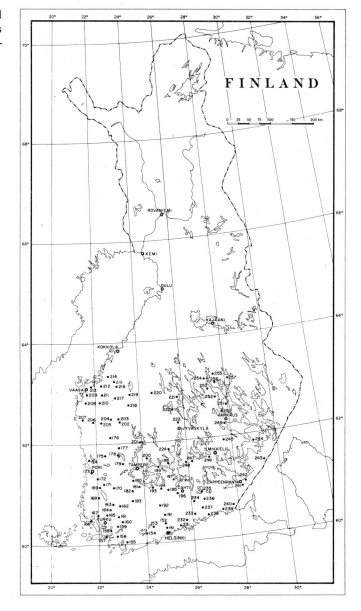

Fig. 69. Sampling sites in Finland (wheat fields). The last three numerals of each sample pair number (48151—48257) are given.

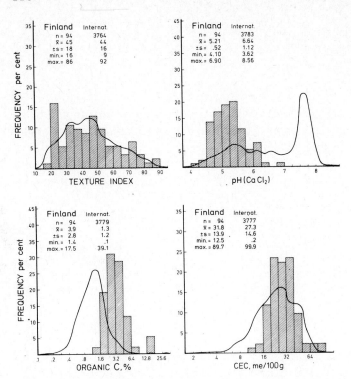

Fig. 70. Frequency distributions of texture, pH, organic carbon content, and cation exchange capacity in soils of Finland (columns). Curves show the international frequency of the same characters.

The average texture of the Finnish soils is very similar to that of all the soils studied and the range of variation is exceptionally wide. In general, the finest textured soils occur in the coastal areas of southern and western Finland and the coarsest in the central parts of the country. Most of the fine textured soils were classified as Cambisols and the coarse textured as Podsols. The soils are generally acid. The pH(CaCl$_2$) varies from 4.1 to 6.9 and the national average (5.2) is among the lowest found in this study. The average organic matter content (org. C 3.9 %) of Finnish soils (including five peat soils) is three times as high as the international mean (1.3 %). Only the soils of New Zealand are higher in organic matter. Compared to all soils studied, the cation exchange capacity of Finnish soils is slightly on the high side and the range of variation is relatively wide. Electrical conductivity, CaCO$_3$ equivalent and sodium contents of soils are low compared to most other countries (Appendixes 2 and 4).

5.2.2 Macronutrients

The frequency distributions of nitrogen, phosphorus, potassium, calsium and magnesium contents in original Finnish wheat samples and in respective soils are shown in Figures 71—75. To compare the average macronutrient situation in Finland to that in other countries, see Figs 6—10 (Section 2.2.3).

The mean **nitrogen** content of the soils (0.288 %) is more than twice as high as the international mean (0.133 %) and it is exceeded only by that of New Zealand (0.340 %; Appendix 2). Only two countries, Hungary and the Republic of Korea, have higher national averages for the N content of wheat (5.46 and 5.27 %, respectively) than Finland (5.03 %). The higher rates of nitrogen applied to the sampled wheat crops in Hungary

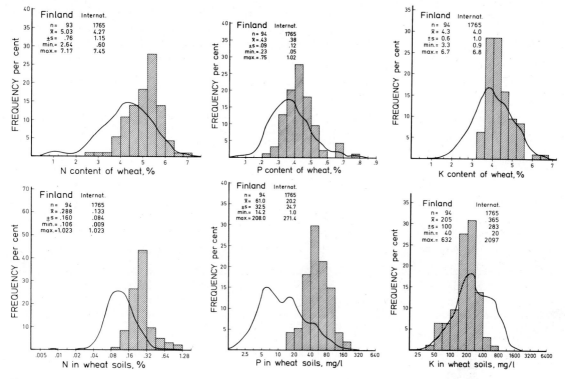

Fig. 71. Nitrogen, Finland.

Fig. 72. Phosphorus, Finland.

Fig. 73. Potassium, Finland.

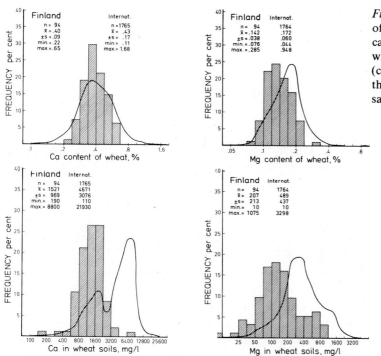

Fig. 74. Calcium, Finland.

Fig. 75. Magnesium, Finland.

Figs 71–75. Frequency distributions of nitrogen, phosphorus, potassium, calcium and magnesium in original wheat samples and respective soils (columns) of Finland. Curves show the international frequency of the same characters.

(140 ± 74 kg N/ha) and Korea (100 ± 25 kg) compared to Finland (76 ± 24 kg) and New Zealand (9 ± 16 kg) are likely to explain these differences (Appendix 5). Peat soils and other soils of high organic matter content were, in general, higher in N than other Finnish soils. See also Fig. 6.

The mean **phosphorus** ($NaHCO_3$ sol.) content of the soils is the third highest (after Belgium and Malta) in this study, and the average P content of wheat is exceeded only by those of Belgium, Malta and Hungary (Appendix 2). In all these countries applications of phosphorus fertilizer to the sampled wheat crops were above the international average (Belgium 42 ± 9, Finland 43 ± 26, Hungary 52 ± 55, Malta 26 ± 0, international mean 21 ± 27 kg P/ha, Appendix 5).

In general, the highest soluble soil P and plant P contents in Finnish material were found in coarse and medium textured soils and in plants grown in those soils, apparently because both the applied and the native soil P were less firmly fixed in these soils than in finer textured soils. See also Figs 7 and 12 in Sections 2.2.3—2.2.4.

The mean **potassium** content of the original Finnish wheats is at the international average in spite of the relatively low exchangeable K of the soils (Figs 73 and 8). Evidently, the high potassium application (mean 63 ± 19, international mean 21 ± 44 kg K/ha) compensates for the low contents of native soil K. The generally low pH of the soils may also contribute to the relatively high K content of the plants (See Fig. 11, Section 2.2.4).

The **calcium** contents of the wheats are only slightly, but the exchangeable Ca of soils are considerably below the international average (Fig. 74, Appendix 2). See also Fig. 9 and related text.

The **magnesium** content of the soils varies widely from one soil to another, but the national average is one of the lowest among the countries in this study. The low soil Mg contents are also reflected in plants. Only five countries have lower national mean Mg contents of wheat than Finland (Appendix 2), but when maize growing countries are included in the comparison (pooled plant Mg contents, Fig. 10) Finland's position is the ninth lowest. Response to Mg is likely in several of the sites sampled.

5.2.3 Micronutrients

Boron. The B contents of Finnish soils and pot-grown wheats are given in Figures 76 and 77 and show that both soils and plants contain slightly less than the international average (see also Figures 22 and 25). A great majority of the sample pairs are within the normal B range (Zone III) and only a few in the two lowest or the two highest B Zones. No clear geographical differences between the high B and low B soils can be distinguished. Low and high B values can often be found in fields located relatively close to each other (e.g. sampling sites 48159 and -160, 48177 and -178, 48199 an -200, 48249 and -250).

The limited number of samples in this study may give too high an estimate of the general B status of Finnish soils since the average hot water soluble soil B content (0.55 mg/l) is somewhat higher than those (based on more numerous samples) published earlier by Sippola and Tares (1978) and Kurki (1979), 0.38 and 0.48 mg/l, respectively. The latter data, based on the results of general soil testing, showed also that the average B content of Finnish soils has been improving from the years 1966—1970 (0.41 mg/l) to 1976—77 (0.48 mg/l). The improvement was attributed to increased B fertilization following soil testing and to the addition of some B to most of the fertilizers generally used in Finland.

Fig. 76. Regression of B content of pot-grown wheat (y) on hot water soluble soil B (x), Finland. For details of summarized international background data, see Chapter 4.

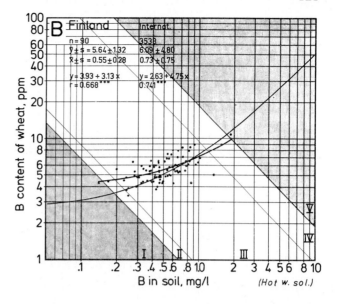

Fig. 77. Regression of B content of pot-grown wheat (y) on CEC-corrected (hot water soluble) soil B (x), Finland.

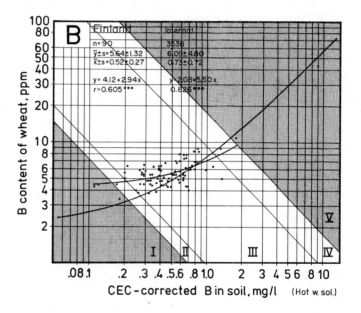

Copper. The national averages for Cu contents in wheat and soils are slightly on the low side compared to the whole international range (Figs 78, 79, 27 and 29). Because of the relatively high organic matter contents of Finnish soils, the correction of AAAc-EDTA extractable soil Cu values for organic carbon reduces the soil Cu values somewhat in relation to other countries and so lowers their position in the "international Cu field". Although most of the Cu values remain within the normal range, the relative number of Cu values falling in the low Cu Zones (I and II) becomes more than doubled and indicates that the possibility of response to copper fertilization is somewhat more likely than in most other countries.

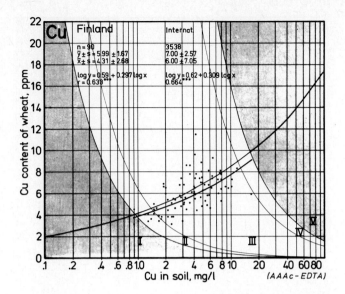

Fig. 78. Regression of Cu content of pot-grown wheat (y) on acid ammonium acetate-EDTA extractable soil Cu (x), Finland.

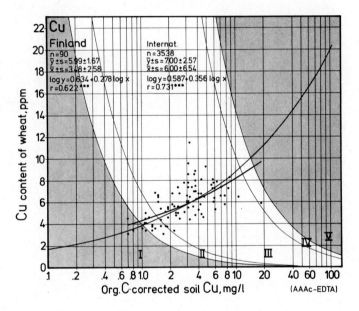

Fig. 79. Regression of Cu content of pot-grown wheat (y) on organic carbon-corrected (AAAc-EDTA extr.) soil Cu (x), Finland.

Most of the samples within the low Cu Zones I and II represent relatively coarse textured soils in the central parts of the country. Previous investigations (e.g. Tainio, 1960; Tähtinen, 1971) indicate that Cu deficiency occurs quite frequently in Finland, especially in the peat soils and in the coarse textured mineral soils of the northern parts of the country (for reasons explained earlier, the present material does not include samples from the northern parts). The average AAAc-EDTA extractable Cu contents of Finnish soils, based on a substantially larger number of soil samples (n = 2015) including also northern Finland, is 2.83 ± 2.90 (Sippola and Tares, 1978). This suggests that Cu deficiency may be more common in Finland than is indicated by the limited data of this study.

According to soil testing data for the period 1955—1977 (Kurki, 1979) the Cu status of Finnish soils was declining until about 1970, but since then a slight improvement has been observed. This was attributed to the increased use of fertilizers containing Cu.

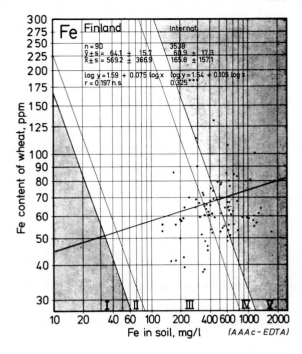

Fig. 80. Regression of Fe contents of pot-grown wheat (y) on acid ammonium acetate-EDTA extractable soil Fe (x), Finland.

Iron. The Fe contents of plant and soil samples are plotted in Fig. 80. The plant Fe—soil Fe correlation does not reach the 5 percent but only the 10 percent significance level. Both the average Fe contents of plants and soils exceed the respective international means, and Finland is the only country in which the national plant-soil average falls in the highest Fe Zone (V) (Fig. 31). Almost half of the samples fall within Zone V and less than one-third within the normal range (Zone III) (Fig. 80). The soils having the highest extractable Fe contents (> 1000 mg/l) include the five peat soils and soils with high clay and organic matter contents, high cation exchange capacity and low pH. The average AAAc-EDTA extractable Fe content of 2015 Finnish soils (677 ± 656 mg/l) presented by Sippola and Tares (1978) was still slightly higher than the national average recorded in this study (569 ± 367 mg/l).

Manganese. The regressions of Mn contents of pot-grown wheat on the DTPA extractable soil Mn without and with pH corrections are given in Figures 81 and 82. Correcting DTPA Mn for pH changes the distribution of points in the Figure, improves the correlation from being non-significant (r =0.104 n.s.) to a highly significant level (r = 0.639***), and moves the Finnish material to the right in the "international Mn field" (see Figures 33 and 37). Notwithstanding that the Finnish soils are generally acid with high availability of Mn to plants (Fig. 34), the national mean values for plant Mn and extractable soil Mn are lower than the respective mean values for the whole international material. This is an indication of relatively low total Mn contents of Finnish soils. However, cases of "primary" deficiency (due to low total Mn content) are very rare compared to "secondary" deficiency (low availability mainly due to high soil pH).

Most of the Finnish samples fall in the normal Mn Zone (III), only a couple of sample pairs are in the two lowest and three in the two highest zones (Fig. 82). The highest Mn contents of plants and soils were found in samples collected from sites where acid soils (pH $<$ 5) prevailed and the lowest from sites having soils of only moderate acidity.

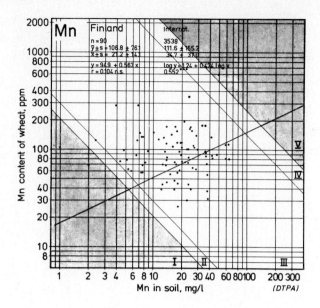

Fig. 81. Regression of Mn content of pot-grown wheat (y) on DTPA extractable soil Mn (x), Finland.

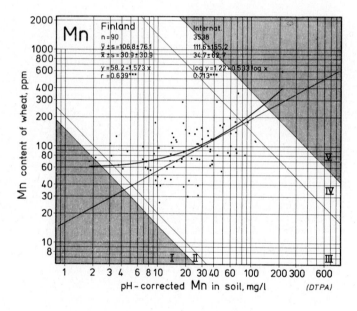

Fig. 82. Regression of Mn content of pot-grown wheat (y) on DTPA extractable soil Mn corrected for pH (x), Finland.

Molybdenum. The average Mo content of wheat grown on Finnish soils is somewhat lower than the international average, while the mean of ammonium oxalate-oxalic acid extractable soil Mo exceeds the international mean by a factor of 2.6 (Fig. 83). This inconsistency is eliminated by correcting the soil Mo values for pH (Fig. 84). In consequence, points in the Figure took up new positions; the plant Mo—soil Mo correlation improved substantially and the whole Finnish material is moved to the left in the "international Mo field" (see also Figures 15 and 20). After pH correction only four points remain in the two highest Mo Zones (IV and V), one only is in Zone II and most of the remainder are within the normal range.

Fig. 83. Regression of Mo content of pot-grown wheat (y) on ammonium oxalate-oxalic acid extractable soil Mo (x), Finland.

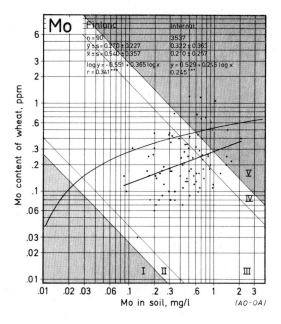

Fig. 84. Regression of Mo content of pot-grown wheat (y) on pH-corrected AO-OA extractable soil Mo (x), Finland.

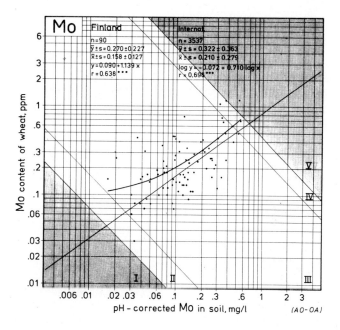

The highest Mo values were found in samples from sites where the soils have pH above the average and medium texture, while the lowest values were recorded from sites of low soil pH and either coarse (TI 16—30) or fine (TI 50—80) soil texture. In coarse textured soils the low Mo values are apparently due to low total contents and in fine textured soils due to low plant availability (see also Fig. 16). No clear geographical distinction can be made between the high and low Mo areas. The present material does not indicate any serious Mo problems in Finland. However, heavy nitrogen applications (300—450 kg N/ha/yr) commonly practised by fodder producers may induce Mo deficiency on acid soils if the pH is not kept at a reasonable level by liming (Rinne *et al.*, 1974 b).

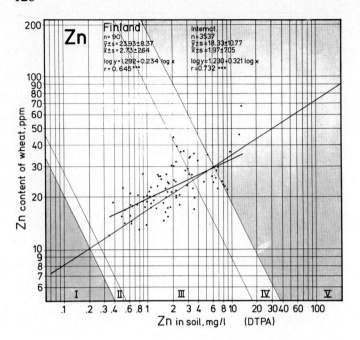

Fig. 85. Regression of Zn content of pot-grown wheat (y) on DTPA extractable soil Zn (x), Finland.

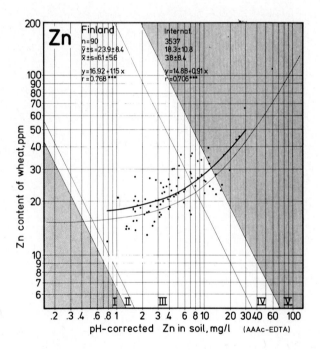

Fig. 86. Regression of Zn content of pot-grown wheat (y) on AAAc-EDTA extractable soil Zn corrected for pH (x), Finland.

Zinc. The national averages for both plant Zn and extractable soil Zn are distinctly above the respective international averages (Figures 85 and 86). Irrespective of the extraction method used (DTPA or pH-corrected AAAc-EDTA), the correlations between plant and soil Zn are high. Comparison of Figures 85 and 86 shows that the locations of individual points have changed little. With only a few exceptions, the same points falling in Zones V and IV when DTPA was used as an extractant are within the same zones when the soil Zn contents were determined by AAAc-EDTA method and corrected for pH.

No Zn values fall in the two lowest Zn zones, and although almost one-third of the Zn points are within the two highest zones no extremely high Zn contents were found (see e.g. Figs 67 and 68). Geographically, the lowest and highest Zn values are quite evenly distributed all over the southern part of the country.

5.2.4 Summary

The Finnish soils are generally acid with widely varying texture, high organic matter content and relatively high cation exchange capacity.

The P and N contents of soils and original plants are among the highest in this study and those of Mg among the lowest. The K and Ca contents of wheat are average but those of soils are low.

The Mn and Mo contents of Finnish soils and pot-grown wheat are, with few exceptions, within the normal international range. The Cu and B contents are somewhat below but those of Zn and Fe distinctly above those of most other countries. It appears that no extensive, but some local (Cu and B), micronutrient problems are to be expected in Finland.

5.3 Hungary

5.3.1 General

The sampling sites in Hungary were well distributed across the country (Fig. 87) and soil properties therefore varied widely. About half of the soils sampled were classified as Phaeozems (125). Other soils commonly found were Chernozems (46), Luvisols (25), Vertisols (15) and Cambisols (10).

The frequency distribution graphs for soil texture, pH and cation exchange capacity given in Fig. 88 show that on average, the soil properties in the Hungarian sample material correspond closely with the whole international material and the ranges of variation are almost as wide as those internationally. Organic matter contents in Hungarian soils are relatively high and quite uniform. The average electrical conductivity and $CaCO_3$ equivalent values of Hungarian wheat soils are lower than those internationally (Appendix 2), but in the case of maize soils the Hungarian averages exceed the respective international means (Appendix 3). With a few exceptions (mainly Halosols) the sodium contents of the soils are relatively low.

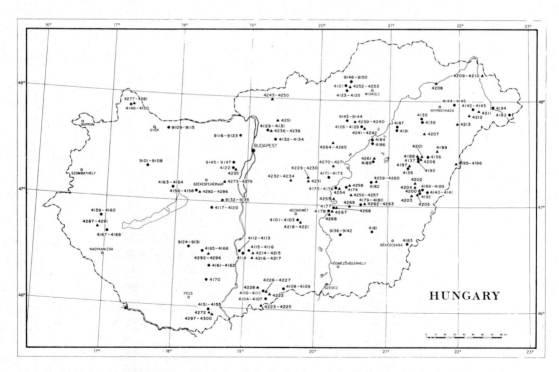

Fig. 87. Sampling sites in Hungary (points = wheat fields, triangles = maize fields). The last four numerals of each sample pair number (44101—44300 and 49101—49150) are given.

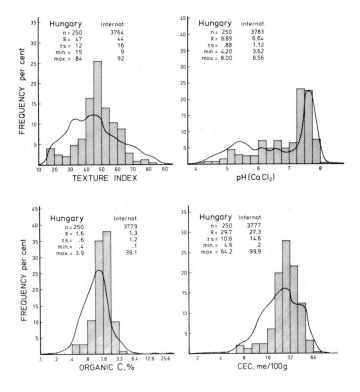

Fig. 88. Frequency distributions of texture, pH, organic carbon content, and cation exchange capacity in soils of Hungary (columns). Curves show the international frequency of the same characters.

5.3.2 Macronutrients

The frequency distributions of macronutrient (N, P, K, Ca and Mg) contents of the original wheat and maize samples and respective soils are expressed in Figures 89 to 93.

The Hungarian averages for **nitrogen** contents of soils are high compared to most other countries in this study (Appendixes 2 and 3, Figs 6 and 89). Hungarian wheats contain on average more N than those of any other country, and only Belgian maize samples have higher mean N contents than Hungarian maize. It is apparent that apart from the inherent high N contents of these soils, the liberal applications of nitrogen fertilizers to the sampled wheat and maize crops (140 ± 74 and 136 ± 58 kg N/ha, respectively) are further responsible for the high N contents of Hungarian plants.

The average **phosphorus** contents of the original wheat and maize (0.49 and 0.50 %, respectively) are among the three highest national mean values recorded in this study (Figs 90 and 7). The contents of soluble P in Hungarian soils are also well above the average

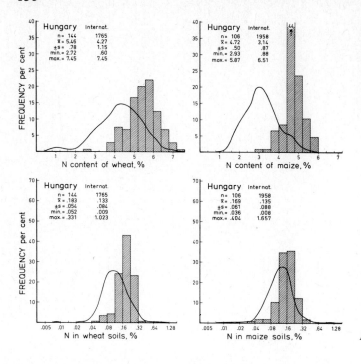

Figs 89–93. Frequency distributions of nitrogen, phosphorus, potassium, calcium and magnesium in original wheat and maize samples and respective soils (columns) of Hungary. Curves show the international frequency of the same characters.

Fig. 89. Nitrogen, Hungary.

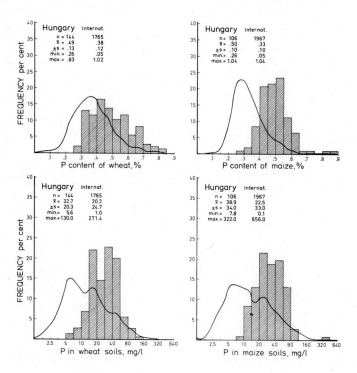

Fig. 90. Phosphorus, Hungary.

international level. The very high and largely varying rates of phosphorus fertilizers applied to the sampled crops (52 ± 55 and 52 ± 33 mg P/ha, respectively) also appear to have affected the P contents of Hungarian crops both by increasing the average plant P level and by widening the variation ranges.

Fig. 91. Potassium, Hungary.

Fig. 92. Calcium, Hungary.

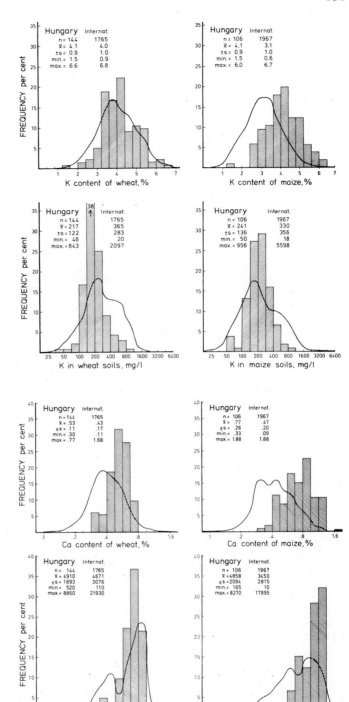

The exchangeable **potassium** contents of most Hungarian soils are well below the international average. However, the average K content of maize is considerably above the international mean and that of wheat slightly so (Figs 91 and 8). Both crops had received generally high but widely varying rates of potassium fertilizer (98 ± 96 and 104 ± 56 kg

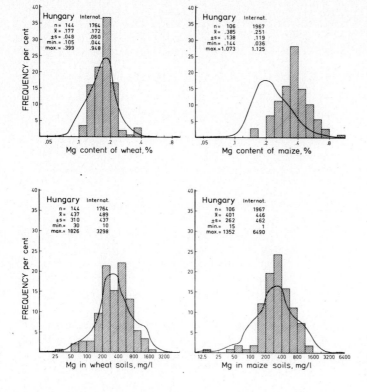

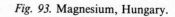
Fig. 93. Magnesium, Hungary.

K/ha, respectively) which is likely to explain the dissimilarity between the K contents of soils and plants as well as the wide variation of K contents in plants. The average level of potassium fertilization in Hungary was exceeded only by that of Belgium.

Calcium contents of both indicator crops as well as of the respective soils are high (Figs 92 and 9).

The Hungarian mean for Ca contents of maize is the highest and that of wheat is exceeded only by Turkey (Appendixes 2 and 3).

Exchangeable **magnesium** contents of Hungarian soils as well as the Mg contents of wheat are very close to the international average but high Mg contents are typical of Hungarian maize (Fig. 93, Appendixes 2 and 3). According to these data a response to Mg fertilization seems unlikely. See also Fig. 10.

Fig. 94. Regression of B content of pot-grown wheat (y) on hot water soluble soil B (x), Hungary. For details of summarized international background data, see Chapter 4.

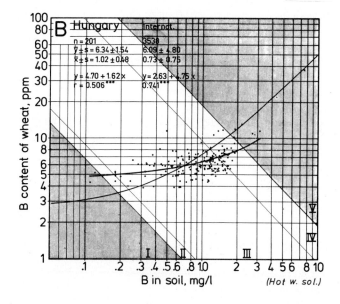

Fig. 95. Regression of B content of pot-grown wheat (y) on CEC-corrected (hot water soluble) soil B (x), Hungary.

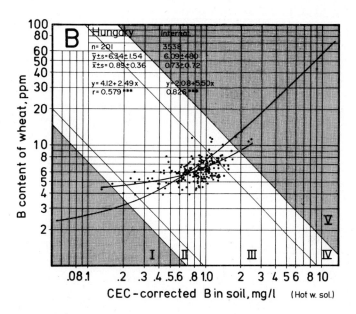

5.3.3 Micronutrients

Boron. The B status of the soils and the B content of pot-grown wheat are, in general, at a satisfactory level (Figures 94 and 95). Only four out of 201 sample pairs fall in the two lowest Zones (I and II). Although about 12 percent of the samples are within the two highest B Zones (Fig. 95), there were no extremely high plant or soil B values. No geographical areas of low or high B soils can be clearly defined although most of the high B values were measured in samples from sites east of Donau, either near the southern border or in the middle eastern part of the country on Phaeozem and Chernozem soils.

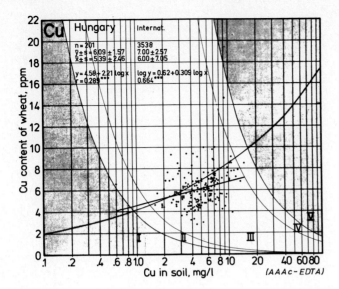

Fig. 96. Regression of Cu content of pot-grown wheat (y) on acid ammonium acetate-EDTA extractable soil Cu (x), Hungary.

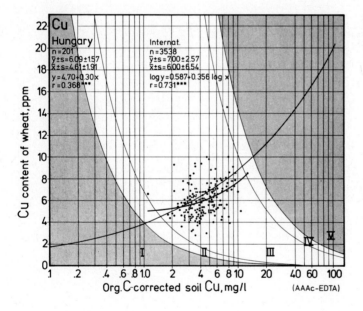

Fig. 97. Regression of Cu content of pot-grown wheat (y) on organic carbon-corrected (AAAc-EDTA extr.) soil Cu (x), Hungary.

Copper. Hungarian mean values for plants and soil Cu correspond closely to the respective international values (Figures 27, 29, 96 and 97). Since the variation of Cu contents is rather narrow and no extremely low or high Cu values were recorded, the Hungarian sample material falls almost completely in to the normal copper Zone (III). According to these analytical data, Cu problems seem unlikely.

Fig. 98. Regression of Fe contents of pot-grown wheat (y) on acid ammonium acetate-EDTA extractable soil Fe (x), Hungary.

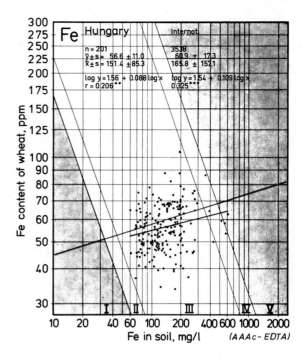

Iron. The Hungarian material includes some plant samples with fairly low Fe contents (Fig. 98), but, in general, the Fe status of Hungarian soils and plants seems to be quite "normal". Fe data are distributed within the normal range (Zone III) with only six exceptions, four in Zone IV and two just slightly inside Zone V. In the "international Fe field" (Fig. 31), Hungary stands at the centre.

Manganese. The correlation between the plant Mn and DTPA extractable soil Mn is high and because of the widely varying pH of Hungarian soils the correlation is further improved by correcting the soil Mn values for pH (Figs 99 and 100). The averages for plant and soil Mn contents deviate less from the respective international means than those of the most other countries (Figs 33 and 37). The Mn variation in Hungary, however, is very wide, and some very high as well as quite low plant and soil Mn values are included. Most of the high Mn values (falling in Zones IV and V) originated from sites in the northeastern part of the country or northeast of Budapest on soils classified as Cambisols, Arenosols or Luvisols. With several exceptions the low Mn values originate from southern areas east of Donau or from the middle eastern part of the country. These were often recorded on Phaeozem and Vertisol soils.

Molybdenum. The initially low correlation between plant Mo and AO-OA extractable soil Mo (Fig. 101) improves and becomes highly significant when correction is made for pH (Fig. 102) owing to the wide variation in pH of Hungarian soils. Internationally, the Hungarian plant Mo and soil Mo contents are slightly on the low side but about 93 percent of Mo values are within the normal Mo range (Zone III, Fig. 102). See also Fig. 20. Indications of quite severe Mo deficiency are shown by samples from two sites (44133 and 44134) located near Budapest.

Zinc. The Hungarian mean plant Zn content is the seventh lowest among 29 countries, and the mean soil Zn is the ninth lowest regardless of the extraction method used (Figures

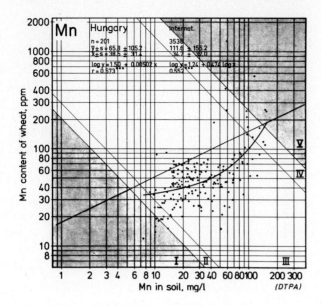

Fig. 99. Regression of Mn content of pot-grown wheat (y) on DTPA extractable soil Mn (x), Hungary.

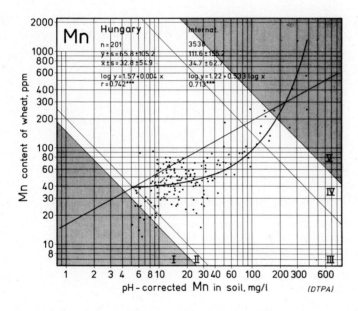

Fig. 100. Regression of Mn content of pot-grown wheat (y) on DTPA extractable soil Mn corrected for pH (x), Hungary.

41 and 43). The correlations between plant Zn and soil Zn, irrespective of the extraction method, are good and both methods give a similar general picture of the Zn status of Hungarian soils (Figures 103 and 104).

It should be noted that when the DTPA extraction method was used more samples fall in the two highest Zn Zones (IV and V) than when the soils were extracted with AAAc-EDTA and the results corrected for pH, the latter method giving relatively more low Zn values (Zones I and II). The differences, however, are not marked. At some locations response to Zn fertilization can be expected. No extreme Zn values were recorded.

Fig. 101. Regression of Mo content of pot-grown wheat (y) on ammonium oxalate-oxalic acid extractable soil Mo (x), Hungary.

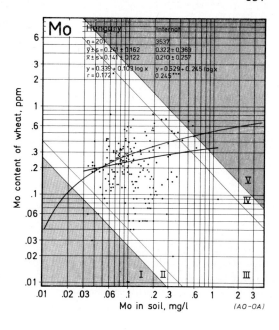

Fig. 102. Regression of Mo content of pot-grown wheat (y) on pH-corrected AO-OA extractable soil Mo (x), Hungary.

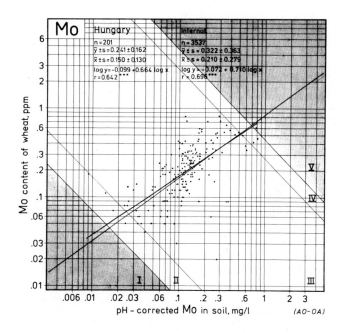

5.3.4 Summary

The Hungarian soils included in this study vary widely in regard to texture, pH, and cation exchange capacity and somewhat less in regard to organic matter content.

The P, Ca and N contents of soils and original plants are generally very high. Most of the K and Mg contents of soils are below the international average but those of plants are at an average level or above. The soil and plant contents of most macronutrients vary

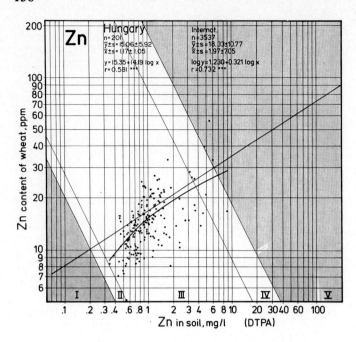

Fig. 103. Regression of Zn content of pot-grown wheat (y) on DTPA extractable soil Zn (x), Hungary.

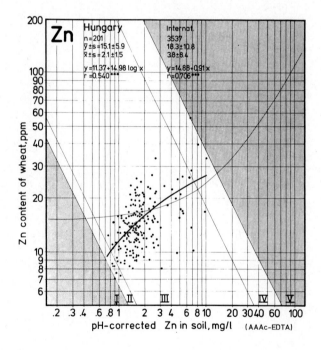

Fig. 104. Regression of Zn content of pot-grown wheat (y) on AAAc-EDTA extractable soil Zn corrected for pH (x), Hungary.

widely, partly because of generally high though varying N, P, K fertilizer application.

The micronutrient contents of Hungarian soils and plants are commonly at the normal international level; B is slightly on the high side, but Cu, Fe, Mn, Mo and Zn are on the low side. Compared to other micronutrients the variation ranges for B and Mn are wide: both low and high B and Mn values were recorded.

5.4 Italy

5.4.1 General

The most important agricultural areas in Italy are relatively well represented by the soil and plant samples collected from the country (Fig. 105). About half of the sampled soils

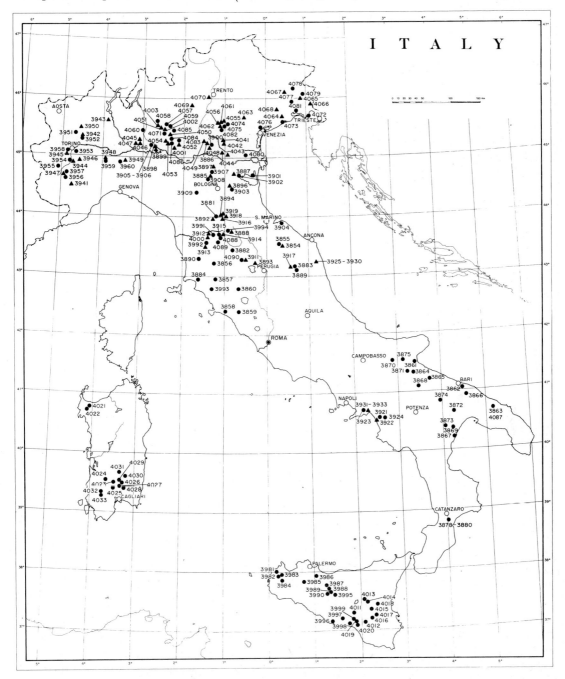

Fig. 105. Sampling sites in Italy (points = wheat fields, triangles = maize fields). The last four numerals of each sample pair number (43854—44090) are given.

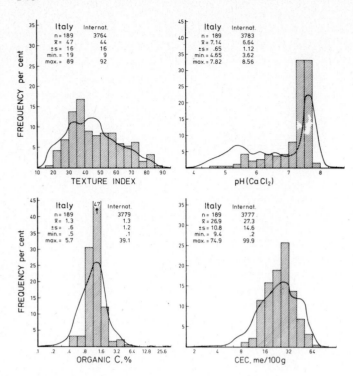

Fig. 106. Frequency distributions of texture, pH, organic carbon content, and cation exchange capacity in soils of Italy (columns). Curves show the international frequency of the same characters.

were classified according to the FAO/Unesco classification system. The frequency distribution of these was: Fluvisols (33 %), Luvisols (20 %), Cambisols (18 %), Regosols (13 %), Vertisols (13 %), Andosols (3 %) and Arenosols (1 %).

The mean and frequency distribution of texture (TI) of the soils studied correspond closely to those of all the international samples (Fig. 106). The majority of soils have an alkaline pH(CaCl$_2$) and their organic matter contents and cation exchange capacity values are at or close to the international average. Electrical conductivity and CaCO$_3$ equivalent values in Italian soils vary greatly and are generally higher than in soils of most other countries (Appendixes 2—4).

5.4.2 Macronutrients

The average **nitrogen** contents of Italian wheat and maize soils and crops are clearly above the respective international means (Figs 6 and 107), the ranges of variation are wide. In part, the high contents and wide variation of plant N are obviously due to the generally high, but varying, rates of nitrogen fertilizer applications (Appendix 5).

The **phosphorus** contents of the original Italian wheat and maize plants vary greatly but on average are above the international levels (Figs 7 and 108). Wide variations in the P contents of soils are also noticeable. For wheat soils the national mean is somewhat lower than for maize soils. Phosphorus fertilizer applications to the sampled crops (42 $\pm$ 29 and 51 $\pm$ 44 kg P/ha, respectively) are among the highest recorded in this study (Appendix 5).

Figs 107–111. Frequency distributions of nitrogen, phosphorus, potassium, calcium and magnesium in original wheat and maize samples and respective soils (columns) of Italy. Curves show the international frequency of the same characters.

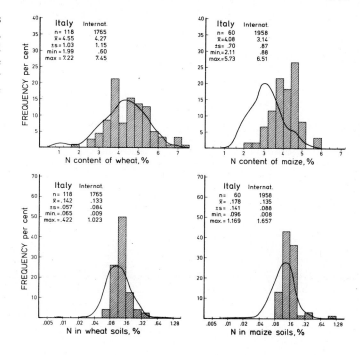

Fig. 107. Nitrogen, Italy.

Fig. 108. Phosphorus, Italy.

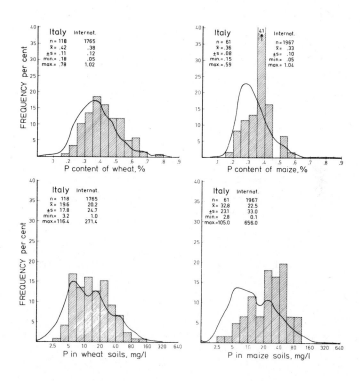

There are wide variations in the **potassium** contents of Italian plants and soils (Fig. 109) and in the rates of application of potassium fertilizer to the sampled wheat and maize crops (29 ± 31 and 80 ± 98 kg K/ha, respectively). In general, the K contents of Italian plants are at the international average or slightly lower. The low K content of soils,

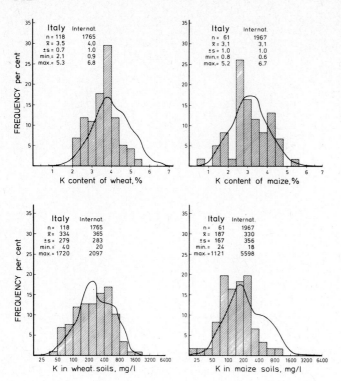

Fig. 109. Potassium, Italy.

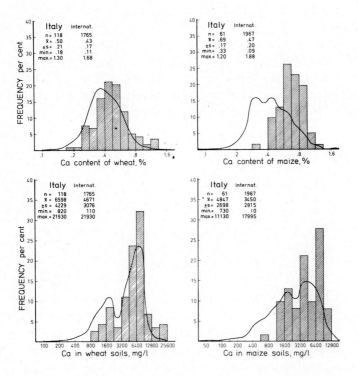

Fig. 110. Calcium, Italy.

Fig. 111. Magnesium, Italy.

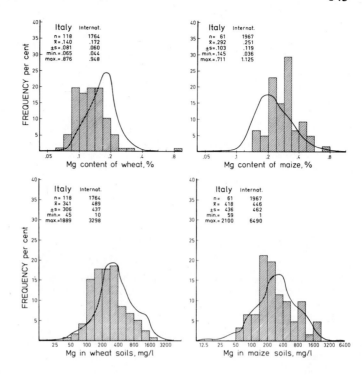

especially maize soils, has been compensated for by heavy potassium fertilizer applications (Appendix 5).

The average **calcium** contents of both the plants and the soils of Italy are among the highest national averages recorded in this study (Figs 9 and 110), while those of **magnesium** are at an average international level (Fig. 10). However, there is a difference between the two crops (Fig. 111). The Mg contents of maize and maize soils, representing the northern part of Italy (Fig. 105), are relatively much higher than the Mg contents of wheat and associated soils. About half the samples of wheat and wheat soils come from the southern part of the country including Sardinia and Sicilia, and usually have a lower Mg content than those from northern Italy.

5.4.3 Micronutrients

Boron. The B situation of Italian soils and plants is normal. In both the "international B fields" (Figs 22 and 25) Italy stands at the centre and has national mean B values near the respective international averages. However, owing to the relatively wide variation of Italian B values about eight percent of those ones show quite high B contents (Zones IV and V, Figs 112 and 113). These are not restricted to any specific geographical area but are scattered in various locations from Sicilia to Trieste. According to these data, deficiency of B is not as common in Italy as in many other countries but response to B can be expected at several locations (see also Section 2.7).

Copper. Italian soils are exceptionally rich in Cu. With the Philippines and Brazil, Italy occupies one of the highest positions in the "international Cu fields" (Figs 27 and 29). The

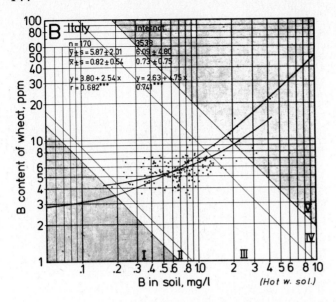

Fig. 112. Regression of B content of pot-grown wheat (y) on hot water soluble soil B (x), Italy. For details of summarized international background data, see Chapter 4.

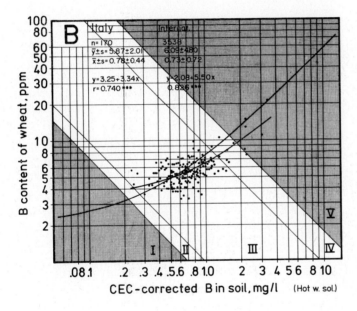

Fig. 113. Regression of B content of pot-grown wheat (y) on CEC-corrected (hot water soluble) soil B (x), Italy.

correlations between plant Cu and soil Cu are very good (Figs 114 and 115). About one-third of Italian Cu values fall in the two highest Cu Zones (IV and V), and of these two-thirds occur in Zone V. The majority of the highest (Zone V) Cu values were found in Bologna, Modena, Firenze, Arezzo, Perugia and Ascoli Piceno Provinces and bordering areas. Many of the high Cu soils were classified as Luvisols. Deficiency of Cu seems unlikely in Italy, since no low Cu values were recorded among the Italian sample material consisting of 170 sample pairs.

Iron. Against the international background the Fe content of Italian plants and soils is normal (Figs 31 and 116). On average, the Fe contents of soils are slightly on the high side, but those of plants remain close to the international average. Although over ten percent of the Fe values fall within the two highest Fe zones none of these occupy any extreme

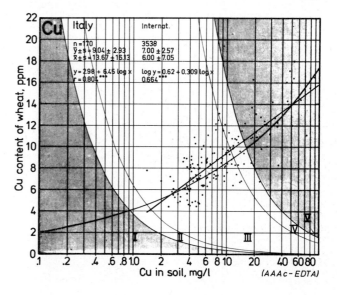

Fig. 114. Regression of Cu content of pot-grown wheat (y) on acid ammonium acetate-EDTA extractable soil Cu (x), Italy.

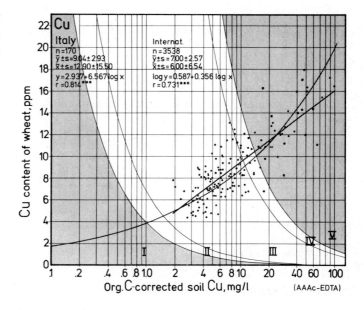

Fig. 115. Regression of Cu content of pot-grown wheat (y) on organic carbon-corrected (AAAc-EDTA extr.) soil Cu (x), Italy.

positions. The analyses of Italian samples do not indicate any severe Fe deficiency.

Manganese. Because most Italian soils are alkaline, mean $pH(CaCl_2)$ 7.1, the availability of Mn to plants is low. The Italian mean values for plant and soil Mn contents are far below the respective international averages and Italy occupies a low position in the international scale (Figs 33, 37, 117 and 118). Correction of the DTPA extractable soil Mn values for pH improves the plant Mn — soil Mn correlation, depresses the soil Mn values and brings the Italian regression line close to the international line. About 22 percent of the Italian Mn values lie within the two lowest Mn Zones (I and II), more than half of them in Zone I, and some are very low (Fig. 118). The majority of low (Zone I) Mn values occur in the Po Valley, but soils of apparent Mn shortage are also found in the Salerno area, in Sicilia and in Sardinia. No high Mn values were found in the Italian samples.

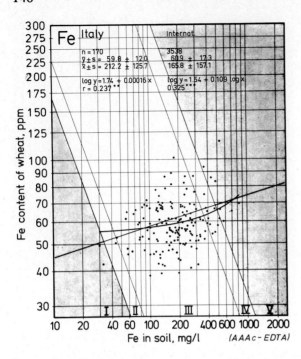

Fig. 116. Regression of Fe contents of pot-grown wheat (y) on acid ammonium acetate-EDTA extractable soil Fe (x), Italy.

Molybdenum. The high soil pH has the opposite effect on Mo to that on Mn and renders Mo readily available to plants. In the "international Mo fields" Italy stands fairly high (Figs 15 and 20). The plant and soil Mo contents in Italy vary widely and the sample material includes several high as well as a few relatively low Mo values (Figs 119 and 120). Although pH correction substantially improves the Italian plant—soil Mo correlation, the improvement is not as great as in many other countries and internationally where the variations in soil pH values are wider than in Italy. The highest (Zone V) Mo values are geographically widely distributed. Many of them are found in the same areas and sites as the low Mn values (see also Mo—Mn correlation in Fig. 46). In all cases the soils are alkaline and their pH is above the Italian average. Mo deficiency is unlikely to occur in Italy except in a few locations where the soils are exceptionally acid.

Zinc. The mean Zn contents of Italian wheats correspond closely to the mean for the whole international material, as do the Italian mean values for soil Zn irrespective of the extraction method used (Figs 41, 43, 121 and 122).

Ranges of variation in both the plant Zn and the soil Zn are wide, though most of the Zn values are "normal". The frequency distribution of high Zn values in the Italian material corresponds to that of the whole international material: about 5 percent of Italian Zn values fall in each of the high Zn Zones, IV and V. Of the 17 high Zn sample pairs falling in Zones V and IV in Fig. 122 (pH-corr. AAAc-EDTA), 16 occur in the same zones when the DTPA method was used (Fig. 121). Many of the high Zn values originate in the same areas where high Cu values are typical (see Copper, above) but a number of samples from the upper Po Valley are also high in Zn. The low Zn soils were found mainly in the southern part of the country, especially in Sicilia where almost all soils and plants are low in Zn. Although the Zn content of most Italian soils seems to be medium to high, shortage of Zn may often be found locally.

Fig. 117. Regression of Mn content of pot-grown wheat (y) on DTPA extractable soil Mn (x), Italy.

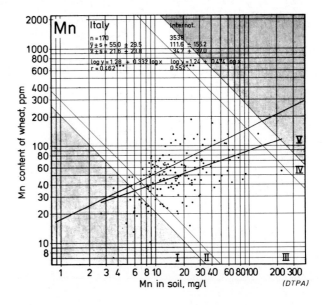

Fig. 118. Regression of Mn content of pot-grown wheat (y) on DTPA extractable soil Mn corrected for pH (x), Italy.

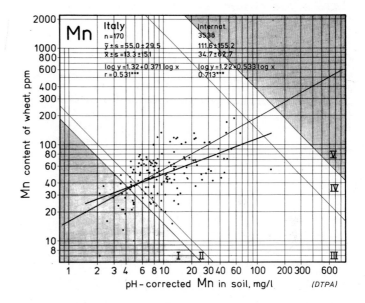

5.4.4 Summary

The Italian soils studied vary greatly in texture and they usually have an alkaline pH, medium content of organic matter and medium CEC. Compared with the whole international range Italian soils and plants are high in N, moderately high in P while K is at or slightly below the international average. For all the above macronutrients the ranges of variation are wide. Generally high, but greatly varying rates of N P K fertilizers were applied. On average, the Ca status of Italian plants and soils is high but that of Mg is only at a medium level.

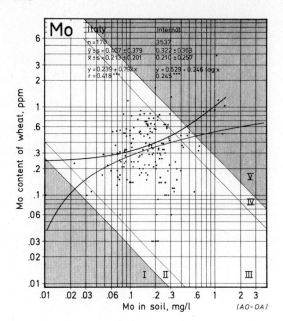

Fig. 119. Regression of Mo content of pot-grown wheat (y) on ammonium oxalate-oxalic acid extractable soil Mo (x), Italy.

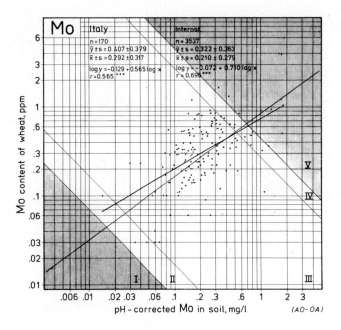

Fig. 120. Regression of Mo content of pot-grown wheat (y) on pH-corrected AO-OA extractable soil Mo (x), Italy.

Among the six micronutrients studied the B and Fe contents of soils and plants are usually within the "normal" international range, but especially in the case of B both low and relatively high values were recorded. Italian soils are exceptionally rich in Cu and Cu deficiency seems very unlikely. However, a great number of samples show alarmingly high contents of Cu. Owing to the high pH of most soils, high Mo and low Mn values are typical in Italy. In many areas shortage of Mn is apparent. Zn contents vary greatly, medium and high values being most common but response to Zn may be obtained locally.

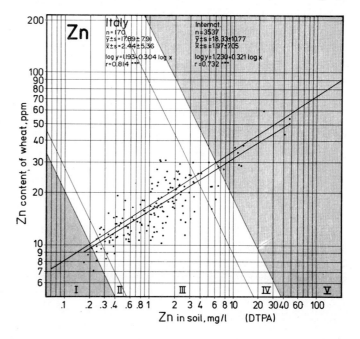

Fig. 121. Regression of Zn content of pot-grown wheat (y) on DTPA extractable soil Zn (x), Italy.

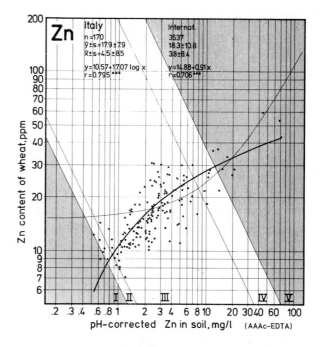

Fig. 122. Regression of Zn content of pot-grown wheat (y) on AAAc-EDTA extractable soil Zn corrected for pH (x), Italy.

5.5 Malta

5.5.1 General

The Maltese material consists of 25 paired wheat and soil samples originating from five different areas: Rabat, Mosta, Zurrieq, Qormi and Zejtun. The wheat samples were all of one variety ('Capelli').

The soils included in this study are medium textured with high pH, having a medium organic matter content and a cation exchange capacity usually somewhat lower than the international average (Fig. 123). The national mean electrical conductivity is among the highest and that of $CaCO_3$ equivalent is the highest in this study (Appendixes 2 and 4). The ranges of variation of all the above soil properties are narrow.

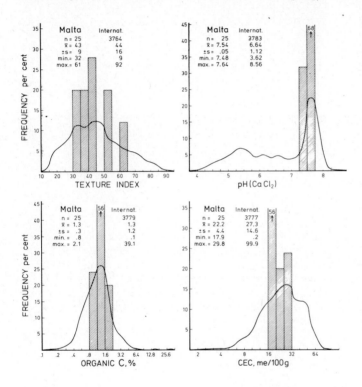

Fig. 123. Frequency distributions of texture, pH, organic carbon content, and cation exchange capacity in soils of Malta (columns). Curves show the international frequency of the same characters.

5.5.2 Macronutrients

The total **nitrogen** contents of Maltese soils are at an average international level but those of the original wheat are low (Fig. 124, Appendix 2, see also Fig. 6). All the wheat had been fertilized with 60 kg N/ha which is slightly less than the average amount applied to the wheat sampled elsewhere in this study (Appendix 5).

The average **phosphorus** content of the original Maltese wheat is the highest among the national mean values and the P content of soils the second highest after Belgium (Fig. 125 and Appendix 2). All soils were fertilized with 26 kg P/ha. See also Fig. 7.

Potassium contents of both wheat and soils are at an average international level (Fig.

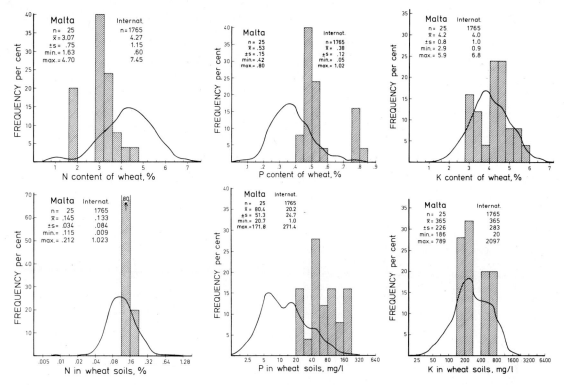

Fig. 124. Nitrogen, Malta.   *Fig. 125.* Phosphorus, Malta.   *Fig. 126.* Potassium, Malta.

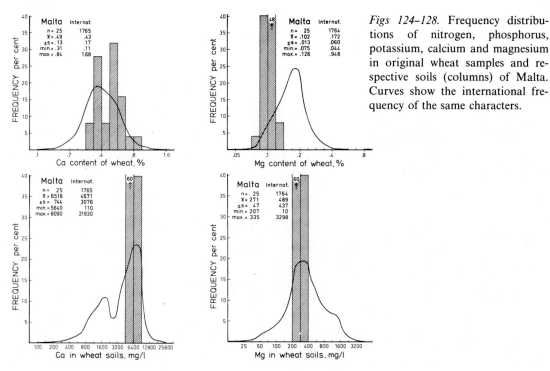

Figs 124–128. Frequency distributions of nitrogen, phosphorus, potassium, calcium and magnesium in original wheat samples and respective soils (columns) of Malta. Curves show the international frequency of the same characters.

Fig. 127. Calcium, Malta.   *Fig. 128.* Magnesium, Malta.

126). High rates of potassium fertilizer (83 kg K/ha) were applied to all soils. See also Figs 8 and 11 and related text.

Calcium contents of Maltese wheat and soils are among the highest while **magnesium** contents, especially those of wheat are among the lowest (Figs 127 and 128, Appendix 2). See also Figs 9 and 10.

5.5.3 Micronutrients

Boron. The national mean values of B are among the highest (Figs 22 and 25) and a considerable percentage of these samples fall in the two highest B Zones (IV and V, Figs 129 and 130). The variation of B, however, is narrow and no low or extremely high B

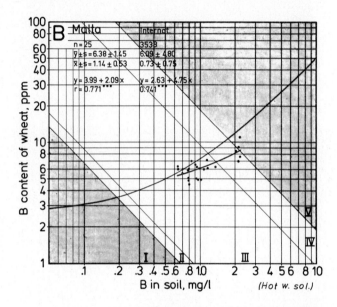

Fig. 129. Regression of B content of pot-grown wheat (y) on hot water soluble soil B (x), Malta. For details of summarized international background data, see Chapter 4.

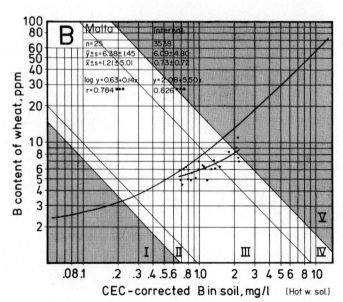

Fig. 130. Regression of B content of pot-grown wheat (y) on CEC-corrected (hot water soluble) soil B (x), Malta.

values were measured from Maltese soils or pot-grown wheat.

Copper. All Cu contents determined from Maltese plants and most soil Cu values are above the international average (Figs 131 and 132), although no sample had an extremely high Cu value.

Iron. Among the 29 countries studied, Malta has the lowest national mean values for plant and soil Fe (Fig. 31). More than half of the Maltese samples are within the lowest Fe Zone (I) and only one-third lie within the "normal" range (Fig. 133). On the basis of this finding, although from a limited number of samples, it seems that the application of iron in Malta would be beneficial for crops sensitive to iron deficiency.

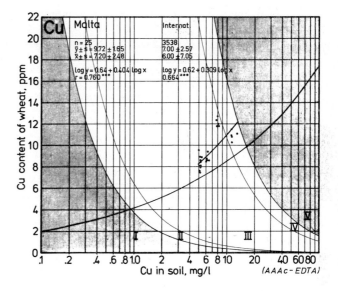

Fig. 131. Regression of Cu content of pot-grown wheat (y) on acid ammonium acetate-EDTA extractable soil Cu (x), Malta.

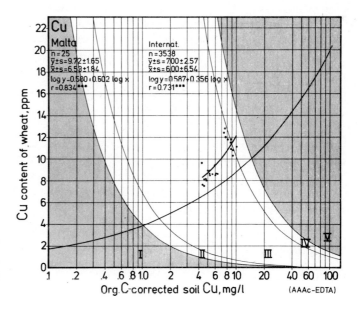

Fig. 132. Regression of Cu content of pot-grown wheat (y) on organic carbon-corrected (AAAc-EDTA extr.) soil Cu (x), Malta.

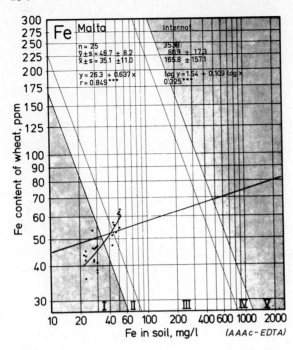

Fig. 133. Regression of Fe contents of pot-grown wheat (y) on AAAc-ammonium acetate-EDTA extractable soil Fe (x), Malta.

Manganese. Eighty percent of the Maltese soil and plant samples is so low in Mn that the samples fall in the lowest Mn Zone (Figs 33, 37, 134 and 135). The national mean values for soil and plant Mn are far below those of any other country included in this study. It seems that a response to manganese fertilization would frequently be obtained in Malta, providing that the Mn status of other Maltese soils is not much better than that of those studied here.

Molybdenum. In Malta where all the soils included in this study were alkaline, $pH(CaCl_2)$ 7.48—7.64, the availability of Mo to plants was high (Fig. 16) although the extractability of Mo by the AO-OA extractant was low. The plant Mo values given in Fig. 136 are therefore far above and soil Mo values far below the irrespective international means. Correcting soil Mo values for pH raises these values to the international average (Fig. 137). See also Figures 15 and 20. In spite of the high average Mo content of plants a great majority of the Maltese Mo values are within the "normal" range, and since the variations of both plant and soil Mo are narrow no extreme single Mo values were recorded in the Maltese material.

Zinc. Contrary to the results obtained for Fe and Mn, the Zn contents of Maltese plants and soils were very high. Such high national mean values for Zn were otherwise only recorded for Belgium (Figs 41 and 43). Irrespective of the extraction method used, only five out of the 25 Maltese Zn sample pairs are within the normal Zn range (Figs 138 and 139), 15 lie within Zone IV or on the low side of Zone V and five sample pairs lie extremely high in Zone V.

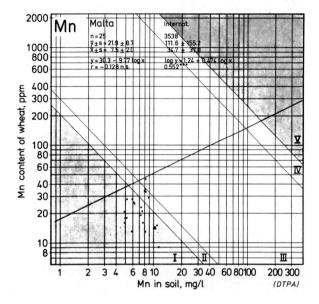

Fig. 134. Regression of Mn content of pot-grown wheat (y) on DTPA extractable soil Mn (x), Malta.

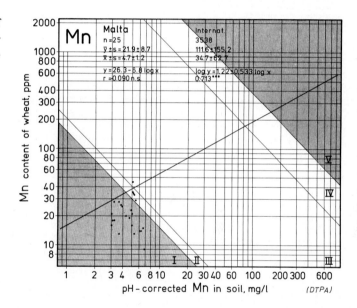

Fig. 135. Regression of Mn content of pot-grown wheat (y) on DTPA extractable soil Mn corrected for pH (x), Malta.

5.5.4 Summary

The soils of Malta included in this study usually have medium texture, organic matter content and CEC. Their pH, electrical conductivity and $CaCO_3$ equivalent values are high.

Many extremes of nutrient content were found. The P and Ca contents are at the highest international level and Mg contents are at a low level. The mean Fe and Mn contents of Maltese soils and plants are the lowest and those of Zn the highest recorded in this study.

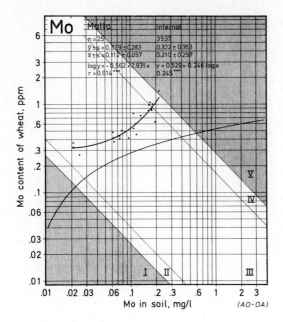

Fig. 136. Regression of Mo content of pot-grown wheat (y) on ammonium oxalate-oxalic acid extractable soil Mo (x), Malta.

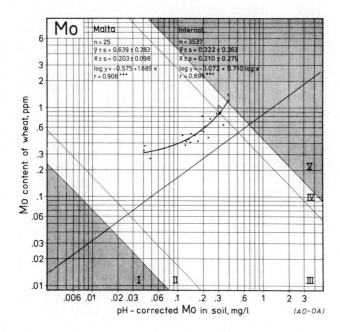

Fig. 137. Regression of Mo content of pot-grown wheat (y) on pH-corrected AO-OA extractable soil Mo (x), Malta.

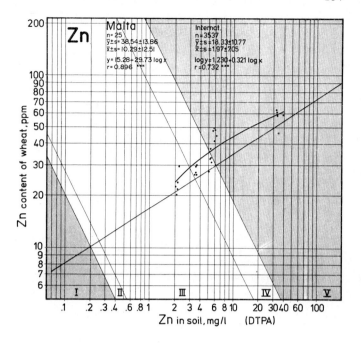

Fig. 138. Regression of Zn content of pot-grown wheat (y) on DTPA extractable soil Zn (x), Malta.

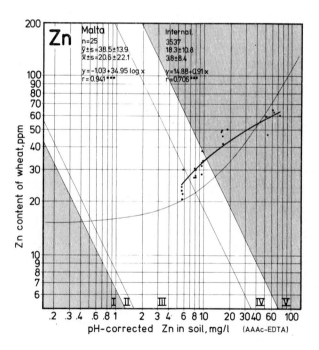

Fig. 139. Regression of Zn content of pot-grown wheat (y) on AAAc-EDTA extractable soil Zn corrected for pH (x), Malta.

5.6 New Zealand

5.6.1 General

In addition to samples collected from the North and South Islands of New Zealand (Fig. 140), eight maize-soil sample pairs (45401—08, not shown in the map) came from Tongatabu, Tonga Islands, and two (45432, 45439) from Rarotonga, Cook Islands, and are included in the sample material received from New Zealand. Out of 29 soils classified, Fluvisols were the most common occurring at 8 sampling sites. Other soils were classified

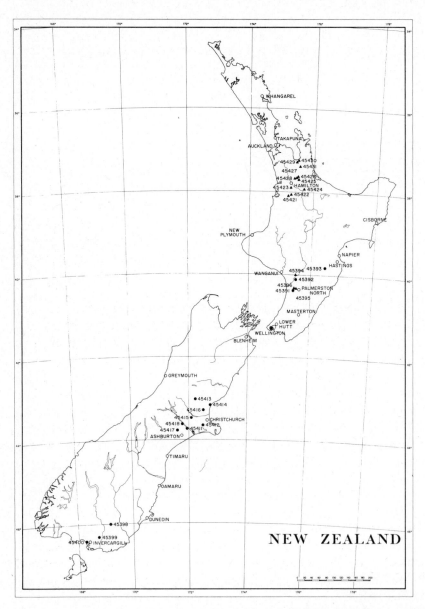

Fig. 140. Sampling sites in New Zealand (points = wheat fields, triangles = maize fields).

as Gleysols (5), Cambisols (4), Andosols (4), Regosols (3), Histosols (2), Luvisols (2) and Phaeozem (1).

The textural distribution of the sampled soils is shown in Fig. 141. The soils from North and South Island are usually medium to heavy textured whilst all Tonga soils are heavy, the texture index ranging from 70 to 82. The soils received from New Zealand are all somewhat acid. The national average pH is one of the lowest and the material includes a Histosol sample (45428) with pH(CaCl$_2$) 3.62 which is the lowest pH of any examined in this study. The latter soil failed to grow a plant sample in the pot experiment. The organic matter contents of all New Zealand soils are above the international mean (Fig. 141). Due in part to the two peat soils, the New Zealand mean (5.1 %) is higher than any other national mean content of organic carbon, and because of the generally heavy texture and high organic matter content of New Zealand soils their average cation exchange capacity is higher than that of any other country. The heavy textured Tonga soils especially show a high CEC (62—74 me/100 g). The electrical conductivity values are at an average international level or below and, with an exception of one sample from Rarotonga, the CaCO$_3$ equivalent values were at zero level.

5.6.2 Macronutrients

The national average for **nitrogen** contents in New Zealand soils and original maize were the highest recorded in this study, and N contents of wheat among the highest (Appendixes 2 and 3). Furthermore, the highest single value for soil total N of this material was measured from a New Zealand peat soil. The sampled wheat crops were fertilized with relatively small amounts of nitrogen fertilizer (9 ± 16 kg N/ha). The higher and more varying rates applied to maize (55 ± 67 kg N/ha) are an apparent reason for the wider variation of maize N than of wheat N contents (Fig. 142). For international comparison see also Fig. 6.

Phosphorus contents of the original New Zealand wheat samples and associated soils (Fig. 143) vary little around the international means. The P contents of maize correspond to the international average but those of maize soils are higher. Both the P contents of maize and of maize soils vary more than those of wheat and wheat soils. The lowest P values in maize and maize soils were obtained from Tonga samples. Phosphorus fertilizer applications to sampled wheat crops were at rates much lower (16 ± 7 kg P/ha) than those applied to maize (30 ± 29 kg P/ha). In spite of this the uptake of P by maize in relation to P in the soil was rather low. A partial reason for this may be the high CEC of maize soils (average CEC of maize soils is 60 and that of wheat soils 35 me/100 g, Appendixes 2 and 3) making P less available to plants. See Fig. 14. The P situation of New Zealand in an international context is seen in Fig. 7.

Potassium content of wheat and the exchangeable K in wheat soils are clearly on the low side, but K contents of maize and maize soils are substantially on the high side compared with respective distributions for the whole international spectrum (Fig. 144, Appendixes 2 and 3). There was also a considerable difference in rates of applied potassium fertilizer to wheat and to maize crops (7 ± 12 and 44 ± 25 kg K/ha, respectively). All the highest soil K values (> 800 mg/l) were obtained from Tonga samples. As a whole (in terms of general mean plant K—soil K) New Zealand stands close to the centre of the "international K field" (Fig. 8).

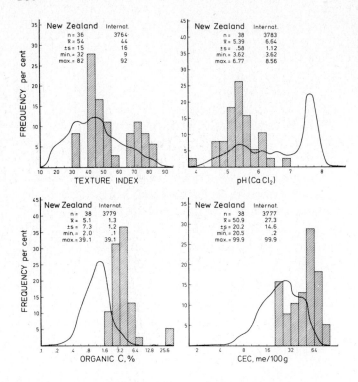

Fig. 141. Frequency distributions of texture, pH, organic carbon content, and cation exchange capacity in soils of New Zealand (columns). Curves show the international frequency of the same characters.

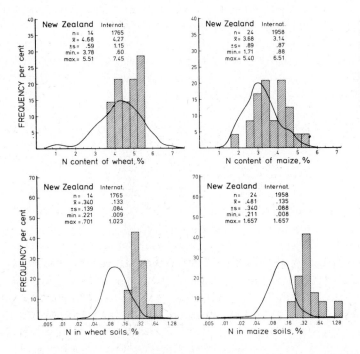

Figs 142–146. Frequency distributions of nitrogen, phosphorus, potassium, calcium and magnesium in original wheat and maize samples and respective soils (columns) of New Zealand. Curves show the international frequency of the same characters.

Fig. 142. Nitrogen, New Zealand.

Calcium contents of both the original plant species are relatively low as are the exchangeable Ca contents of wheat soils (Fig. 145). The relatively high Ca average for maize soils (3215 mg Ca/l) is mainly due to the soils of Tonga and Rarotonga (3370—7120 mg/l). Since the latter soils are heavy textured and have a high CEC, their high Ca contents are not reflected as high Ca contents of maize. See Figs 12 and 14 and related text (Section

Fig. 143. Phosphorus, New Zealand.

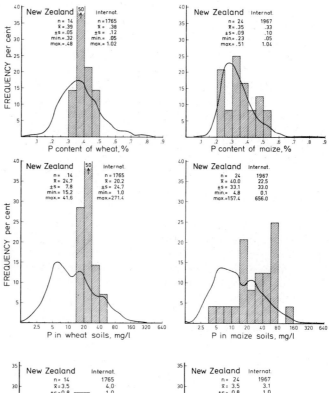

Fig. 144. Potassium, New Zealand.

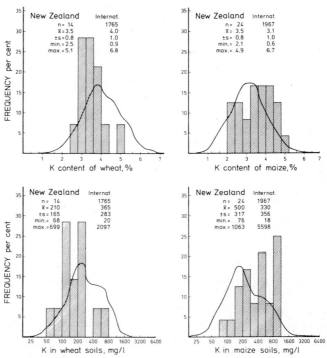

2.2.4). In general, New Zealand stands on the low side in the international comparison (Fig. 9).

As in the case of calcium the **magnesium** contents of original wheat and maize are low (Fig. 146). The same is true for exchangeable Mg contents of wheat soils and of maize soils obtained from the two main islands of New Zealand.

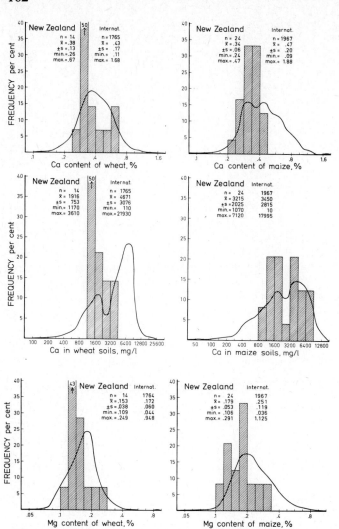

Fig. 145. Calcium, New Zealand.

Fig. 146. Magnesium, New Zealand.

The high mean Mg value of maize soils (518 mg/l) is due to the high Mg contents of the ten Tonga and Rarotonga samples which all exceed 650 mg/l and are among the 13 samples making up the three right-hand side columns in Fig. 146. As in the case of Ca, the uptake of Mg from these soils (heavy texture + high CEC) is relatively low, so explaining the low Mg contents of maize. See Figs 12 and 14 and related text (Section 2.2.4). Internationally New Zealand's Mg status is similar to that of Ca (Fig. 10).

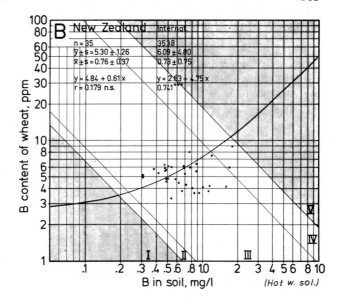

Fig. 147. Regression of B content of pot-grown wheat (y) on hot water soluble soil B (x), New Zealand. For details of summarized international background data, see Chapter 4.

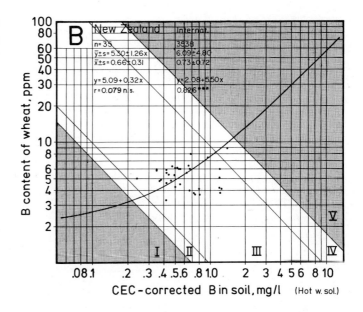

Fig. 148. Regression of B content of pot-grown wheat (y) on CEC-corrected (hot water soluble) soil B (x), New Zealand.

5.6.3 Micronutrients

Boron. The B contents of soils and plants are at an average international level, the variations are narrow and almost all B values (Figs 147 and 148) are within the normal range (Zone III). On the basis of these limited analytical data, response to B may be obtained only occasionally with crops of high B requirement.

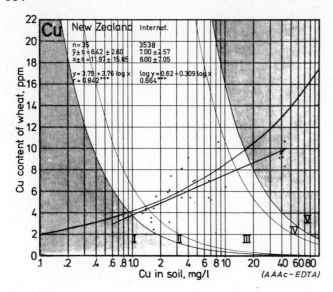

Fig. 149. Regression of Cu content of pot-grown wheat (y) on acid ammonium acetate-EDTA extractable soil Cu (x), New Zealand.

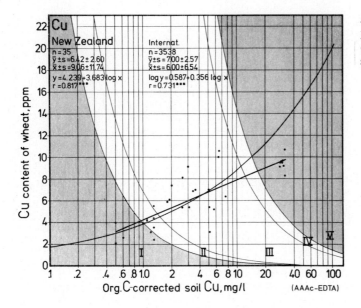

Fig. 150. Regression of Cu content of pot-grown wheat (y) on organic carbon-corrected (AAAc-EDTA extr.) soil Cu (x), New Zealand.

Copper. In spite of the limited number of samples from New Zealand, the variations of both the plant and soil Cu values are very wide (Figs 149 and 150). Only half of the samples is within the normal Cu range, one-fourth is in the two lowest Zones (I and II), indicating possible Cu shortage, and one-fourth is in the highest Zone pointing to a surplus. All low Cu values come from the South and North Islands and the high values from Tonga.

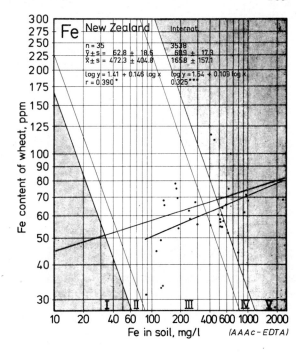

Fig. 151. Regression of Fe contents of pot-grown wheat (y) on acid ammonium acetate-EDTA extractable soil Fe (x), New Zealand.

Iron. The mean Fe contents of plants is only slightly above the international mean, but the mean for soil Fe is the second highest (after Finland) in this study (Fig. 31). More than half of the Fe determinations are within the two highest Fe Zones (Fig. 151). The highest extractable soil Fe value was obtained from a peat (Histosol) soil, the three highest plant Fe contents (> 100 ppm) from plants grown on Fluvisol soils and the three lowest (< 34 ppm) from plants grown on Andosols (see also Fig. Fe in Appendix 7).

Manganese. Because of the generally low soil pH no very low plant or soil Mn values were recorded in the New Zealand material (Figs 152 and 153). The mean values for both the plant and soil Mn are above their respective international means, nevertheless the bulk of the Mn values are within the normal Mn range (Zone III). Although some of the Mn values are within Zones IV and V, there are none in the New Zealand material with alarmingly high Mn contents.

Molybdenum. Owing to the low availability of Mo to plants in acid soils the Mo contents of wheat grown in pots on New Zealand soils are low (Figs 154 and 155). The national mean is the second lowest (after Brazil) among the 29 countries (Figs 15 and 20). The New Zealand mean for AO-OA extractable soil Mo is only slightly below the international mean (Fig. 154) but decreases to less than one-third of it when the soil Mo values were corrected for pH (Fig. 155). The variations of plant and soil Mo values in the New Zealand material are narrow and correlations between plant and soil Mo are not significant, even though pH correction changes the direction of the regression from negative to positive. More than one-third of the Mo determinations (Fig. 155) are within the two lowest Mo Zones. Although this limited material does not include any extremely low Mo values, it is possible that Mo deficiency problems might exist in New Zealand.

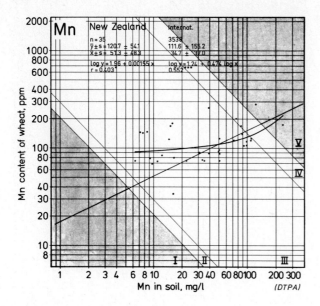

Fig. 152. Regression of Mn content of pot-grown wheat (y) on DTPA extractable soil Mn (x), New Zealand.

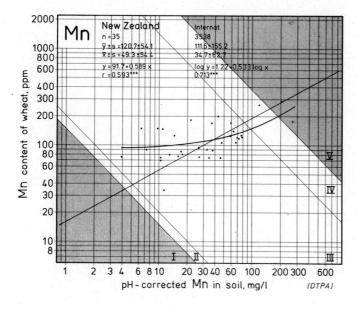

Fig. 153. Regression of Mn content of pot-grown wheat (y) on DTPA extractable soil Mn corrected for pH (x), New Zealand.

Zinc. The mean values of both the plant and extractable soil Zn contents of New Zealand are somewhat above the respective international values, irrespective of the extraction method used (Figs 156 and 157). The variations of both the plant Zn and soil Zn are relatively wide and their mutual correlations are highly significant. Forty percent of the Zn values fall within the two highest Zn Zones (IV and V), but in only one sample pair (45427, Luvisol, Morringsville) are the Zn contents exceptionally high. Most of the other high Zn values occur in Tonga and Rarotonga. The six Zn values within or close to Zones I and II come from the South Island where deficiency of Zn should be suspected.

Fig. 154. Regression of Mo content of pot-grown wheat (y) on ammonium oxalate-oxalic acid extractable soil Mo (x), New Zealand.

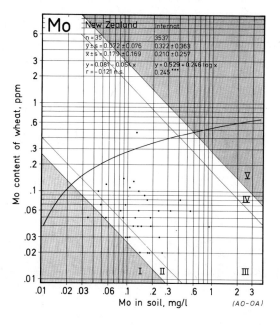

Fig. 155. Regression of Mo content of pot-grown wheat (y) on pH-corrected AO-OA extractable soil Mo (x), New Zealand.

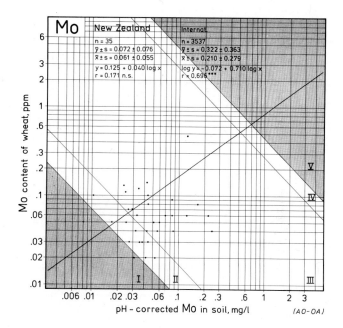

5.6.4 Summary

The soils of New Zealand included in this study are usually medium to heavy textured, severely to moderately acid, have a high organic matter content and high cation exchange capacity. The P and K contents of wheat and wheat soils are generally low while those of maize and maize soils are higher. The lowest P and K values occur in Tonga. The Ca and Mg contents of plants and soils are low, those from Tonga and Rarotonga being exceptions.

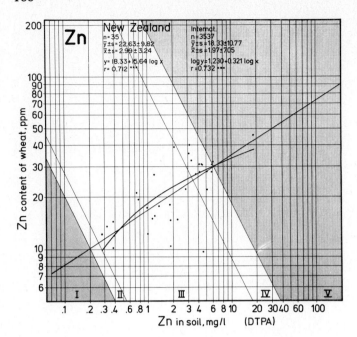

Fig. 156. Regression of Zn content of pot-grown wheat (y) on DTPA extractable soil Zn (x), New Zealand.

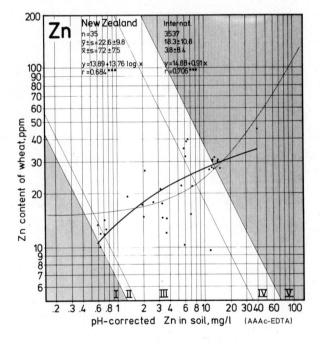

Fig. 157. Regression of Zn content of pot-grown wheat (y) on AAAc-EDTA extractable soil Zn corrected for pH (x), New Zealand.

Among the micronutrients, high Fe, moderately high Mn, normal B and low Mo values were typical. In the cases of Cu and Zn the contents varied widely between low and high values. For both micronutrients the lowest values were obtained from samples originating in both North and South Island and the highest came from Tonga and Rarotonga.

6. Latin America

6.1 Argentina

6.1.1 General

The Argentinian samples were collected from two relatively small, but agriculturally important, areas (Fig. 158). Almost all the soils were classified as Phaeozems.

Fig. 158. Sampling sites in Argentina (points = wheat fields, triangles = maize fields). The last three numerals of each sample pair number are given.

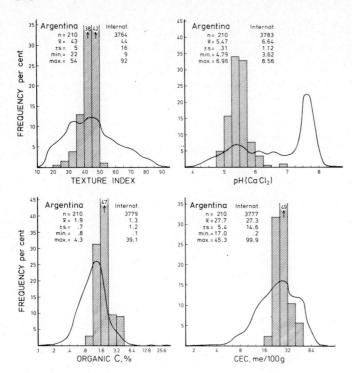

Fig. 159. Frequency distributions of texture, pH, organic carbon content, and cation exchange capacity in soils of Argentina (columns). Curves show the international frequency of the same characters.

The soils sampled are medium textured and relatively acid (Fig. 159). Over 90 percent have a texture index between 30 and 50 and pH(CaCl$_2$) between 5.0 and 6.0. On average, the organic matter contents of the Argentinian soils are appreciably above the international mean but their cation exchange capacities are at a "normal" level. For these soil properties the ranges of variation are narrow. Low electrical conductivity and CaCO$_3$ equivalent values as well as low sodium contents are typical for Argentinian soils. (Appendixes 2—4).

6.1.2 Macronutrients

Owing to the high proportion of organic matter in Argentinian soils, their average total **nitrogen** content is one of the highest in this study (Figs 6, 160, Appendixes 2 and 3). In spite of the low rates of nitrogen fertilizer applied to the sampled wheat and maize crops (3 ± 14 and 6 ± 16 kg N/ha, respectively), the national averages for N contents of these crops are exceeded by a few countries only.

The **phosphorus** contents of Argentinian soils and original indicator plants vary widely. In general, they are at the average international level or slightly above (Figs 7 and 161). No phosphorus fertilizer was applied to the sampled wheat crops and the rates applied to maize were relatively low (6 ± 18 kg P/ha).

The contents of exchangeable **potassium** of all sampled Argentinian soils are higher than the international average, the variation range is relatively narrow and the national mean values for soils under wheat and maize (784 and 790 mg K/l, Figs 162 and 8) are the highest recorded in this study. Although no potassium fertilizer was applied the average K

Figs 160–164. Frequency distributions of nitrogen, phosphorus, potassium, calcium and magnesium in original wheat and maize samples and respective soils (columns) of Argentina. Curves show the international frequency of the same characters.

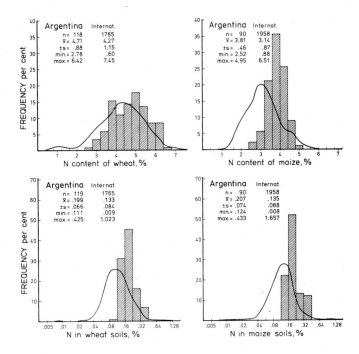

Fig. 160. Nitrogen, Argentina.

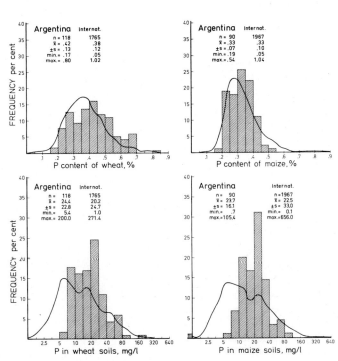

Fig. 161. Phosphorus, Argentina.

contents of wheat and maize are among the highest (Appendixes 2 and 3, Fig. 8). See also footnote on page 173.

The contents of exchangeable **calcium** in Argentinian soils are below the international average, as are the Ca contents of the original wheats (Fig. 163). The Ca contents of maize clearly show a bimodal frequency distribution. Since there is no similar distribution con-

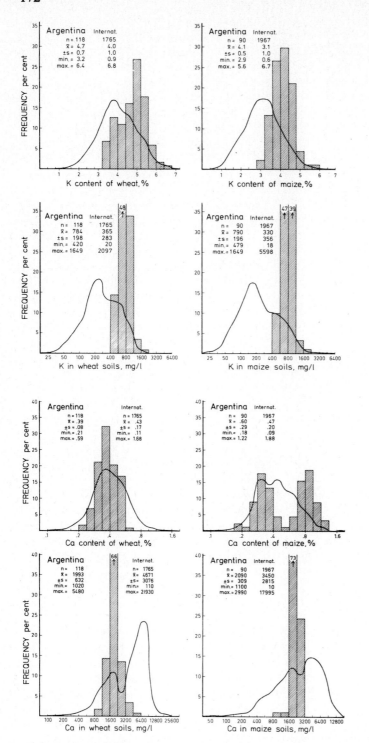

Fig. 162. Potassium, Argentina.

Fig. 163. Calcium, Argentina.

cerning exchangeable Ca in maize soils and no lime applications to maize were reported, this bimodality cannot be attributed to soils or liming. The only marked difference between the maize samples falling into the lower Ca group (Ca < 0.6 %) and the higher group (Ca > 0.6 %) that could be detected from the available data was that the low Ca maize samples were collected at later stages of growth. The age difference (41 and 54 days

Fig. 164. Magnesium, Argentina.

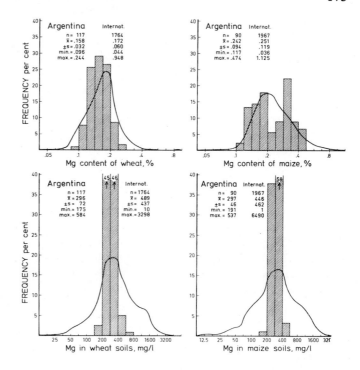

from planting) was 13 days on average. Although the Ca content of a plant may decrease considerably in two weeks when growing vigorously, it is doubtful that it can explain the whole difference between the two Ca content groups; for example, the data of Sippola *et al.* (1978) show a decrease of about 30—35 percent in the Ca content of wheat during much the same period of growth. See also Section 1.2.2. For the international standing of Ca in Argentinian soils and plants see Fig. 9.

The **magnesium** contents of original Argentinian wheat and maize, as well as the exchangeable Mg of soils, are somewhat below the respective international averages (Fig. 164). Again there is very slight variation in the exchangeable Mg contents of maize soils but the Mg contents of maize vary widely. The lower and the higher maize Mg groups consist of almost the same maize samples as the respective low and high Ca groups, but the difference is not as pronounced[1]. For the international rating of Mg in Argentina see Fig. 10.

---

[1] Comparisons of N, P and K contents of maize between the two sample groups (with Ca contents < 0.6 and > 0.6 %) showed that there was a statistically significant difference in the K contents of maize (means of 3.85 and 4.45 %, respectively) with no significant difference in the echangeable soil K. In the cases of N and P the comparisons were obscured by differences in soil N and P contents and N and P fertilizers applied.

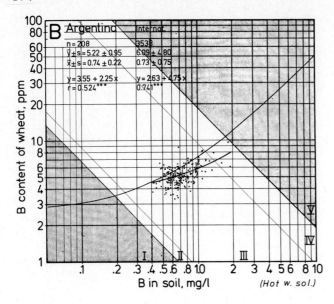

Fig. 165. Regression of B content of pot-grown wheat (y) on hot water soluble soil B (x), Argentina. For details of summarized international background data, see Chapter 4.

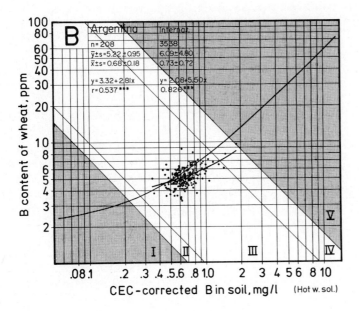

Fig. 166. Regression of B content of pot-grown wheat (y) on CEC-corrected (hot water soluble) soil B (x), Argentina.

6.1.3 Micronutrients

Boron. Because of the homogeneity of Argentinian soils the ranges of variation in B contents of soils and plants are unusually narrow (Figs 165 and 166). The Argentinian national mean values for both the plant and the soil B are close to their international means and no extremely low or high B contents were recorded. See also Section 2.7.

Fig. 167. Regression of Cu content of pot-grown wheat (y) on acid ammonium acetate-EDTA extractable soil Cu (x), Argentina.

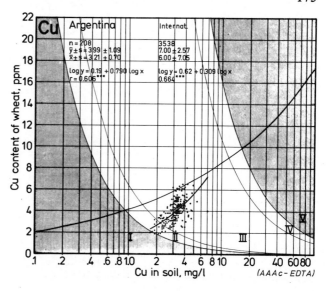

Fig. 168. Regression of Cu content of pot-grown wheat (y) on organic carbon-corrected (AAAc-EDTA extr.) soil Cu (x), Argentina.

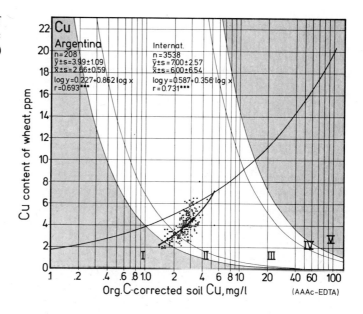

Copper. The Argentinian national mean values for plant Cu and soil Cu are the second and the fifth lowest, respectively, recorded in this study (Figs 27 and 29). Of the 208 sample pairs, only one plant Cu and one soil Cu value exceed the respective international mean (Fig. 167). After correction for soil organic carbon content (Fig. 168) no soil Cu value exceeds the international mean and the number of Cu values falling in Zones I and II is more than doubled. Although no extremely low soil Cu values are recorded in this material, the Cu status of Argentinian soils and especially that of plants is generally low. It is possible that crops sensitive to Cu deficiency would respond to applied Cu.

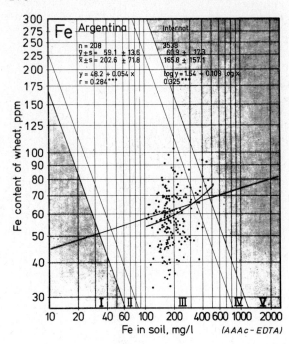

Fig. 169. Regression of Fe contents of pot-grown wheat (y) on acid ammonium acetate-EDTA extractable soil Fe (x), Argentina.

Iron. As is the case for most other nutrients in the Argentinian material, the variation of soil Fe is narrow. In general, the soil Fe values tend to be high, relative to international levels while plant Fe contents are average (Figs 31 and 169). Despite a few plant samples having quite low Fe contents, none of the Fe sample pairs is within the two lowest Fe Zones. No extremely high Fe values were recorded, although 13 sample pairs fall within the two highest Zones.

Manganese. The Mn situation in Argentina is very similar to that of Fe (Figs 169—171) as are the locations of Argentina in the "international Mn and Fe fields" (Figs 31, 33 and 37).

Molybdenum. The discordance between generally low plant Mo and high AO-OA extractable soil Mo contents (Fig. 172) can be largely eliminated by correcting the soil Mo values for pH (Fig. 173). This also improves the plant Mo — soil Mo correlation. Argentina stands close to the centre of the "international Mo field" (Fig. 20). Despite the relatively wide variation in plant Mo contents, no Mo values lie within the two lowest Mo Zones and only a few fall in Zones IV and V (Fig. 173). A single sample pair (44653) from Cordoba province yielded exceptionally high Mo contents.

Zinc. Generally, the Zn contents of Argentinian soils and plants are at an average international level (Figs 41 and 43). The mean values for plant Zn content are slightly below and those for soil Zn a little above the respective international means (Figs 174 and 175). Irrespective of the method used to extract Zn from soils, the correlations are high and both methods give a very similar picture of the Zn levels in the country. No sample had a low value (Zones I and II) but about ten percent of the samples fall in the high Zones (IV and V). The Zn contents of some of these, mainly from the Gral Pueyrredon area, were relatively high in both plant and soil. Zn deficiency in Argentina is possible but less likely than in most other countries. See also Section 2.7.

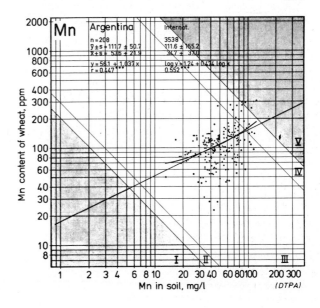

Fig. 170. Regression of Mn content of pot-grown wheat (y) on DTPA extractable soil Mn (x), Argentina.

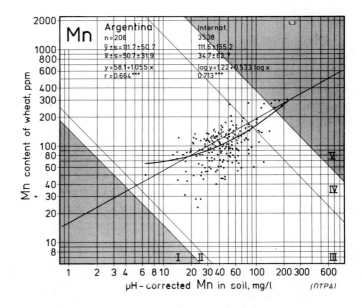

Fig. 171. Regression of Mn content of pot-grown wheat (y) on DTPA extractable soil Mn for pH (x), Argentina.

6.1.4. Summary

The Argentinian soil samples (mainly Phaeozems) were rather homogeneous with medium texture, medium cation exchange capacity, low pH and high organic matter content. The K contents of soils and plants are the highest and N contents among the highest in this study. P contents are of an average international level. The Ca and Mg contents of soils are on the low side but those of maize vary widely. Among the micronutrients only some Cu and Zn contents are relatively low. No low values were recorded for B, Fe, Mn, or Mo.

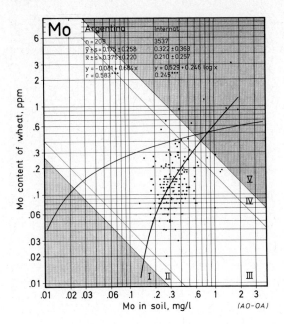

Fig. 172. Regression of Mo content of pot-grown wheat (y) on ammonium oxalate-oxalic acid extractable soil Mo (x), Argentina.

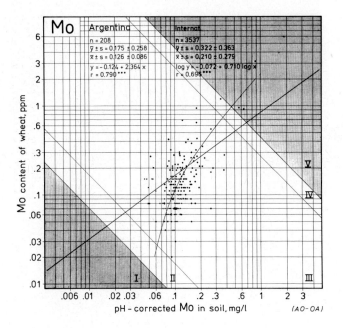

Fig. 173. Regression of Mo content of pot-grown wheat (y) on pH-corrected AO-OA extractable soil Mo (x), Argentina.

The great majority of values for these elements are within the normal international range but, depending on the element and extraction method used, one to ten percent of the values fall within Zones IV and V. This study suggests that, in the Argentinian areas sampled, micronutrient deficiencies are less likely than in most other countries.

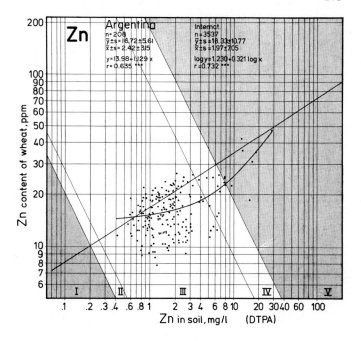

Fig. 174. Regression of Zn content of pot-grown wheat (y) on DTPA extractable soil Zn (x), Argentina.

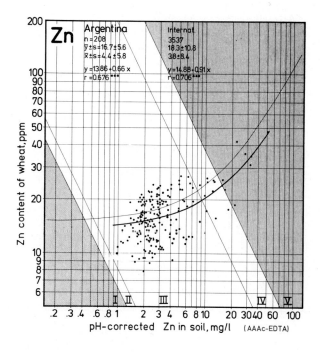

Fig. 175. Regression of Zn content of pot-grown wheat (y) on AAAc-EDTA extractable soil Zn corrected for pH (x), Argentina.

6.2 Brazil

6.2.1 General

The Brazilian soil and original wheat samples for this study come from the southern part of the country (Fig. 176). No maize fields were sampled. The majority (69 %) of the sampled soils were classified as Ferralsols. Other soils were Nitosols (14 %), Luvisols (9 %) and Acrisols (8 %).

Fig. 176. Sampling sites in Brazil. The last three numerals of each sample pair number are given.

Fig. 177. Frequency distributions of texture, pH, organic carbon content, and cation exchange capacity in soils of Brazil (columns). Curves show the international frequency of the same characters.

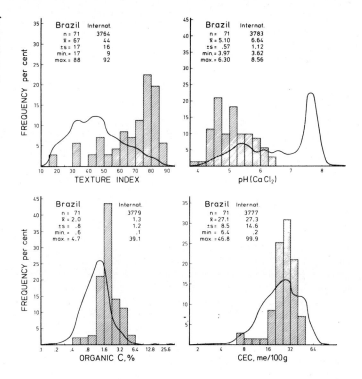

Although some light textured soils are represented the majority of soils are very heavy textured (Fig. 177). The national mean texture index is the highest recorded in this study (Appendix 4). All Brazilian soils are severely or moderately acid, pH(CaCl$_2$) 3.97—6.30, and a lower average soil pH occurs only in two African countries. The contents of soil organic matter are high compared to most other countries. In spite of this and the heavy texture of most soils, the cation exchange capacity values are not very high but usually at an average international level.

6.2.2 Macronutrients

The total **nitrogen** contents of Brazilian soils are high, the applied rates of nitrogen fertilizer low (19 ± 11 kg N/ha) and the N contents of wheat at an average international level (Figs 178 and 6; Appendix 2).

In spite of a fairly low **phosphorus** content in Brazilian soils the P contents of original wheats are at an average international level or even slightly above (Figs 179 and 7). The sampled wheat crops were heavily fertilized with phosphorus (49 ± 30 kg P/ha) which apparently increased the P contents of plants. The variations in the plant P, soil P and applied fertilizer P are wide.

The **potassium** contents of Brazilian soils and wheats are similar to those of phosphorus: low exchangeable soil K, fairly high plant K and relatively high K fertilization (33 ± 18 kg K/ha), all varying widely (Fig. 180).

The **calcium** contents of wheats and the contents of exchangeable Ca in Brazilian soils are low; the national mean values for both are among the lowest recorded in this study (Figs 181 and 9; Appendix 2). Many of the sampled fields had been limed during recent

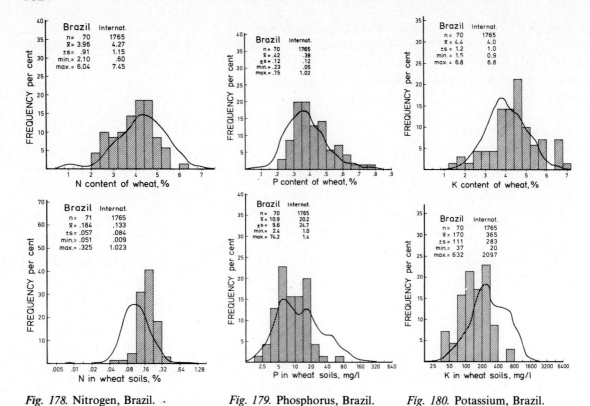

Fig. 178. Nitrogen, Brazil.   Fig. 179. Phosphorus, Brazil.   Fig. 180. Potassium, Brazil.

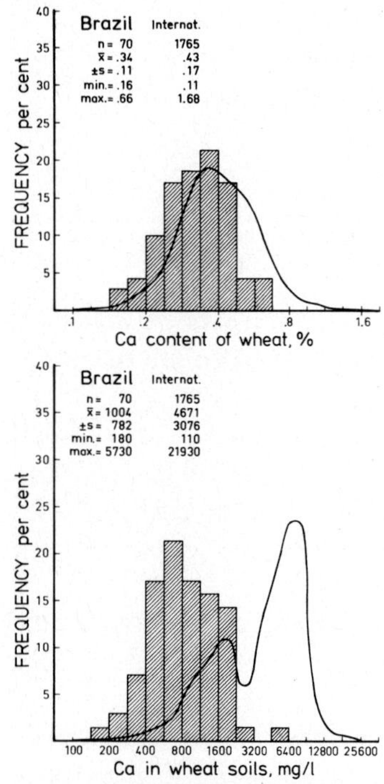

Fig. 181. Calcium, Brazil. 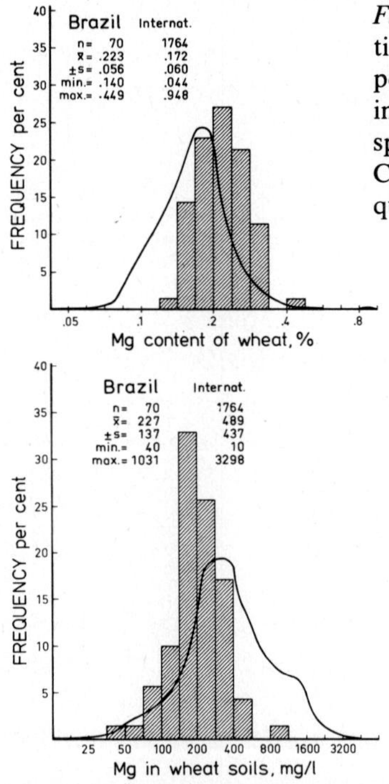   Fig. 182. Magnesium, Brazil.

Figs 178–182. Frequency distributions of nitrogen, phosphorus, potassium, calcium and magnesium in original wheat samples and respective soils (columns) of Brazil. Curves show the international frequency of the same characters.

years with small to moderate rates of application.

In spite of the low exchangeable **magnesium** contents of soils the Mg contents of wheat were high (Figs 182 and 10). No information was obtained on possible magnesium fertilization or on the quality (Mg content) of applied lime.

6.2.3 Micronutrients

Boron. On average the B contents of soils and plants are slightly low in the "international B fields" (Figs 22 and 25). The ranges of variation for both the soil B and the plant B are exceptionally narrow (Figs 183 and 184). The whole of the Brazilian sample material is

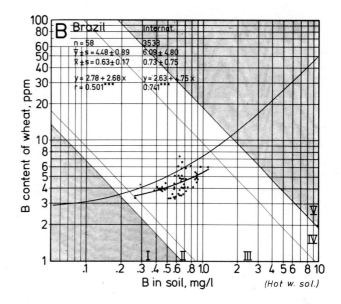

Fig. 183. Regression of B content of pot-grown wheat (y) on acid ammonium soluble soil B (x), Brazil. For details of summarized international background data, see Chapter 4.

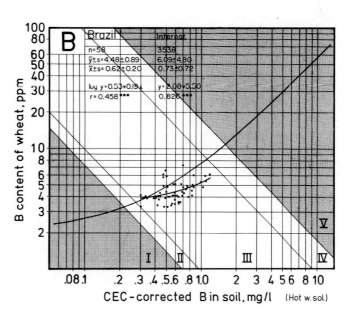

Fig. 184. Regression of B content of pot-grown wheat (y) on CEC-corrected (hot water soluble) soil B (x), Brazil.

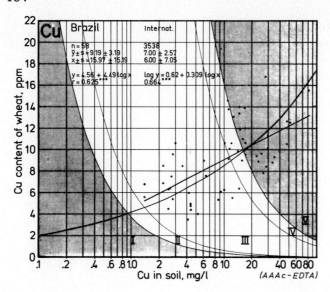

Fig. 185. Regression of Cu content of pot-grown wheat (y) on acid ammonium acetate-EDTA extractable soil Cu (x), Brazil.

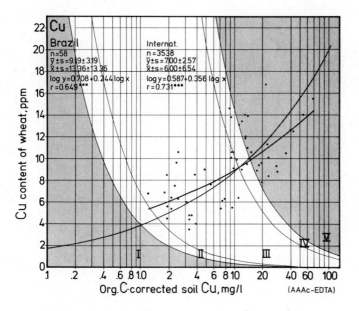

Fig. 186. Regression of Cu content of pot-grown wheat (y) on organic carbon-corrected (AAAc-EDTA extr.) soil Cu (x), Brazil.

within the "normal" B range (Zone III). There is not indication of B excess but at some sites response to B, depending on the crop, could be obtained (see also Section 2.7).

Copper. In spite of the generally high organic matter contents and low pH of the soils (see Figs 28 and 30) Brazil stands in the "international Cu fields" at the very highest level on a par with the Philippines and Italy (Figs 27 and 29). This indicates high total Cu contents in Brazilian soils. The Cu contents of plants as well as the contents of extractable soil Cu vary largely. One half of the Cu values are within the normal Cu range and the other half in the two highest Cu Zones (IV and V); some would therefore indicate at least excess of Cu if not Cu toxicity (Figs 185 and 186). The distribution of the highest Brazilian Cu values is not limited to any specific geographical area but they are relatively more common in Luvisols than in other soils (see also Fig. Cu in Appendix 7).

Fig. 187. Regression of Fe contents of pot-grown wheat (y) on acid ammonium acetate-EDTA extractable soil Fe (x), Brazil.

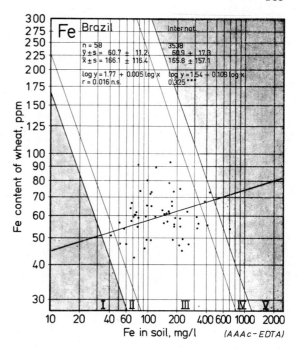

Iron. The Brazilian averages for the Fe contents of plants and soils are almost the same as those for the whole international material (Fig. 187). Since the variations are relatively small most of the Brazilian material is within the normal Fe range and there are no extreme Fe values.

Manganese. The mean Mn contents of Brazilian soils and plants are about three times as high as the respective general international means in this study (Fig. 188). Owing to the low pH of Brazilian soils, correction of soil Mn values for pH not only improves the plant Mn — soil Mn correlation but also increases the difference between the Brazilian and international mean values for soil Mn (Fig. 189). The Brazilian mean for extractable soil Mn is the highest recorded in this study and the national mean for plant Mn is exceeded only by three African countries (Figs 33 and 37). Although widely varying plant and soil Mn values are included in the Brazilian material only one Mn sample pair falls in the lowest Mn Zone. More than half the material is within Zones IV and V, including many samples with extremely high Mn contents. The highest Mn values were relatively more typical for Nitosols and plants grown on these than for other soils, although many extremely high Mn contents were found in Ferralsols and associated plants. Mn toxicity to soybean in Brazil was reported e.g. by Almeida and S'fredo (1979). The risk of Mn toxicity could be reduced or even eliminated by raising the soil pH through liming (see also Fig. 34).

Molybdenum. Unlike Mn, the Mo contents of Brazilian soils and plants are low (Figs 190 and 191). On average, the pot-grown wheat contains ten times as much Mo as the wheat grown on Brazilian soils (0.322 and 0.032 ppm, respectively) and the Brazilian mean is the lowest among the 29 countries investigated (Figs 15 and 20). The apparently medium high AO-OA extractable soil Mo values become much lower when corrected for pH, and would be even lower if corrected for texture (Fig. 21). The effect of texture must therefore

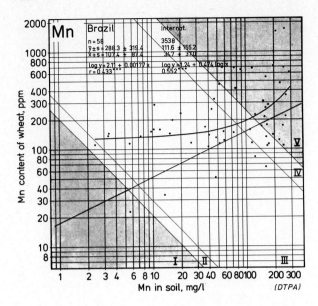

Fig. 188. Regression of Mn content of pot-grown wheat (y) on DTPA extractable soil Mn (x), Brazil.

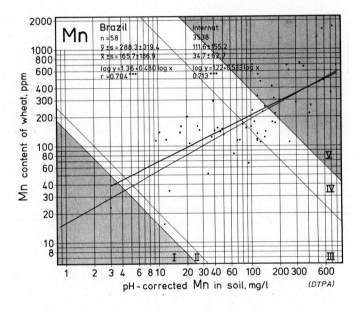

Fig. 189. Regression of Mn content of pot-grown wheat (y) on DTPA extractable soil Mn corrected for pH (x), Brazil.

be kept in mind when interpreting the Brazilian Mo data. Probably because of analytical inaccuracy in determining very low and only slightly varying Mo contents, the correlation between plant Mo and soil Mo for Brazilian material is not significant except when corrected for both pH and texture (Table 8). Less than one-third of the Brazilian Mo values are within the lower half of the normal range, and almost half lie in the lowest Mo Zones (Fig. 191). Low Mo contents are perhaps more typical of the Ferralsols and associated plants than of other Brazilian soils. To recapitulate, it is evident that the basic reason for the low Mo contents of Brazilian wheat is the low plant availability of Mo due to two typical characteristics of the Brazilian soils, namely the acidity and the heavy texture (see Figs 16 and 19). The great majority of Brazilian wheats with the lowest Mo contents (0.02

Fig. 190. Regression of Mo content of pot-grown wheat (y) on ammonium oxalate-oxalic acid extractable soil Mo (x), Brazil.

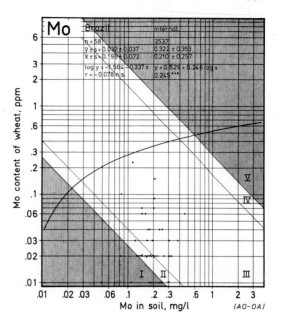

Fig. 191. Regression of Mo content of pot-grown wheat (y) on pH-corrected AO-OA extractable soil Mo (x), Brazil.

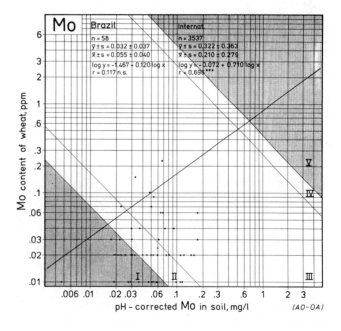

ppm or less) had grown on very heavy textured soils (TI > 70) usually accompanied by low pH (< 5.0). Obviously, in the long run, raising the soil pH by liming would be the best means of increasing the availability of Mo.

Zinc. The average Zn content of Brazil is high in the "international Zn fields" (Figs 41 and 43). The standard deviations of both plant Zn and soil Zn, however, are larger than in

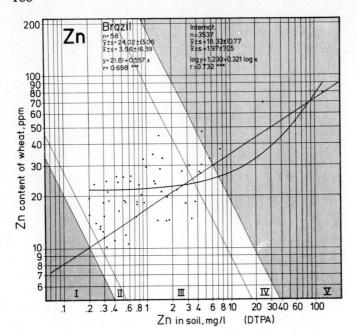

Fig. 192. Regression of Zn content of pot-grown wheat (y) on DTPA extractable soil Zn (x), Brazil.

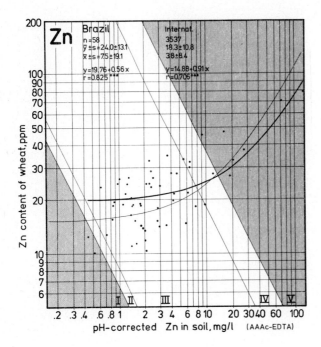

Fig. 193. Regression of Zn content of pot-grown wheat (y) on AAAc-EDTA extractable soil Zn corrected for pH (x), Brazil.

most other countries and irrespective of the method of extracting soil Zn the correlations are good (Figs 192 and 193). A relatively high percentage (16—21 % depending on extraction method) of the Zn sample pairs are within the two highest Zn Zones and two of these represent very high values. Although no sample was extremely low in Zn, a response to Zn at several sites is possible (see also Section 2.7).

6.2.4 Summary

Most of the Brazilian soils studied, mainly Ferralsols, have heavy texture, low pH, high organic matter content and medium cation exchange capacity.

The macronutrient (P, K, Ca and Mg) content of soils is usually low but that of N high. The P, K and N contents of original Brazilian wheat are at an average international level, but whereas the Ca contents tend to be low, the Mg contents tend to be high. The sampled crops were fertilized with P and K at high rates but with low rates of N.

The standing of micronutrients in Brazilian soils and plants varies considerably from element to element. Only the contents of Fe are at the "normal" international level. The content of B is somewhat low, that of Mo is very low while those of Cu and Mn are very high. The contents of Zn vary widely between high and low.

6.3 Ecuador

6.3.1 General

The wheat and soil samples were collected from Cañar, Carchi, Chimborazo, Imbabura and Pichincha Provinces. Maize soils were not sampled. Because of the small quantity of soil in the samples received from Ecuador only a part of the analytical programme could be carried out, and therefore only incomplete data can be presented.

The pH of soils vary greatly, from 4.7 to 7.7, the majority being on the acid side (Fig. 194). The organic matter contents are at an average international level and cation exchange capacity somewhat below that. Very high electrical conductivity is typical for soils sampled from Ecuador (Appendix 2).

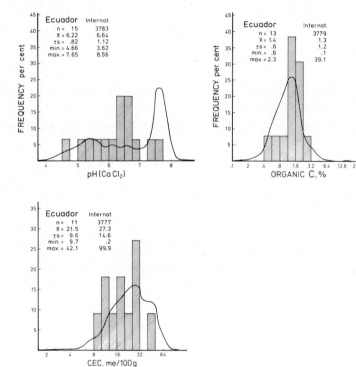

Fig. 194. Frequency distributions of pH, organic carbon content, and cation exchange capacity in soils of Ecuador (columns). Curves show the international frequency of the same characters.

6.3.2 Macronutrients

In spite of the relatively high total **nitrogen** contents of soils and some nitrogen fertilization (44 ± 13 kg N/ha) the N contents of the original wheats are low by international standards (Figs 195 and 6; Appendix 2). In spite of the small amount of material submitted, the ranges of plant and soil N variation are quite wide. One wheat sample especially (No. 47356) differs from the others due to its high N content. This sample had grown on a soil with the highest total N content recorded in any Ecuador soil, and had also been fertilized

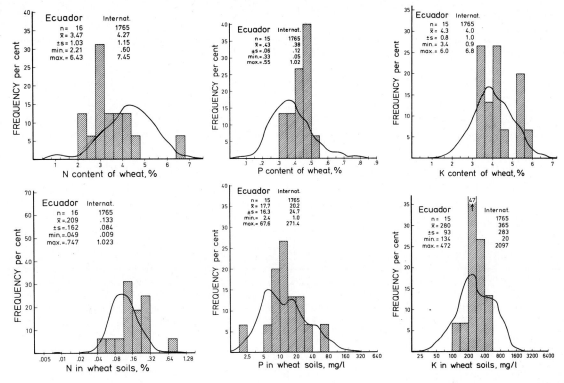

Fig. 195. Nitrogen, Ecuador.   *Fig. 196.* Phosphorus, Ecuador.   *Fig. 197.* Potassium, Ecuador.

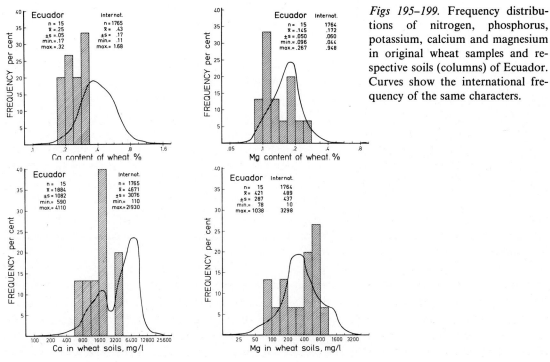

Fig. 198. Calcium, Ecuador.   *Fig. 199.* Magnesium, Ecuador.

Figs 195–199. Frequency distributions of nitrogen, phosphorus, potassium, calcium and magnesium in original wheat samples and respective soils (columns) of Ecuador. Curves show the international frequency of the same characters.

with the highest amount of N (60 kg N/ha) applied to the plants sampled from that country.

The **phosphorus** and **potassium** contents of soils in Ecuador are slightly below, and the P and K contents of wheats somewhat above the respective international means (Figs 7, 8, 196 and 197). The rates of phosphorus fertilizers (23 ± 8 kg P/ha) applied to the sampled crops were higher than those in most other countries in this study but somewhat less potassium fertilizers were used (13 ± 6 kg K/ha, Appendix 5).

The frequency distributions presented in Fig. 198 show that the occurrence of **calcium** in the soils and wheat of Ecuador is low in an international context. In fact, the mean Ca content of wheat in Ecuador is the lowest national mean obtained in this study (Appendix 2). The **magnesium** contents are slightly lower than average (Figs 199 and 10; Appendix 2).

6.3.3 Micronutrients

The small size of soil samples received from Ecuador precluded pot experiments, and the value of having an indicator plant of one species and one variety in the international context was lost (see Section 1.2). It is therefore attempted to give a general picture of the micronutrient content of plants in Ecuador by comparing analytical data of the original wheat samples with the respective data from other countries given in Appendix 2.

Boron. The B contents of the sampled soils and plants of Ecuador vary very little. All plant B values are below and only one hot water extractable soil B value exceeds the respective international means (6.56 ppm and 0.81 mg/l). The national mean for soil B (0.42 mg/l) is the fifth lowest and that of plant B (2.33 ppm) the second lowest in the entire study (Appendix 2). From these limited data, B deficiency seems likely in many locations in Ecuador. Its deficiency for several crops and response to B fertilization have been reported by Tollenar (1966) and Mestanza and Lainez (1970).

Copper. The Cu content of plants in Equador, based on the results of plant analyses, seems to be generally low. However, although the national mean (7.7 ± 1.7 ppm) is the seventh lowest (Appendix 2), the Ecuador material does not include any extremely low values.

Manganese. Mn contents determined from the Ecuador wheats are, with one exception, below the international average and the national mean (34.3 ± 18.8 ppm) is the very lowest recorded for original wheats (Appendix 2). The uncorrected DTPA extractable soil Mn in Ecuador soils varies from 1.2 to 76.7 mg/l and the mean (24.1 mg/l) is only slightly below the international mean. If these soil values were corrected (see Fig. 34 and related text) it would lower the mean by about one-third and greatly reduce many of the single values. It is possible that a response to Mn fertilizer would be obtained especially in soils where the soil pH is relatively high (e.g. sites 47353—54 in Imbabura and 47362—65 in Chimborazo).

Molybdenum. Although the average Mo content of original Ecuador wheats (0.40 ± 0.40 ppm) is far below the international mean (0.94 ± 1.03 ppm), the mean for the uncorrected AO-OA extractable soil Mo (0.249 ± 0.232 mg/l) exceeds the respective international mean (0.204 ± 0.209 mg/l, Appendix 2). Correcting the soil Mo values for pH would largely eliminate this contradiction (see Fig. 16 and related text) by lowering the soil Mo values. This limited material from Ecuador does not include any exceptionally high Mo values, and although many Mo values are low there is no indication of any severe Mo deficiency.

Zinc. The plant and soil Zn values determined from the Ecuador sample material are usually at or somewhat below the average international level (Appendix 2). No extremes were recorded.

6.3.4 Summary

Because of the limited material and incomplete analytical data only a vague picture of the soil nutrient content in Ecuador can be deduced. Most of the sampled soils have pH ($CaCl_2$) below 7, a medium content of organic matter and low to medium cation exchange capacity. The macronutrient contents of soils are internationally low to medium, N being an exception. For micronutrients some very low values, especially of Mn and B, were recorded but, in general, low to medium micronutrient contents are typical for soils sampled from Ecuador.

6.4 Mexico

6.4.1 General

The geographical distribution of sampling sites in Mexico is given in Fig. 200, separately for wheat and maize fields. The most important wheat and maize production areas were quite well covered by sampling.

Out of the 214 soils classified, Xerosols (28 %) were the most common, followed by Kastanozems (21 %), Yermosols (16 %) and Fluvisols (14 %). Both wheat and maize were grown on these soils. The rest of the soils were used mainly for growing maize: Regosols (5 %), Vertisols (4 %), Phaeozems (3 %), Andosols (3 %), Arenosols (2 %), Chernozems

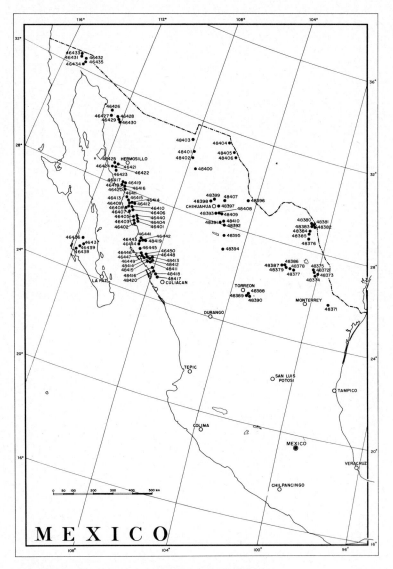

Fig. 200 a. Sampling sites in Mexico (wheat fields).

(1 %), Cambisols (1 %) and Ferralsols (1 %).

As illustrated in Fig. 201 the Mexican soils vary greatly in texture, from very coarse to very fine. The great majority of soils are alkaline but some with quite low pH are included. Most of the soils are low in organic matter and although soils with medium or high cation exchange capacity dominate there are also some with very low CEC. Electrical conductivity and $CaCO_3$ equivalent values vary greatly (Appendixes 2—4). Every sixth soil contains more than 500 mg/l of CH_3COONH_4 extractable sodium. These have $pH(CaCl_2)$ usually around 8, high electrical conductivity and high B content.

In general, a wide variety of soils, with extensively varying characteristics are represented in the Mexican material. Consequently, the nutrient contents of these soils differed greatly.

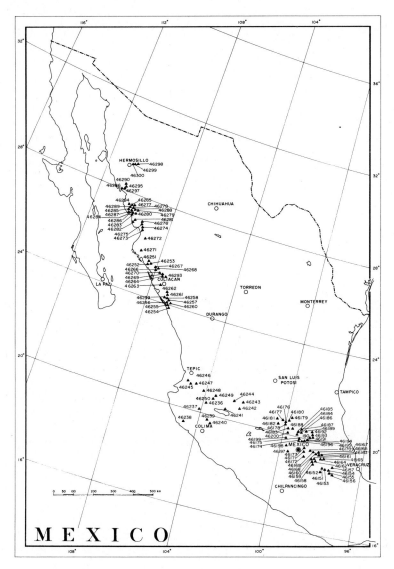

Fig. 200 b. Sampling sites in Mexico (maize fields).

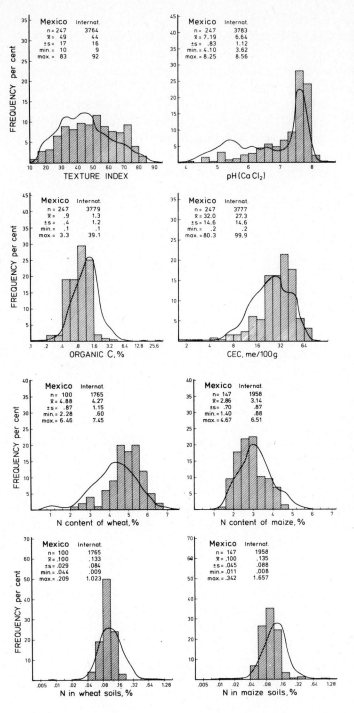

Fig. 201. Frequency distributions of texture, pH, organic carbon content, and cation exchange capacity in soils of Mexico (columns). Curves show the international frequency of the same characters.

Figs. 202–206. Frequency distributions of nitrogen, phosphorus, potassium, calcium and magnesium in original wheat and maize samples and respective soils (columns) of Mexico. Curves show the international frequency of the same characters.

Fig. 202. Nitrogen, Mexico.

6.4.2 Macronutrients

The total **nitrogen** contents of Mexican soils are relatively low from the international viewpoint (Fig. 202, Appendixes 2 and 3). The N contents of both wheat and maize vary widely. This is apparently due to greatly varying rates of nitrogen fertilizer application

Fig. 203. Phosphorus, Mexico.

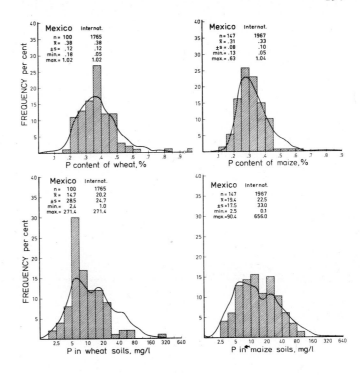

(wheat: 122 ± 54 and maize: 40 ± 49 kg N/ha). The average rate of N applied to wheat is one of the highest recorded in this study and seems to be the main reason for the high N contents of wheat. For example, all wheat samples with a N content less than 3.6 percent had received no N fertilizer at all while those with N ≥ 6.0 % were fertilized with 100 to 153 kg N/ha. While the N contents of wheat were high those of maize are low, and when the plant N contents are pooled (Fig. 6) the Mexican average equalled the international mean.

The **phosphorus** content of Mexican wheat soils is relatively low while the P content of wheat is at an average international level (Fig. 203). In both cases the range of variation is wide. The P contents of maize and maize soils are at an average international level but also vary widely. The sampled wheat crops were fertilized with phosphates at somewhat higher rates than the maize (9 ± 13 and 4 ± 7 kg P/ha, respectively). For the international rating of P in Mexico, see Fig. 7.

The national mean contents of **potassium** in wheat and wheat soils are among the highest in this study (Fig. 204, Appendix 2). The mean K content of maize soils is also high but that of maize itself is below the international average. Relatively wide variations were typical of all K values. Practically no potassium fertilizer was applied to either of the sampled crops. A good response to potassium fertilizer could only be expected at the lowest soil and maize K levels. See also Fig. 8.

The content of exchangeable **calcium** in Mexican soils, especially in wheat soils, is high (Fig. 205). In wheat the Ca content is, on average, at the normal international level but the maize samples contain relatively little Ca. In all cases there are wide variations.

The **magnesium** content of Mexican soils is generally high but again with rather wide

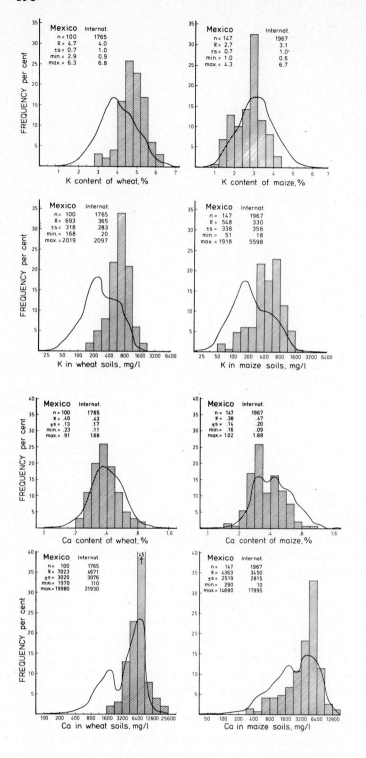

Fig. 204. Potassium, Mexico.

Fig. 205. Calcium, Mexico.

variations (Fig. 206). The two indicator crops have almost equal mean Mg contents while the wheat Mg content is above the respective international average and maize below that. Some quite low Mg contents of maize are included. See also the position of Mexico in the "international Mg field" (Fig. 10).

Fig. 206. Magnesium, Mexico.

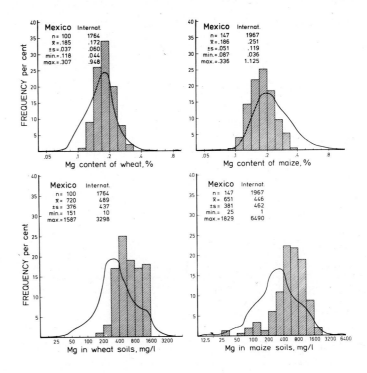

6.4.3 Micronutrients

Boron. Owing to the heterogeneity of the Mexican soil material the B contents of soils and plants vary greatly from one sample to another (Figs 207 and 208). High B values dominate and in the "international B fields" (Figs 22 and 25) Mexico stands high. The frequency of high B values (points in Zones IV and V) is more than twice as high as in the whole international material. There seem to be no distinct geographical areas of high B but high values were recorded from several states including Hidalgo, Mexico and Tlaxcala in the central part of the country, as well as Sonora, Sinaloa and Coahuila in the west and north. High B contents were relatively more common for Regosols, Yermosols, Fluvisols and Xerosols and plants growing in such soils than for other Mexican soils (see also Fig. B in Appendix 7). Some of the high B values may be due to B contained in irrigation water.

Although only a few B sample pairs are within the two lowest B Zones and no extremely low B values were found, response to B fertilization may be obtained at several locations, especially if plants with a high B requirement are grown.

Copper. Mexico stands close to the centre of the "international Cu fields" (Figs 27 and 29). In only one sample pair (46225) are the Cu contents exceptionally high but low values are more frequent, two of them (46249—50) being very low (Figs 209 and 210). Many of the low Cu values came from the western states of Jalisco, Nayarit and Sinaloa, but low values are also found in the central states. All four Arenosols included in the Mexican soil material were low in Cu as were many Andosols, Regosols and Fluvisols.

Iron. Excluding Malta, where the number of samples was small, Mexico has the lowest

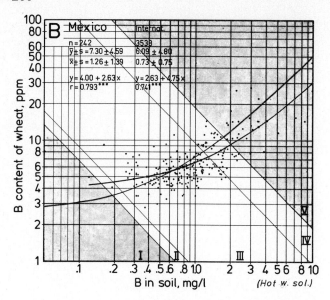

Fig. 207. Regression of B content of pot-grown wheat (y) on hot water soluble soil B (x), Mexico. For details of summarized international background data, see Chapter 4.

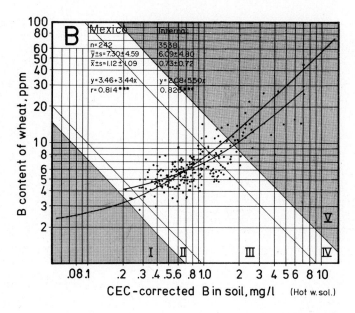

Fig. 208. Regression of B content of pot-grown wheat (y) on CEC-corrected (hot water soluble) soil B (x), Mexico.

national mean values for soil and plant Fe (Fig. 31). More than 25 % of the Fe values fall in the lowest Fe Zone and almost half of the Mexican material falls in the two lowest Zones (I and II, Fig. 211). Even the Fe values within Zone III are predominantly clustered toward the lower boundary. It seems likely that in many places crops sensitive to Fe deficiency would respond to iron fertilization. Soils low in Fe seem to be more common in the northern states, Nuevo León, Coahuila, Chihuahua and Sonora than in the southern parts of the country. Low Fe values were relatively more common in Yermosols and Kastanozems, but also occurred in the Regosols, Xerosols and Fluvisols.

Manganese. Mexico is close to the centre of the "international Mn field" (Figs 33 and

Fig. 209. Regression of Cu content of pot-grown wheat (y) on acid ammonium acetate-EDTA extractable soil Cu (x), Mexico.

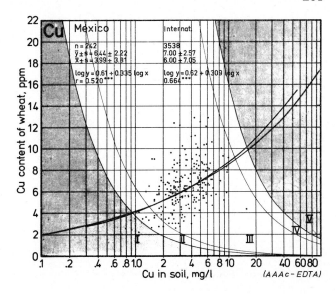

Fig. 210. Regression of Cu content of pot-grown wheat (y) on organic carbon-corrected (AAAc-EDTA extr.) soil Cu (x), Mexico.

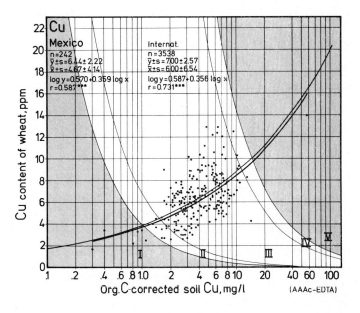

37). Although the bulk of Mexican material lies within the "normal" range for Mn (Figs 212 and 213) the variation is wide and with both high and more frequently, low values being recorded. Over ten percent of the Mn values fall in the two lowest Mn Zones. The lowest Mn contents were from alkaline soils and plants grown on alkaline soils except for sample pairs No. 46157 and 46246 (co-ordinates in Fig. 212: plant Mn 45, soil Mn 3.6; and plant Mn 33, soil Mn 0.9) where the soil pH(CaCl$_2$) values were 6.27 and 5.40, respectively. These soils were exceptionally coarse textured, low in organic carbon and apparently low in total Mn. The highest Mn Zones (IV and V) include samples only from sites with low soil pH, usually below 5.0. The geographical distribution of low and high Mn sites is

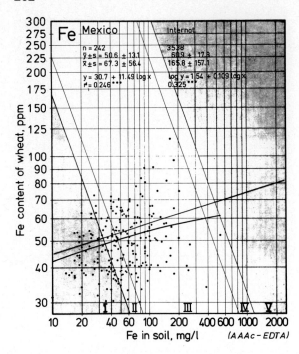

Fig. 211. Regression of Fe contents of pot-grown wheat (y) on acid ammonium acetate-EDTA extractable soil Fe (x), Mexico.

obscure. The low Mn values may be more typical of the soils of Baja California, Sonora and Sinaloa and high Mn of the Mexico and Jalisco States.

Molybdenum. As in case of other nutrients the contents of Mo in Mexican soils and plants vary considerably. As a whole, Mexico stands high in the "international Mo fields" (Figs 15, 20 and 21). Because of widely varying pH and texture in Mexican soils the relatively weak correlation between plant Mo and AO-OA extractable soil Mo is substantially improved by correction for pH and texture (Figs 214 and 215; Table 8). Almost without exception plants with low Mo content (< 0.1 ppm) have been grown on soils with low pH combined with either very fine or very coarse texture. Correspondingly, plants with high Mo content (> 1.0 ppm) have grown on alkaline, medium textured soils (see also Figs 16 and 19 and related text). The highest Mo values were recorded from Coahuila and Sonora, often in samples from irrigated sites.

Zinc. Regarding average Zn contents of both soils and plants, Mexico stands close to the centres of the "international Zn fields" (Figs 41 and 43) but samples vary widely in Zn content between low and high (Figs 216 and 217). About six percent of the Zn values lie in the two highest Zn Zones irrespective of the extraction method used, some of them relatively far into Zone V. However, low Zn values are more common. Depending on extraction method, 8 (DTPA) to 17 (AAAc-EDTA, pH-corr.) percent of the Zn values fall within the two lowest Zn Zones with a number of samples close to these. Although the most extreme values are lacking, shortage of Zn is likely in many Mexican soils. Most of the high Zn values were measured from the Hidalgo and Jalisco samples, while shortage of Zn may be more common of Queretaro, Sonora and Sinaloa than elsewhere.

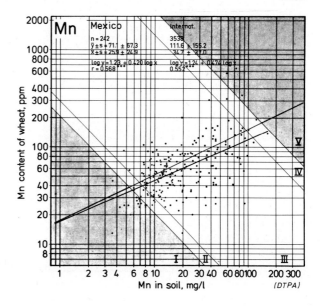

Fig. 212. Regression of Mn content of pot-grown wheat (y) on DTPA extractable soil Mn (x), Mexico.

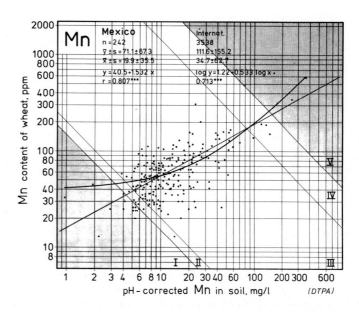

Fig. 213. Regression of Mn content of pot-grown wheat (y) on DTPA extractable soil Mn corrected for pH (x), Mexico.

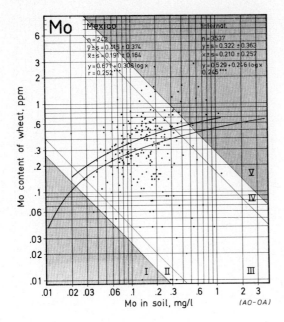

Fig. 214. Regression of Mo content of pot-grown wheat (y) on ammonium oxalate-oxalic acid extractable soil Mo (x), Mexico.

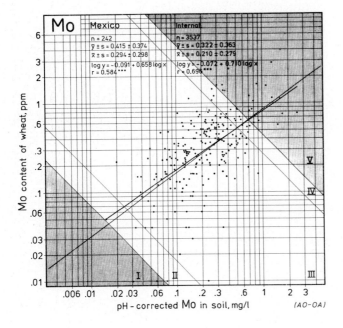

Fig. 215. Regression of Mo content of pot-grown wheat (y) on pH-corrected AO-OA extractable soil Mo (x), Mexico.

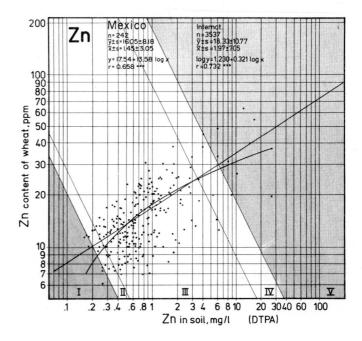

Fig. 216. Regression of Zn content of pot-grown wheat (y) on DTPA extractable soil Zn (x), Mexico.

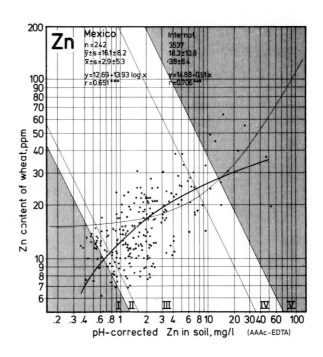

Fig. 217. Regression of Zn content of pot-grown wheat (y) on AAAc-EDTA extractable soil Zn corrected for pH (x), Mexico.

6.4.4 Micronutrient contents of original plants with special reference to varieties

Since the original Mexican wheat material consists only of high yielding varieties, comparisons with local varieties are not possible. Variety data were obtained for 71 of the 147 maize samples supplied, of which 24 were classified as HYV. This group consisted of varieties H 407, H 412, H 505 and H 517. The other variety group was more heterogeneous comprising seven different varieties. Since the majority (29 samples) of these were of a local variety, 'Criollo', comparison in Table 24 is made between the HYV and 'Criollo'. On average, the maize plants of the two variety groups were sampled at the same physiological age. Differences between these variety groups were found in their uptake of several micronutrients, but these were most marked for B, Mn and Zn.

Although the HYV maize plants were grown on soils lower in extractable B, Mn and Zn than plants of cv. 'Criollo', their average B, Mn and Zn contents were considerably higher.

Table 24. Comparison of micronutrient contents between high yielding maize varieties and local variety 'Criollo', respective soils and wheat (cv. 'Apu') grown on the same soils in pots. Differences between the mean contents of the two groups followed by the same index letters (a—a) are not statistically significant. Letters a—b indicate significant differences at 5, a—c at 1 and a—d at 0.1 percent level.

Micro-nutrient	Average Micronutrient Contents of:					
	original maize (ppm)		resp. soils (mg/l)[1]		wheats grown on same soils in pots	
	HYV (n = 24)	Criollo (n = 29)	HYV (n = 24)	Criollo (n = 29)	(n = 24)	(n = 29)
Boron	17.9[a]	7.8[d]	0.87[a]	1.01[a]	7.2[a]	7.0[a]
Manganese	76[a]	62[c]	9[a]	66[d]	53[a]	150[d]
Zinc	41[a]	28[a]	0.6[a]	4.4[d]	14[a]	20[b]

[1] CEC-corrected hot water sol. B; pH-corrected DTPA extr. Mn; DTPA extr. Zn.

When the different maize varieties were replaced with one variety of wheat, the plant and soil data for the three micronutrients were in relatively good accordance.

Although factors other than the genetic differences between the varieties, such as irrigation practices, may have affected the above results obtained for original maize, it would seem that much of the apparent contradiction between the results of plant and soil analyses is due to plant variety and that the HYV maize plants are more efficient at absorbing the micronutrients in question than the local variety, 'Criollo'. See also Section 1.2.4.

6.4.5 Summary

Great variations in general soil characteristics are typical of the Mexican soil material in this study. These variations are also reflected in the macronutrient contents of soils and plants.

With regard to micronutrients the concentration of Fe is exceptionally low from the international point of view. Although Cu, Mn and Zn values are generally somewhat low and those of B and Mo tend to be high, high Mn and Zn and low Mo and B contents were frequently recorded. At many locations responses to Fe and Zn fertilization would seem most likely.

6.5 Peru

6.5.1 General

Most of the sampling sites in Peru (Fig. 218) are located in the agriculturally important Sierra and Ceja Selva regions, where the annual rainfall varies from 200 to over 1000 mm. In many cases the sampled fields were irrigated.

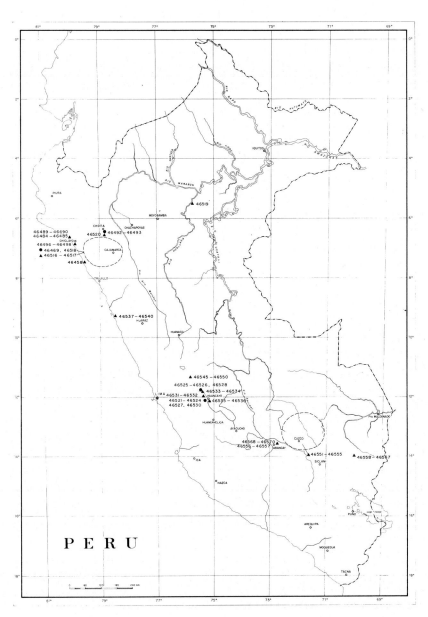

Fig. 218. Sampling sites in Peru (points = wheat fields, triangles = maize fields).

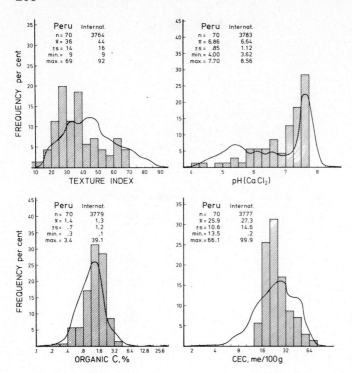

Fig. 219. Frequency distributions of texture, pH, organic carbon content, and cation exchange capacity in soils of Peru (columns). Curves show the international frequency of the same characters.

The textural variation of Peruvian soils is wide (Fig. 219), with coarse soils dominating though many soils with a high texture index are included. The majority of soils are alkaline or moderately acid but the material encompasses a few samples with low pH as well. The organic matter content, cation exchange capacity, electrical conductivity and $CaCO_3$ equivalent values are at an average international level, with fairly wide variations (Fig. 219, Appendixes 2—4).

6.5.2 Macronutrients

The mean contents of total **nitrogen** in Peruvian soils as well as N contents of maize are somewhat above the respective international averages but N contents of original wheats are low (Fig. 220). On average the N status is close to the international mean (Fig. 6). The sampled wheats were fertilized with relatively low rates of nitrogen fertilizers (average 38 ± 35 kg/ha; international average 66 ± 61 kg N/ha, Appendix 5). Wheats with a N content of less than 2.7 % N had not received any nitrogen fertilizers. The comparatively higher N contents of maize may be due to more abundant applications of nitrogen fertilizers (average 46 ± 58 kg N/ha; international average 27 ± 37 kg N/ha).

The **phosphorus** contents of maize and maize soils are at an average international level but those of wheat and wheat soils are somewhat lower (Fig. 221). In general, the average P content of Peruvian soils and crops corresponds closely to the average international level (Fig. 7). Only low to moderate rates of phosphates were applied to the sampled crops

Figs 220–224. Frequency distributions of nitrogen, phosphorus, potassium, calcium and magnesium in original wheat and maize samples and respective soils (columns) of Peru. Curves show the international frequency of the same characters.

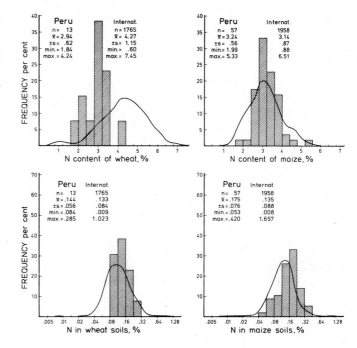

Fig. 220. Nitrogen, Peru.

Fig. 221. Phosphorus, Peru.

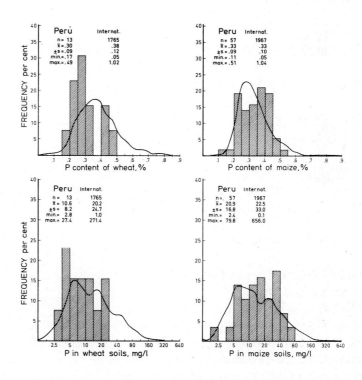

(maize: 8 ± 13 kg P/ha and wheat: 10 ± 13 kg P/ha). A quarter of the Peruvian sampled wheat and maize crops had a P content of less than 0.25 percent. With only two exceptions, these crops had not been fertilized with phosphates.

The generally low contents of **potassium** in Peruvian soils is reflected in the K contents

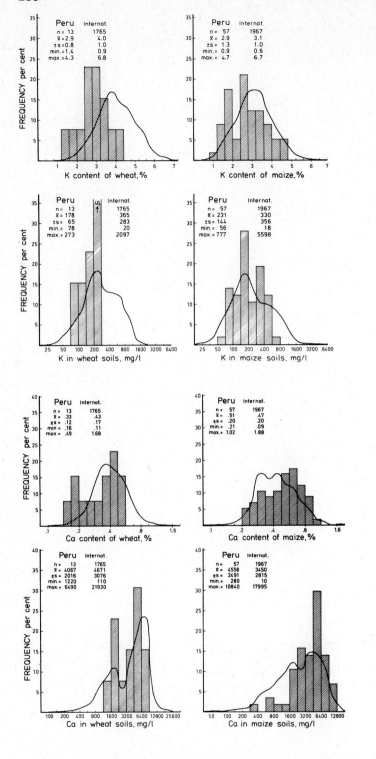

Fig. 222. Potassium, Peru.

Fig. 223. Calcium, Peru.

of plants, especially in the case of wheat (Fig. 222). Rates of applied potassium fertilizer were relatively low (wheat: 12 ± 19 and maize: 9 ± 17 kg K/ha). All sampled crops with low K content (< 2.4 %) were grown on low-K soils (< 200 mg/l) and/or received no potassium fertilizer. See also Fig. 8.

Fig. 224. Magnesium, Peru.

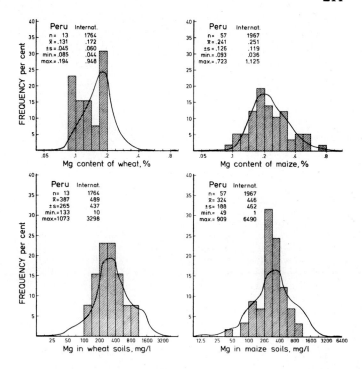

The **calcium** and **magnesium** contents of maize and maize soils were at an average international level or slightly above it, while wheat and wheat soils were somewhat lower in these elements (Figs 223 and 224). Consequently, Peru lies close to the centres of the "international Ca and Mg fields" (Figs 9 and 10).

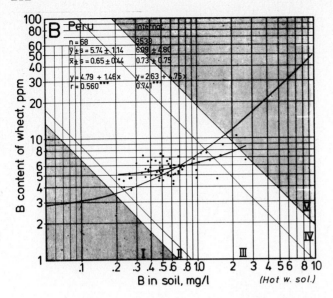

Fig. 225. Regression of B content of pot-grown wheat (y) on hot water soluble soil B (x), Peru. For details of summarized international background data, see Chapter 4.

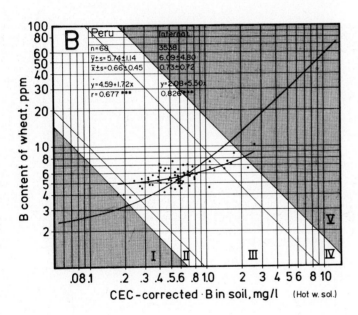

Fig. 226. Regression of B content of pot-grown wheat (y) on CEC-corrected (hot water soluble) soil B (x), Peru.

6.5.3 Micronutrients

Peru is usually located close to the centres of the "international micronutrient fields" (Figs 15, 20, 22, 25, 27, 29, 31, 33, 37, 41 and 43). Although none of the **boron** values measured in the sample material from Peru showed very low B contents (Figs 225 and 226), several of them are low enough to indicate the possibility of B deficiency, if crops with high B requirement are grown (see also Section 2.7). Excess of B seems unlikely.

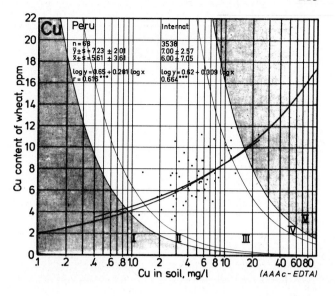

Fig. 227. Regression of Cu content of pot-grown wheat (y) on acid ammonium acetate-EDTA extractable soil Cu (x), Peru.

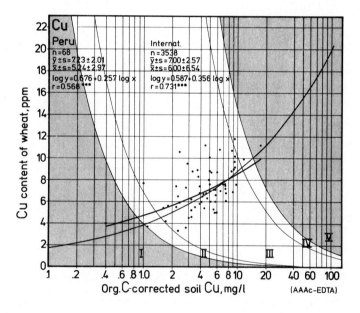

Fig. 228. Regression of Cu content of pot-grown wheat (y) on organic carbon-corrected (AAAc-EDTA extr.) soil Cu (x), Peru.

For the **copper, iron** and **manganese** values recorded from the sample material, the ranges of variation are relatively narrow (Figs 227—231), and only a few sample pairs fall outside the "normal" range (Zone III), and seldom at extreme locations within Zones I or V. Only at one site (No 46546 in Tarma) were very low Cu values measured and quite high plant and soil Mn contents were recorded at three sites (46549, -528 and -565). Two of the latter originated from Tarma in acid soils with pH(CaCl$_2$) 4.00 and 5.28 and one from Paucartambo, soil pH 4.27.

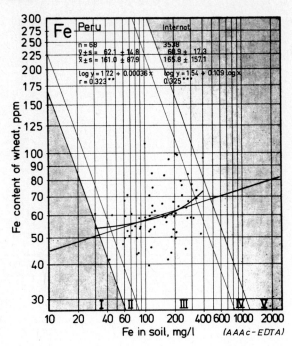

Fig. 229. Regression of Fe contents of pot-grown wheat (y) on acid ammonium acetate-EDTA extractable soil Fe (x), Peru.

The contents of **molybdenum** were more variable (Figs 232 and 233). In addition to four relatively high plant and soil Mo values, one especially low one was recorded. This originated from the same acid soil in Tarma (46549) as the highest Mn value. Also other low Mo values were measured from samples of high Mn content.

Although none of the **zinc** values falls in the lowest Zn Zone, many of them show quite low Zn contents of both soils and plants (Figs 234 and 235) indicating the possibility of Zn shortage (see Section 2.7). The majority of Zn values are normal but in samples from a couple of sites originating in San Lorenzo, Janja, quite high Zn contents were measured.

6.5.4 Summary

Most of the Peruvian soils are coarse textured, slightly alkaline, have medium organic matter content and medium cation exchange capacity. The macronutrient situation is variable, with low to medium contents dominating. For micronutrients normal values are typical. However, the limited data give some indications of possible B, Mo and Zn shortages. In a few samples Mn, Mo and Zn contents were high.

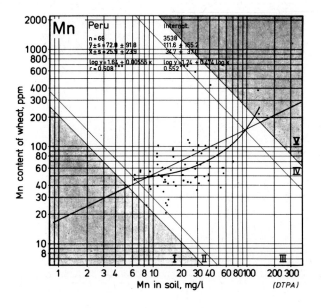

Fig. 230. Regression of Mn content of pot-grown wheat (y) on DTPA extractable soil Mn (x), Peru.

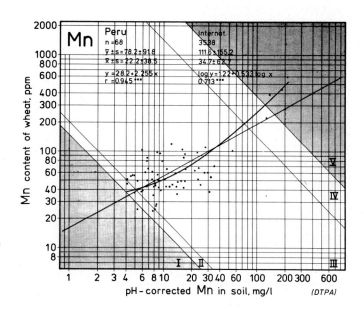

Fig. 231. Regression of Mn content of pot-grown wheat (y) on DTPA extractable soil Mn corrected for pH (x), Peru.

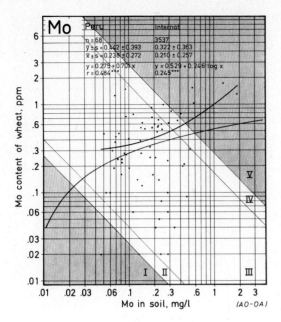

Fig. 232. Regression of Mo content of pot-grown wheat (y) on ammonium oxalate-oxalic acid extractable soil Mo (x), Peru.

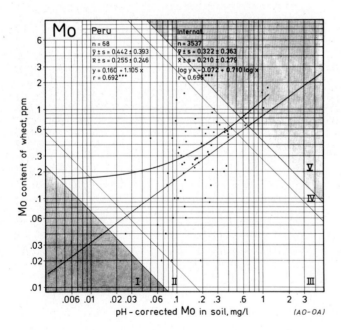

Fig. 233. Regression of Mo content of pot-grown wheat (y) on pH-corrected AO-OA extractable soil Mo (x), Peru.

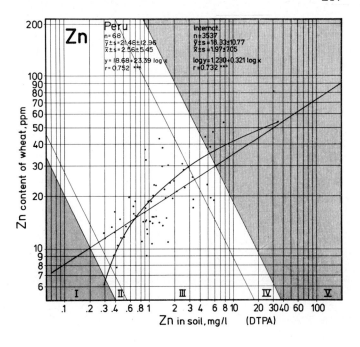

Fig. 234. Regression of Zn content of pot-grown wheat (y) on DTPA extractable soil Zn (x), Peru.

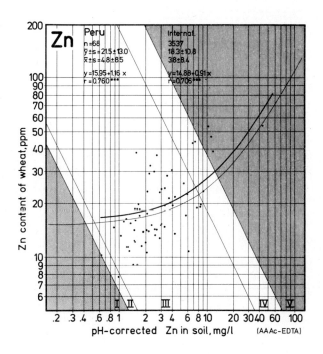

Fig. 235. Regression of Zn content of pot-grown wheat (y) on AAAc-EDTA extractable soil Zn corrected for pH (x), Peru.

7. Far East

7.1 India

7.1.1 General

Soil and original wheat and maize samples received from India for this study came from five States: Haryana (100 sample pairs), Punjab (70), Bihar (50), Delhi (45) and Madhya Pradesh (30). The approximate sampling sites are given in Figure 236. Of the 160 soils

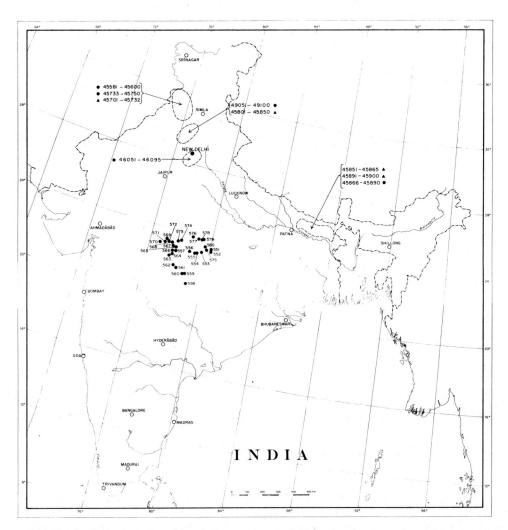

Fig. 236. Sampling sites in India (points = wheat fields, triangles = maize fields). In Madhya Pradesh area only the last three numerals of each sample pair number (45551—45580) are given.

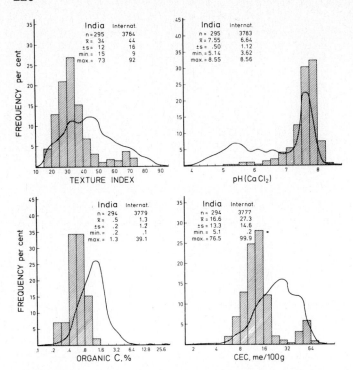

Fig. 237. Frequency distributions of texture, pH, organic carbon content, and cation exchange capacity in soils of India (columns). Curves show the international frequency of the same characters.

classified according to the system adopted by FAO, 55 were Cambisols located mostly in the Punjab, 40 Fluvisols (Bihar and Delhi), 34 Luvisols (Bihar) and 28 Vertisols (Madhya Pradesh). Three Haryana soils were classified as Kastanozems.

Texturally the Indian soils of this study are coarse (Fig. 237). The national mean texture index is among the lowest (Appendixes 2—4). The soils of Madhya Pradesh, however, differ distinctly from other Indian soils. All 30 of them are fine textured (TI = 49 — 73). Equally fine textures (TI 49 and 54) were measured only in two other Indian soils, both from Haryana.

Almost all Indian soils are alkaline, contain little organic matter and have low cation exchange capacity. The national mean value for soil pH is one of the highest and those for organic carbon content and CEC among the lowest in this study. With regard to CEC, the Madhya Pradesh soils differ again from the others; all have a CEC of 30.0 me/100 g or more and other Indian soils from 5.1 to 26.0 me/100 g. Electrical conductivity values of Indian soils (Appendixes 2—4) vary considerably but are generally well below the international average. With the exception of 25 soils from Bihar (45851—75) the $CaCO_3$ equivalent values of Indian soils are relatively low. The sodium contents are usually of the average international level.

7.1.2 Macronutrients

The Indian national mean values for total soil **nitrogen** in both wheat soils and maize soils are the lowest recorded in this study (Appendixes 2—4, Fig. 6). The ranges of variation are also exceptionally narrow (Fig. 238) and only in three Indian soil samples N values exceeding the international mean (0.133 %) were recorded. Unlike soil N, the N contents of original wheat and maize samples vary widely but are usually low. In part, the wide variation is apparently due to largely varying rates of nitrogen fertilizer applications (wheat:

Figs 238–242. Frequency distributions of nitrogen, phosphorus, potassium, calcium and magnesium in original wheat and maize samples and respective soils (columns) of India. Curves show the international frequency of the same characters.

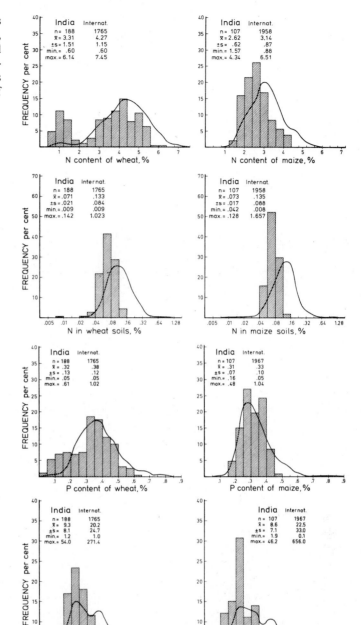

Fig. 238. Nitrogen, India.

Fig. 239. Phosphorus, India.

75 ± 51 and maize: 106 ± 57 kg N/ha). In the case of wheat the effects of varieties might seem to be very clear. Nine out of ten wheats with a N content less than 2.0 percent were classified as local varieties, but the majority of wheats having a N content over 2.0 percent are HYV wheats. For more details, see Section 7.1.4. In the case of maize the varietal effects are less clear, the average N contents of HYV and other varieties were 2.92 % and 2.48 %, respectively.

Less than ten percent of Indian soils have a **phosphorus** content exceeding the international mean (Fig. 239) and the Indian national mean values for soil P are among the very

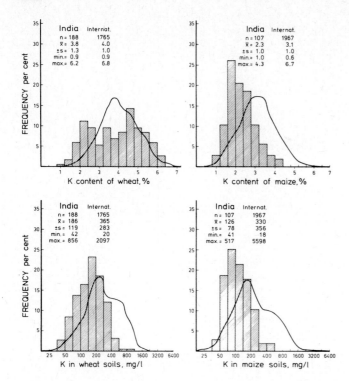

Fig. 240. Potassium, India.

lowest (Appendixes 2—4, Fig. 7). There seems to be no distinct difference in the P contents of soils from the five States. Although relatively heavy dressings of phosphates were applied to the sampled Indian crops (wheat: 15 ± 15 and maize: 18 ± 9 kg P/ha, Appendix 5) the P contents of both indicator plants (especially wheat) are low. There is no clear difference in the P contents of maize between the two variety groups but in the case of wheat it seems to be substantial. See Section 7.1.4.

The average **potassium** status of Indian soils is internationally low (Figs 240 and 8). Low soil K values were most frequently measured from the samples of Bihar and Punjab but were uncommon in the Madhya Pradesh soils. In general, the K contents of maize are low, but those of wheat are at the average international level. In both crops, especially in wheats, the variations of K contents are exceptionally wide. A part of this may be due to varying applications of potassium fertilizers to the sampled crops (wheat: 11 ± 20 and maize: 9 ± 19 kg K/ ha). Another source of variation seems to lie in plant varieties. Again the HYV wheats contain considerably more K than the local varieties (see Section 7.1.4).

In India, as in many other developing countries, attention should be paid to preventing the present low potassium content of the soils falling any further. Especially when higher yields are sought by using improved varieties and heavier applications of nitrogen, it must be realized that (owing to the high K contents of crops) considerable amounts of K are removed from soils. If the removed K is not replaced by fertilizer or manure, the K status of soil is likely to become critically low in a relatively short period of time. This is more obvious in areas where the soils are coarse textured with insignificant K reserves.

In spite of the generally high soil pH, the exchangeable **calcium** contents of most Indian soils remain at or below the international average (Figs 241 and 9). High Ca contents, however, were recorded from almost all Madhya Pradesh soils which, different from other Indian soils, had fine texture and high CEC (see Figs 12 and 14 in Section 2.2.4). On average, the Ca contents of original Indian wheats are internationally slightly on the low

Fig. 241. Calcium, India.

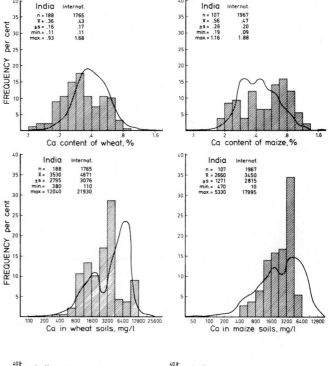

Fig. 242. Magnesium, India.

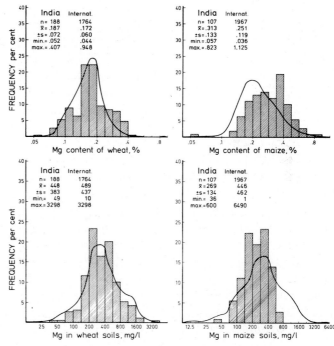

side but those of maize somewhat higher. For both crops the variations in Ca contents are exceptionally wide. Part of it is obviously due to different ability of different plant varieties to absorb Ca from soil (see Section 7.1.4).

In general the **magnesium** contents of Indian soils and plants are good but vary widely from one site to another (Figs 242 and 10). The lowest exchangeable Mg contents were

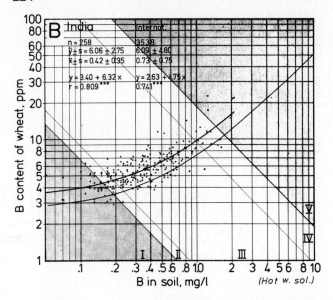

Fig. 243. Regression of B content of pot-grown wheat (y) on hot water soluble soil B (x), India. For details of summarized international background data, see Chapter 4.

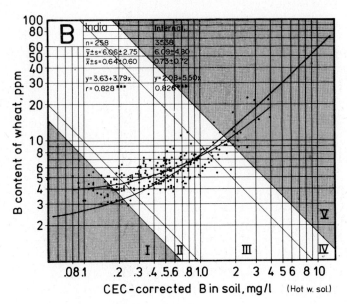

Fig. 244. Regression of B content of pot-grown wheat (y) on CEC-corrected (hot water soluble) soil B (x), India.

most frequently recorded from Bihar soils and the highest from Madhya Pradesh soils. For effects of plant varieties see Section 7.1.4.

7.1.3 Micronutrients

Boron. The average B content of wheat grown in pots of Indian soils corresponds closely to the international mean. Due to the generally low CEC of the soils, correction of soil B values for CEC also brings the Indian average for soil B close to that of the whole international material (Figs 243, 244, 22 and 25). Because of wide variation, both high and low B values occur. The frequency of high B values (Zones IV and V) is about 10 percent and corresponds to that in the whole material, but low values (Zones I and II) are more than

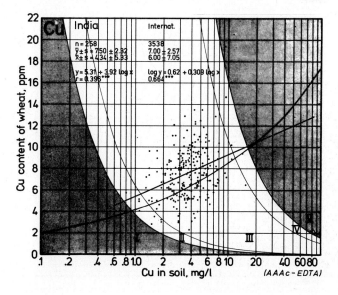

Fig. 245. Regression of Cu content of pot-grown wheat (y) on acid ammonium acetate-EDTA extractable soil Cu (x), India.

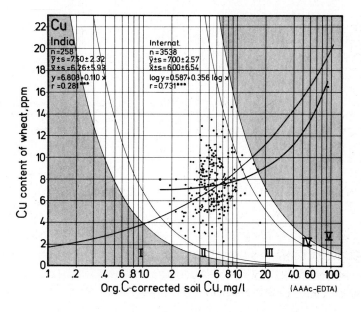

Fig. 246. Regression of Cu content of pot-grown wheat (y) on organic carbon-corrected (AAAc-EDTA extr.) soil Cu (x), India.

twice as frequent. Although low B values occur in other States, most of these originate in Bihar, where four-fifths of the plant and soil B values fall in Zones I and II. Even the remaining Bihar B values are comparatively low. If soil B values 0.3—0.5 mg/l are considered as a critical limit for B deficiency (see Section 2.7) almost half of Indian soils can be suspected of various degrees, from hidden to severe, of B deficiency. According to the present data, response to boron fertilization could be expected especially in Bihar and is least likely in Delhi. The highest B values were recorded from the Punjab and Delhi but only a few of them reach a critically high level. The above data are in relatively good agreement with those presented by Kanwar and Randhawa (1974).

Copper. India's location in the "international Cu field" is central (Figs 27 and 29). Only a few Cu sample pairs fall outside the "normal" Cu range (Figs 245 and 246) and only in one sample pair (45828) from Haryana were extremely high Cu contents measured.

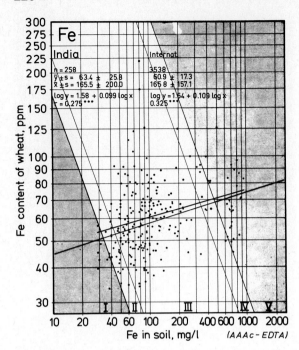

Fig. 247. Regression of Fe contents of pot-grown wheat (y) on acid ammonium acetate-EDTA extractable soil Fe (x), India.

Iron. On average the Fe contents of Indian soils and plants correspond closely to the international mean values for Fe (Figs 31 and 247). The variation, however, is wide and ranges from low to high. About 12 percent of the Indian plant and soil sample pairs are within the two lowest Fe Zones, mainly in Zone II. Even those in Zone I are not in extreme positions. Judging by the present results, there is no severe Fe deficiency at sites where the Indian samples were taken but at some locations plants, sensitive to Fe deficiency, may respond to Fe fertilizer. Samples in the two highest Fe Zones represent about 15 percent of the Indian material. High Fe values were much more typical of Bihar than other States, while most of the low Fe contents were found in samples from the Punjab, Haryana and Delhi.

Manganese. As in other countries with alkaline soils the Mn status in India is very low (Figs 33 and 37). Every seventh plant and soil sample pair falls in the lowest and every fourth in the two lowest Mn Zones (I and II, Figs 248 and 249). The ranges of variation of plant and soil Mn are exceptionally narrow and no high Mn values were found in the Indian material. Low Mn values (Zones I and II) are most frequent in Bihar and Delhi, quite common in the Punjab but rare in Haryana and Madhya Pradesh. Shortage of plant available Mn is apparent in many Indian soils.

Molybdenum. Owing to the high pH of Indian soils Mo is usually readily available to plants. The average Mo content of wheat grown in pots on Indian soils is higher than the average for the whole international material in this study (Figs 250 and 251). The national mean for ammonium oxalate-oxalic acid extractable soil Mo is the sixth lowest among the 29 countries (Fig. 15). Correction of AO-OA Mo values for soil pH improves the plant Mo — soil Mo correlation, raises the Indian mean above the international mean for soil

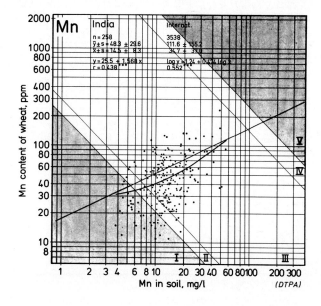

Fig. 248. Regression of Mn content of pot-grown wheat (y) on DTPA extractable soil Mn (x), India.

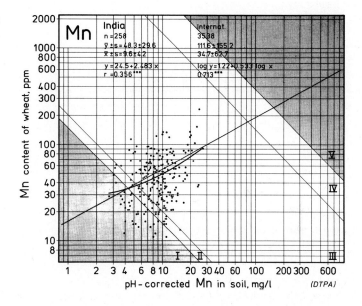

Fig. 249. Regression of Mn content of pot-grown wheat (y) on DTPA extractable soil Mn corrected for pH (x), India.

Mo, and moves India from the sixth lowest to eleventh highest place in the "international Mo field" (Fig. 20). In other words, the apparent Mo status of India is substantially changed. The low Mo values in Zones I and II have almost disappeared and the number of high values in Zones IV and V is increased. According to these results problems of Mo deficiency in India are much less probable than those of excess. High Mo values occurred most frequently in samples from the Punjab and Haryana. In some cases where excess of Mo is combined with low Mn availability, it may be useful to apply manganese sulphate alone or mixed with ammonium sulphate or sulphur to obtain more lasting effects through lowering the soil pH.

Zinc. In the "international Zn fields" (Figs 41 and 43), India occupies one of the very

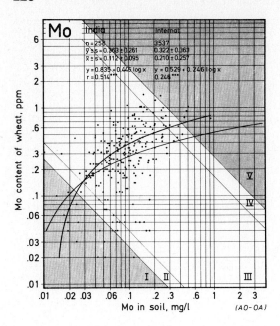

Fig. 250. Regression of Mo content of pot-grown wheat (y) on ammonium oxalate-oxalic acid extractable soil Mo (x), India.

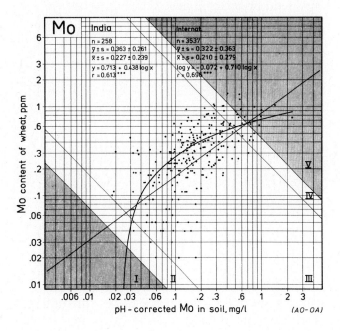

Fig. 251. Regression of Mo content of pot-grown wheat (y) on pH-corrected AO-OA extractable soil Mo (x), India. India.

lowest positions. Depending on the extraction method 18 (DTPA) to 28 (pH-corr. AAAc-EDTA) percent of the Indian Zn values fall in the two lowest Zn Zones (Figs 252 and 253) and even most of the rest are comparatively low. The analyses frequently indicate quite severe Zn deficiency. Only in one sample pair from Haryana (45828) were very high Zn contents measured. The frequency of low Zn values is highest in Bihar, where almost

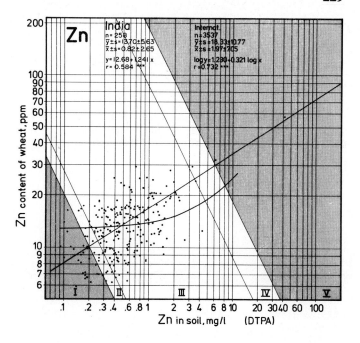

Fig. 252. Regression of Zn content of pot-grown wheat (y) on DTPA extractable soil Zn (x), India.

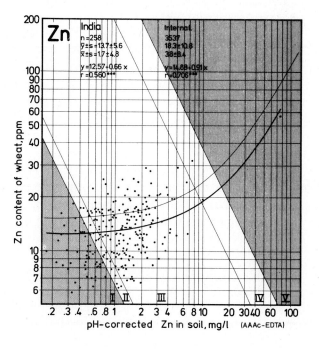

Fig. 253. Regression of Zn content of pot-grown wheat (y) on AAAc-EDTA extractable soil Zn corrected for pH (x), India.

every second sample pair falls in the two lowest Zn Zones. Low Zn values occur also in Haryana, the Punjab and Madhya Pradesh but in Delhi soils the Zn content seems to be somewhat better. In general, Zn deficiency, acute or hidden, seems to be much more common in India than in most other countries. Many investigators have reported the response of various crops to Zn in India.

7.1.4 Nutrient contents of original Indian plants with reference to varieties and other factors

Since the differences in the nutrient contents between the high yielding varieties and local Indian varieties were considerably more pronounced in the case of wheat than in maize, the varietal differences among the original Indian wheats warrant discussion.

Of 188 wheats sampled from India, 82 could be classified with relatively high certainty as HYV and 59 as local varieties. Neither group was homogeneous. With a few exceptions the group of HYV wheats consisted of cultivars HD-1553, WG-357, 'Kalyansona', N 4 and 'Sonalika'. Most of local varieties were identified in the "Field information forms" plainly as "Local".

The average macronutrient contents of the two variety groups are compared in Table 25 where the analytical data of respective soils and fertilizer application are also given. These data indicate that the N, P, K, Ca and Mg contents of HYV wheats are 134, 109, 56, 49 and 68 percent higher than those of the local varieties, respectively. For all five macronutrients these differences are statistically highly significant. In the cases of N and P some of the differences can be accounted for by the larger amounts of N and P fertilizers applied to HYV wheats and, compared with the local varieties, the HYV wheats were grown on soils somewhat richer in P.

The differences in B, Cu, Mn and Mo contents between the original HYV and local wheats were significant without similar (significant) differences in the respective soil data (Table 26). When wheat (cv. 'Apu') was grown in pots on the same soils the differences became statistically non-significant.

According to the data in Tables 25 and 26 there seem to be considerable differences between the two variety groups in their ability to absorb nutrients, especially K, Ca, Mg, B, Cu, Mn and Mo from the soil. However, there are other factors accounting for these differences. The estimated yields of HYV wheats were about 3000 kg/ha and those of local varieties about 2400 kg/ha. On the other hand, the local varieties were sampled at considerably later stages of growth; the difference was 26 days on average. These two factors, as explained in Sections 2.5 and 1.2.2, affect the nutrient contents in opposite directions but the effects of physiological age are likely to be much more pronounced. If we could take into account both these factors, the differences between the nutrient contents of the two variety groups would be much less.

On the basis of the above data and discussion the striking differences in the nutrient contents of the two variety groups obviously are partly due to genetic varietal differences, partly to fertilization (N and P), partly to soil nutrient content (P) and partly to differences in yield level and physiological age of plants at sampling. Not enough is known of the **quantitative** effects of these factors, obviously variable from one nutrient to another, to draw distinct conclusions on their **relative** efficacy. However, these data give further evidence of the fact that estimation of the nutrient status of soils founded on analyses of heterogeneous plant material may lead to misleading conclusions.

Table 25. Comparison of macronutrient contents of high yielding and local varieties of original Indian wheats, respective soils and fertilizer applications. Differences between the mean contents of the two groups followed by the same index letter are not statistically significant. Letters a—b indicate significant differences at 5, a—c at 1, and a—d at 0.1 percent level.

Nutrient	Average Macronutrient Content				Average N, P and K application kg/ha	
	Original wheat %		Wheat soils N %, others mg/l			
	HYV (n = 82)	Local (n = 59)	HYV (n = 82)	Local (n = 59)	HYV (n = 82)	Local (n = 59)
Nitrogen	4.32a	1.85d	0.071a	0.071a	100a	63a
Phosphorus	0.399a	0.191d	10.8a	7.6b	26a	6d
Potassium	4.60a	2.94d	177a	180a	11a	7a
Calcium	0.453a	0.305d	3820a	3883a	—	—
Magnesium	0.229a	0.136d	475a	451a	—	—

Table 26. Comparison of micronutrient contents of high yielding and local varieties of original Indian wheats, respective soils and pot-grown wheats[1] grown on the same soils. For index letters indicating statistical differences, see caption for Table 25.

Micro-nutrient	Average Micronutrient Contents					
	Original wheat (ppm)		Wheat soils[2] (mg/l)		Pot grown wheat (ppm) grown on the same soils	
	HYV (n = 68)	Local (n = 57)	HYV (n = 68)	Local (n = 57)	(n = 68)	(n = 57)
Boron	5.7a	3.4c	0.67a	0.49a	6.4a	5.4a
Copper	11.0a	6.0d	6.2a	5.7a	8.2a	8.1a
Manganese	70a	44d	8.3a	9.7b	46a	48a
Molybdenum	1.38a	0.65d	0.190a	0.204a	0.37a	0.35a

[1] Compared to Table 25 the number of samples is smaller; see Section 1.2.6.
[2] Extraction methods used:
    B: hot water + CEC correction   Mn: DTPA + pH correction
    Cu: AAAc-EDTA + org. C correction   Mo: AO-OA + pH correction

7.1.5 Summary

The Indian soils, except the fine textured high CEC soils of Madhya Pradesh, are usually coarse textured, alkaline, low in organic matter and have a low cation exchange capacity. The soils are generally low in N, P and K, but have satisfactory contents of exchangeable Ca and Mg.

The most probable microelement deficiency problems are those of Zn, B and Mn occurring most likely in Bihar. Many soils of Haryana, Punjab and Madhya Pradesh are also short of Zn and B and those of Delhi low in Mn. Some relatively high B values were recorded in the Punjab and Delhi soils. Many soils of the Punjab and Haryana were low in Fe and high in Mo. Low Fe values were found especially in Delhi and high values in Bihar samples. Of the five States represented in this study, Bihar seems to have more and Delhi and Madhya Pradesh fewer micronutrient problems than the other States.

7.2 Republic of Korea

7.2.1 General

The approximate sources of the Korean sample material taken from 50 wheat and 50 maize fields are given in Figure 254. Soils classified as Cambisols (37 %) were the most common, followed by Acrisols (21), Gleysols (17) and Luvisols (12).

Most of the Korean soils are coarse textured (Figs. 255). Only 14 soils out of 100 had a texture index of 50 or higher. With only a few exceptions the soils are acidic but the pH varies very widely, from 4.1 to 7.5. A medium content of organic matter and low cation exchange capacity are typical as are a low electric conductivity, low $CaCO_3$ equivalent and low sodium contents (Appendixes 2—4).

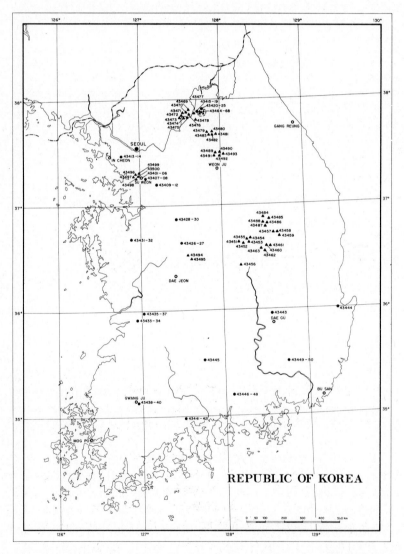

Fig. 254. Sampling sites in the Republic of Korea (points = wheat fields, triangles = maize fields).

Fig. 255. Frequency distributions of texture, pH, organic carbon content, and cation exchange capacity in soils of Korea (columns). Curves show the international frequency of the same characters.

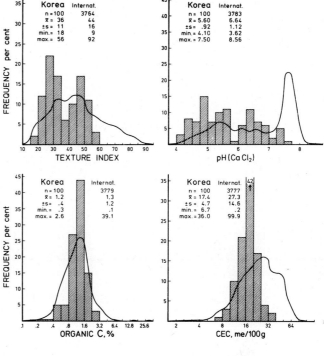

Figs 256–260. Frequency distributions of nitrogen, phosphorus, potassium, calcium and magnesium in original wheat and maize samples and respective soils (columns) of Korea. Curves show the international frequency of the same characters.

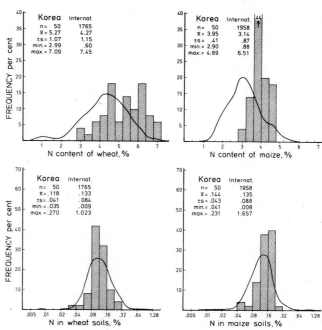

Fig. 256. Nitrogen, Korea.

7.2.2 Macronutrients

The contents of total **nitrogen** in Korean soils are at an average international level but in both original indicator plants the N contents are high (Figs 6 and 256). The average N content of Korean wheat (5.27 %) is the second highest (after Hungary) recorded in this study, but the variation range is wide. Both sampled crops were fertilized with high, but

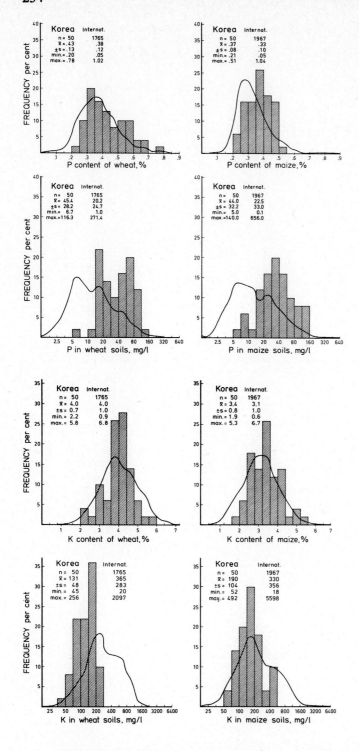

Fig. 257. Phosphorus, Korea.

Fig. 258. Potassium, Korea.

varying doses of nitrogen (wheat: 100 ± 25 and maize: 107 ± 31 kg N/ha, Appendix 5) which is probably the reason for the high N contents of the crops.

On average, the **phosphorus** content of Korean soils is twice as high as the average for the 30 countries of this study (Figs 7 and 257). The P contents of Korean wheat and maize are also high. The crops were fertilized relatively generously with phosphate fertilizers

Fig. 259. Calcium, Korea.

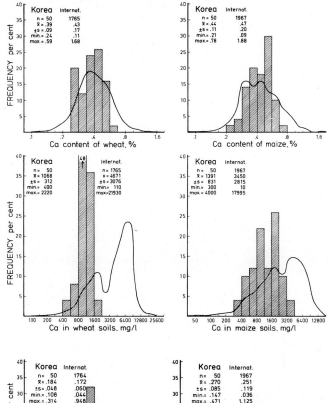

Fig. 260. Magnesium, Korea.

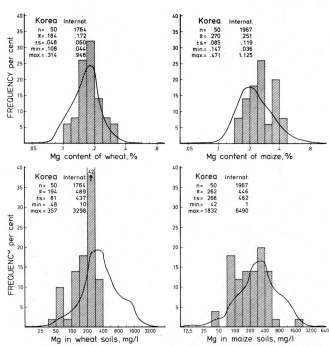

(wheat: 31 ± 7 and maize: 35 ± 14 kg P/ha).

Korean soils are low in exchangeable **potassium** (Figs 8 and 258) but the K content of plants are at an average international level (wheat) or higher (maize). Obviously, this is due to comparatively high potassium fertilization (wheat: 58 ± 10 and maize: 77 ± 80 kg K/ha). See also Fig. 11 and related text in Section 2.2.4.

Korean soils are generally low in exchangeable **calcium** and **magnesium** (Figs 9, 10, 259 and 260) but the Ca and Mg contents of plants are at an average international level. More than one-third of the soils included in this study were limed and on two-thirds farmyard manure was applied during the past few years before sampling. These practices have possibly affected the macronutrient contents of the sampled Korean crops. See also the effect of texture and CEC on the uptake of Ca and Mg, Section 2.2.4.

7.2.3 Micronutrients

Boron. The average B status of Korea is slightly low in an international context (Figs 22 and 25). Despite many B values falling in the low B Zones only a few of these indicate

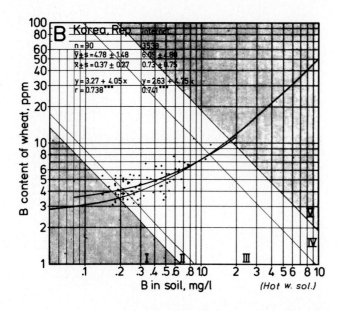

Fig. 261. Regression of B content of pot-grown wheat (y) on hot water soluble soil B (x), Korea. For details of summarized international background data, see Chapter 4.

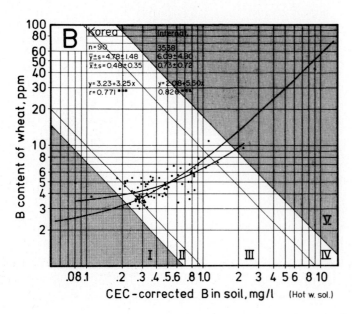

Fig. 262. Regression of B content of pot-grown wheat (y) on CEC-corrected (hot water soluble) soil B (x), Korea.

severe B deficiency (Figs 261 and 262). It is apparent that boron fertilization would be beneficial on many low B soils, especially on crops with a high B requirement such as root crops, legumes, fruits and vegetables. Response of rape to B has been reported, e.g. by Hong (1972) on a soil with 0.15 ppm available B.

Copper. Korea's position in the "international Cu fields" is almost central (Figs 27 and 29). Normal Cu values dominate in the Korean material and only a few fall outside Zone III, none of them showing extreme Cu contents (Fig. 263 and 264). The analytical data from the present material indicate no serious deficiency or toxicity problems in Korea.

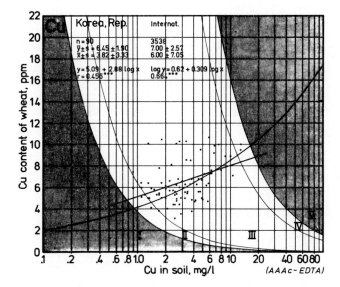

Fig. 263. Regression of Cu content of pot-grown wheat (y) on acid ammonium acetate-EDTA extractable soil Cu (x), Korea.

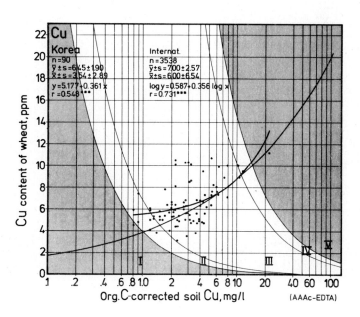

Fig. 264. Regression of Cu content of pot-grown wheat (y) on organic carbon-corrected (AAAc-EDTA extr.) soil Cu (x), Korea.

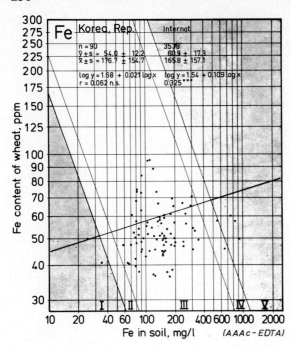

Fig. 265. Regression of Fe contents of pot-grown wheat (y) on acid ammonium acetate-EDTA extractable soil Fe (x), Korea.

Iron. The situation of Fe in Korean soils and plants is very similar to that of Cu (Figs 31 and 265). Normal values dominate and in spite of a few within Zones I and V, none of them is in an extreme position.

Manganese. Korea stands clearly on the high side in the "international Mn fields" (Figs 33 and 37). Owing to substantial pH variation in Korean soils (Fig. 255) the availability of Mn to plants varies, and consequently, the variation range of Mn contents of plants is exceptionally wide (Figs 266 and 267). Correction of DTPA extractable soil Mn values for pH raises the plant Mn — soil Mn correlation to a very high level (r = 0.935***) and more than triples the standard deviation of soil Mn. Both extremely high and low Mn values are included in the Korean material. The highest values come from geographically various locations in the country from sites of low soil pH(CaCl$_2$), usually 4.1—5.0. Correspondingly, the lowest Mn contents were measured in samples from sites where the soil pH was close to 7. See also Figure 34. Evidently, correction of Mn anomalies in several locations would be beneficial, in low Mn sites by using fertilizer containing Mn and in high Mn sites by raising the soil pH through liming.

Molybdenum. The Mo status of Korean soils and plants appears normal (Figs 268 and 269). The varying soil pH affects its availability to plants and the variation range of Mo contents of plants is relatively wide. Correction of AO-OA extractable soil Mo values for pH widens their variation and improves the correlation. However, the material shows neither low nor high extremes of Mo contents. In many cases low Mo contents are combined with high Mn contents (e.g. sample pairs 43436, -461, -463, -495), the common nominator being low soil pH.

Zinc. Only two countries, Belgium and Malta, exceed Korea in the "international Zn

Fig. 266. Regression of Mn content of pot-grown wheat (y) on DTPA extractable soil Mn (x), Korea.

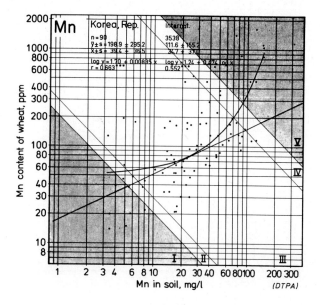

Fig. 267. Regression of Mn content of pot-grown wheat (y) on DTPA extractable soil Mn corrected for pH (x), Korea.

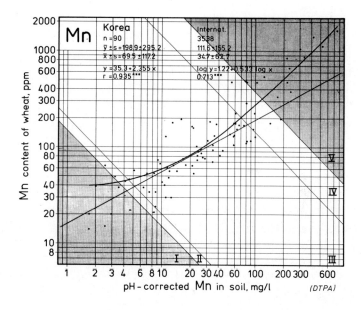

fields" scale (Figs 41 and 43). In spite of large variations in Korean soil and plant Zn contents there are only a few values which would indicate possible shortage of Zn (Figs 270 and 271). The Zn contents of about every fourth sample pair are so high that the samples fall in the two highest Zn Zones. Some of these are so extreme as to include the maximum plant Zn and soil Zn contents recorded for the entire international material submitted in this study. Most of the high Zn sites occur in the Suweon area on soils classified as Luvisols or Cambisols. According to the above results Zn deficiency is rare in Korea but at certain locations problems due to Zn excess may arise.

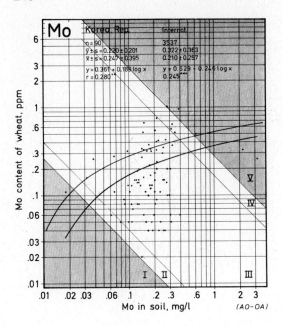

Fig. 268. Regression of Mo content of pot-grown wheat (y) on ammonium oxalate-oxalic acid extractable soil Mo (x), Korea.

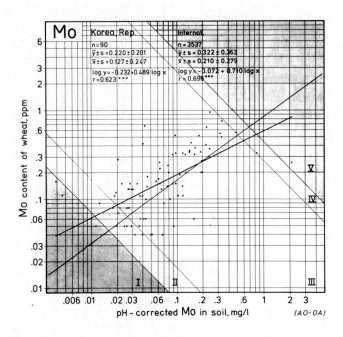

Fig. 269. Regression of Mo content of pot-grown wheat (y) on pH-corrected AO-OA extractable soil Mo (x), Korea.

7.2.4 Summary

The sampled Korean soils are generally coarse in texture, severely to moderately acid, medium in organic matter content and low in cation exchange capacity.

From an international point of view the macronutrient status of Korean soils is relatively low, with the exception of P. The level of macronutrient fertilizer application is relatively high and in addition two-thirds of the soils had received FYM and one-third were limed.

Fig. 270. Regression of Zn content of pot-grown wheat (y) on DTPA extractable soil Zn (x), Korea.

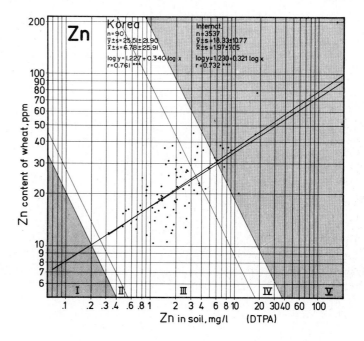

Fig. 271. Regression of Zn content of pot-grown wheat (y) on AAAc-EDTA extractable soil Zn corrected for pH (x), Korea.

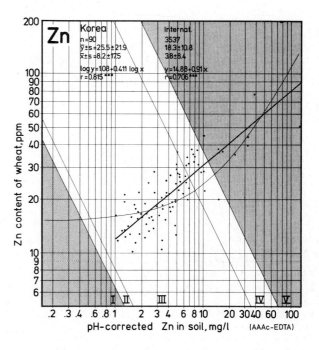

Consequently, the macronutrient contents of plants were at an average international level or above.

Of the micronutrients the levels of Cu, Fe and also Mo are quite normal. In case of B low values dominate and Mn values vary from very low to very high. Low Zn values were rare but some were extremely high. The most likely micronutrient problems in Korea are those of Mn and B deficiency and Mn excess.

7.3 Nepal

7.3.1 General

The sample material received from Nepal represents the Narayani, Seti, Mahakali, Gandaki, Bagmati, Lumbini, Bheri, Koshi, Mechi and Sagarmatha areas. A map indicating the sampling sites is not available.

The great majority of Nepalese soil samples are texturally on the coarse side (Fig. 272). Soil pH varies greatly but the extremes, very acid and very alkaline, are excluded. The organic matter contents of soils as well as their cation exchange capacities are generally low to medium with a relatively narrow range of variation. Values for electrical conductivity, $CaCO_3$ equivalent and sodium content are low for the most part (Appendixes 2 and 4).

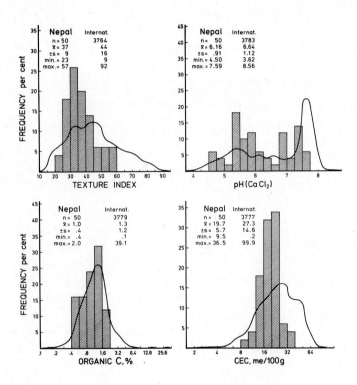

Fig. 272. Frequency distributions of texture, pH, organic carbon content, and cation exchange capacity in soils of Nepal (columns). Curves show the international frequency of the same characters.

7.3.2 Macronutrients

The total contents of **nitrogen** in Nepalese soils and wheats are at an average international level (Figs 6 and 273). For soils, the range of variation is narrow, but wide for the wheats. The use of nitrogen fertilizer (68 ± 70 kg N/ha) corresponds to the average for wheats sampled in this study but the rates varied from 0 to 370 kg N/ha, which partly explains the greatly varying N contents of wheats. The ten HYV wheats included in the Nepalese material had higher N contents (average 4.79 %) than other wheats (4.04 %) but they were

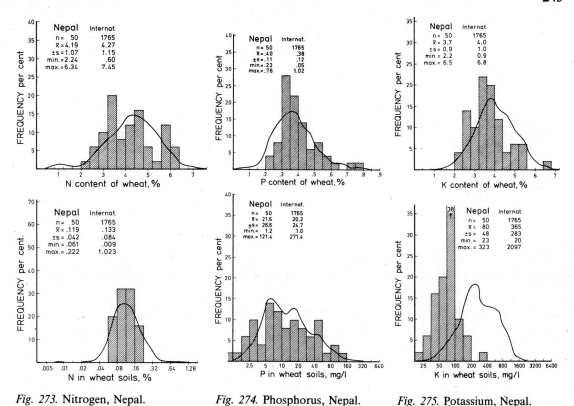

Fig. 273. Nitrogen, Nepal.   *Fig. 274.* Phosphorus, Nepal.   *Fig. 275.* Potassium, Nepal.

Figs 273–277. Frequency distributions of nitrogen, phosphorus, potassium, calcium and magnesium in original wheat samples and respective soils (columns) of Nepal. Curves show the international frequency of the same characters.

grown on soils higher in total N (average 0.160 and 1.109 %, respectively) and were fertilized with larger dressings of N (average 122 and 54 kg N/ha, respectively).

With regard to **phosphorus,** the limited sample material from Nepal very closely matches P conditions in the international material as a whole. The soil and plant P averages are about equal to their respective international averages and the ranges of variation are almost as wide as in the whole material (Figs 7 and 274). The phosphate fertilization level is somewhat lower and the rates vary slightly less (15 ± 20 kg P/ha) than those for the whole material (21 ± 27 kg P/ha).

The extremely low **potassium** contents of Nepalese soils (Figs 8 and 275) are also reflected in the K contents of original wheats, for which the national average is the sixth lowest recorded in this study (Appendix 2). It seems that the moderate dressings of applied potassium fertilizer (16 ± 20 kg K/ha) are sufficient to keep the K contents of plants from dropping very low. Nevertheless, the Nepalese soils are almost without exception deficient in K and a good response to potassium fertilization is likely.

The levels of exchangeable **calcium** and **magnesium** in Nepalese soils are generally low but the Ca and Mg contents of wheats are at or above an average international level with fairly wide variations (Figs 9, 10, 276 and 277). These contradictions may partly be due to relatively high Ca and Mg uptake from coarse textured soils with a low CEC. See Fig. 272 and Section 2.2.4.

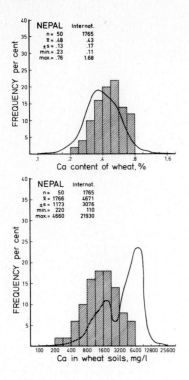

Fig. 276. Calcium, Nepal.

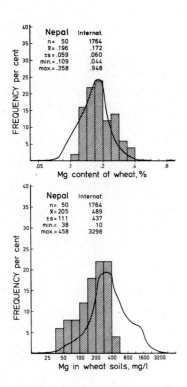

Fig. 277. Magnesium, Nepal.

7.3.3 Micronutrients

Because of a laboratory accident the micronutrient data of 15 samples of wheat grown in pots of Nepalese soil were lost.

Boron. Nepal occupies the very lowest positions in the "international B fields" (Figs 22 and 25). About half of the Nepalese plant and soil samples are within the lowest B Zone and only a third lie in the "normal" range (Zone III), and these are close to its lower limit (Figs 278 and 279). On the basis of the above results, widespread B deficiency, acute or hidden, is likely to exist in Nepal, limiting yields especially of those crops with high B requirement.

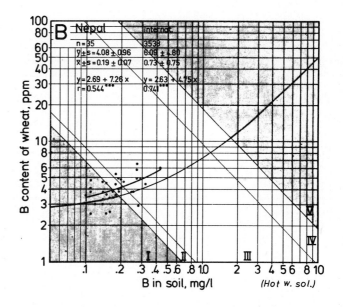

Fig. 278. Regression of B content of pot-grown wheat (y) on hot water soluble soil B (x), Nepal. For details of summarized international background data, see Chapter 4.

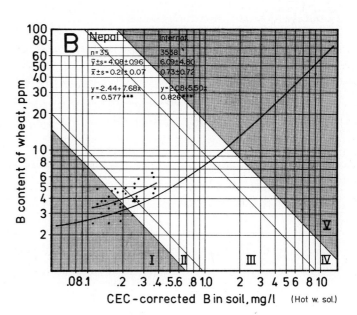

Fig. 279. Regression of B content of pot-grown wheat (y) on CEC-corrected (hot water soluble) soil B (x), Nepal.

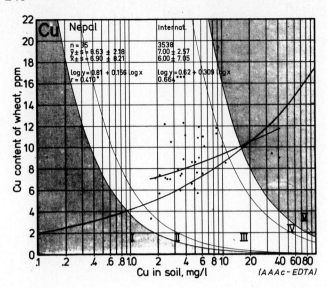

Fig. 280. Regression of Cu content of pot-grown wheat (y) on acid ammonium acetate-EDTA extractable soil Cu (x), Nepal.

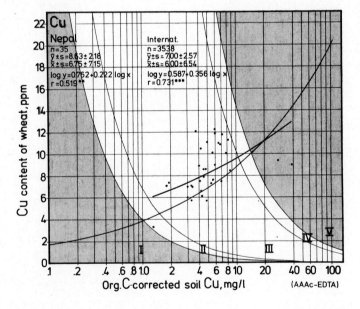

Fig. 281. Regression of Cu content of pot-grown wheat (y) on organic carbon-corrected (AAAc-EDTA extr.) soil Cu (x), Nepal.

Copper. The Cu contents of Nepalese soils seem to be quite satisfactory and Nepal stands relatively high in the "international Cu fields" (Figs 27 and 29). One sample pair falls in Zone II and only three in Zone V, the rest lie within the normal range (Figs 280 and 281). Since no extreme Cu contents were found in the sample material the present results indicate no serious Cu problems in Nepal.

Fig. 282. Regression of Fe contents of pot-grown wheat (y) on acid ammonium acetate-EDTA extractable soil Fe (x), Nepal.

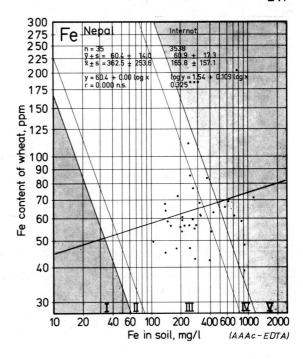

Iron. The mean Fe content of wheat grown in pots of Nepalese soils corresponds to the international mean but the mean for extractable soil Fe contents is exceeded by only two countries, Finland and New Zealand (Figs 31 and 282). There is no correlation between the plant Fe and soil Fe values. Although a few relatively high soil Fe values exist, the availability of Fe to plants in these soils seems to be low. Since no low Fe values occur in the sampled material, Fe deficiency is unlikely in Nepal.

Manganese. According to the analytical data presented in Figures 33, 37, 283 and 284, no Mn deficiency or toxicity problems are to be expected in Nepal, provided that the sample material is representative of the Nepalese soils.

Molybdenum. On average, the Nepalese plant and soil samples contain less than half as much Mo as all samples in this study and Nepal's positions in the "international Mo fields" are clearly low (Figs. 15, 20, 285 and 286). The availability of Mo to plants is highly dependent on soil pH and this varies widely in Nepalese soils. After correction for soil pH the plant Mo—soil Mo correlation coefficient (r) is more than doubled. About one-third of the Nepalese plant—soil Mo points are within the two lowest Mo Zones (I and II), some of them at quite low positions. Mo deficiency is likely to occur in Nepal, most probably in the soils of the Bagmati area but also elsewhere.

Zinc. Nepal occupies one of the low positions in the "international Zn fields" (Figs 41 and 43). The ranges of variation for both the plant and soil Zn contents are comparatively narrow (Figs 287 and 288). Only one sample pair is within the two highest Zn Zones and depending on the extraction method one (DTPA) or seven (pH-corr. AAAc-EDTA) lie in Zone I. According to the latter method Zn deficiency could be more severe than the results of the DTPA method indicate. Although the lowest plant and soil Zn contents are not as low as in some other countries (see e.g. India, Iraq and Turkey), Zn deficiency should be expected at many locations in Nepal. See also Section 2.7.

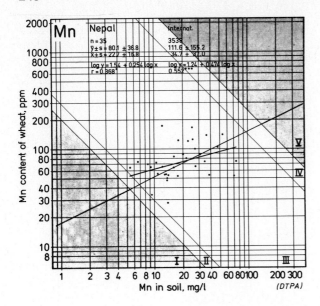

Fig. 283. Regression of Mn content of pot-grown wheat (y) on DTPA extractable soil Mn (x), Nepal.

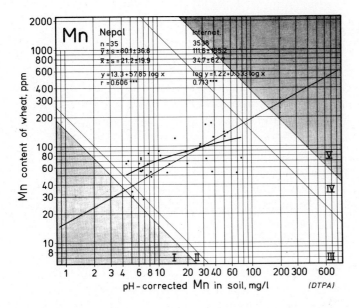

Fig. 284. Regression of Mn content of pot-grown wheat (y) on DTPA extractable soil Mn corrected for pH (x), Nepal.

7.3.4 Summary

The soils of Nepal are relatively coarse textured, widely varying in pH, with low organic matter content and low cation exchange capacity. Of the soil macronutrients the contents of N and P are at an average international level but P contents vary widely. The Ca, Mg and especially K contents of Nepalese soils are low. Micronutrient problems concerning Cu, Fe, and Mn are unlikely. Although the Mo content of many Nepalese soils and plants is rather low, the most evident micronutrient deficiencies in Nepal are those of B and Zn.

Fig. 285. Regression of Mo content of pot-grown wheat (y) on ammonium oxalate-oxalic acid extractable soil Mo (x), Nepal.

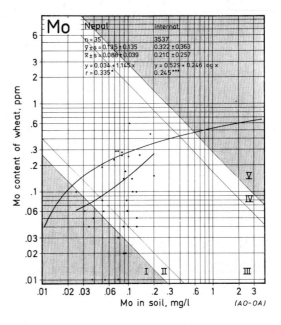

Fig. 286. Regression of Mo content of pot-grown wheat (y) on pH-corrected AO-OA extractable soil Mo (x), Nepal.

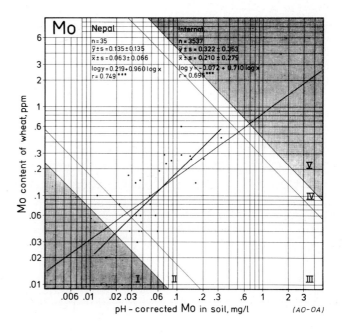

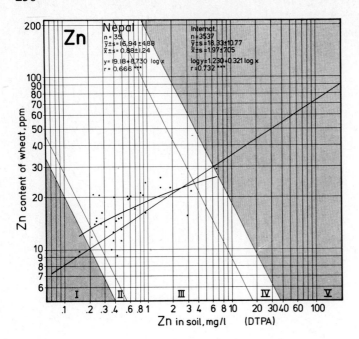

Fig. 287. Regression of Zn content of pot-grown wheat (y) on DTPA extractable soil Zn (x), Nepal.

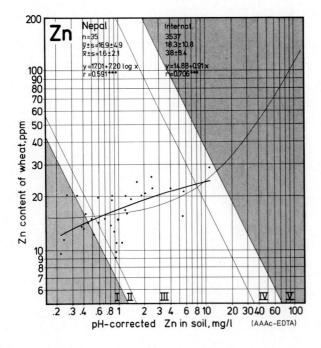

Fig. 288. Regression of Zn content of pot-grown wheat (y) on AAAc-EDTA extractable soil Zn corrected for pH (x), Nepal.

7.4 Pakistan

7.4.1 General

The distribution of the 242 sampling sites in Pakistan (Fig. 289) covers relatively well the most important agricultural areas of the country. Of the 131 soils classified according to the FAO classification system, Yermosols (44 %) and Cambisols (36 %) are far more

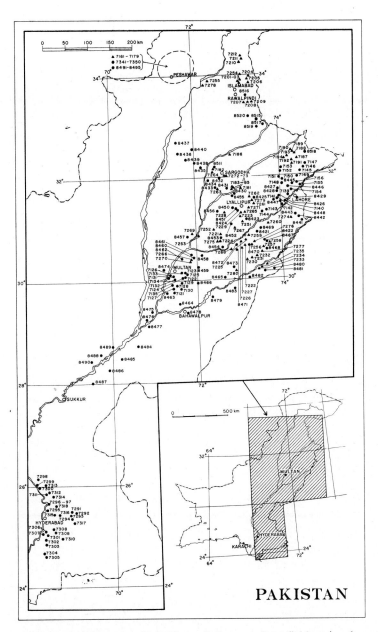

Fig. 289. Sampling sites in Pakistan (points = wheat fields, triangles = maize fields). The last four numerals of each sample pair number are given.

common than others: Fluvisols (9 %), Halosols (7 %), and Luvisols and Xerosols (2 % each).

Most soils are coarse to medium textured and highly alkaline (Fig. 290). The national average pH(CaCl$_2$) is higher than in any other of the countries studied and the standard deviation is exceptionally narrow. Low organic matter content and low cation exchange capacity are typical of Pakistani soils. The values for electrical conductivity and CaCO$_3$ equivalent are relatively high (Appendixes 2—4). In many locations, especially in the Multan area, the alkalinity and high electrical conductivity are combined by high Na contents of soils.

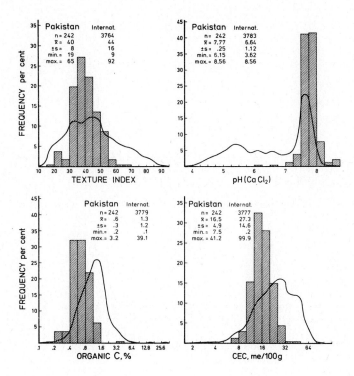

Fig. 290. Frequency distributions of texture, pH, organic carbon content, and cation exchange capacity in soils of Pakistan (columns). Curves show the international frequency of the same characters.

7.4.2 Macronutrients

Because of the relatively high and greatly varying nitrogen fertilizer applications (wheat 58 ± 57 and maize 83 ± 90 kg N/ha), the **nitrogen** contents of original wheat and maize vary widely but are generally at an average international level in spite of the exceptionally low total content of soil N (Figs 6 and 291).

Pakistani soils, on average, are poorer in 0.5 M NaHCO$_3$ extractable **phosphorus** than soils of any other country, but the P contents of plants are only slightly below the international means (Figs 7, 292 and Appendixes 2—4). Greatly varying but mostly only moderate amounts of phosphates had been applied to the sampled crops (wheat, 9 ± 15 and maize, 12 ± 29 kg P/ha, Appendix 5). The discrepancy between low soil P and relatively high plant P contents may partly be due to the effects of CEC (Fig. 14, Section 2.2.4).

Figs 291–295. Frequency distributions of nitrogen, phosphorus, potassium, calcium and magnesium in original wheat and maize samples and respective soils (columns) of Pakistan. Curves show the international frequency of the same characters.

Fig. 291. Nitrogen, Pakistan.

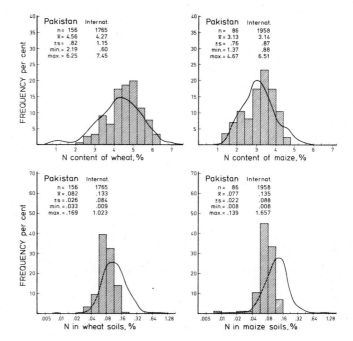

Fig. 292. Phosphorus, Pakistan.

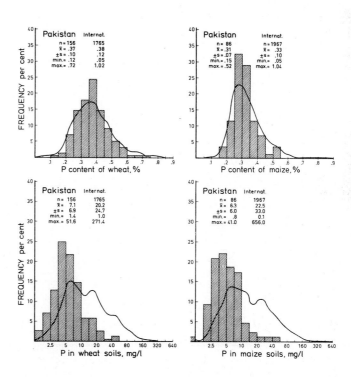

The **potassium** contents of Pakistani soils are slightly low in the international scale (Figs 293 and 8). In spite of this and of the small rates of potassium application (Appendix 5) the K contents of plants are generally high but vary widely from one sample to another. Obviously the high rates of nitrogen application have accelerated the uptake of K from soils (e.g. Rinne *et al.*, 1974 a; Sillanpää, 1974). For example, maize plants containing more

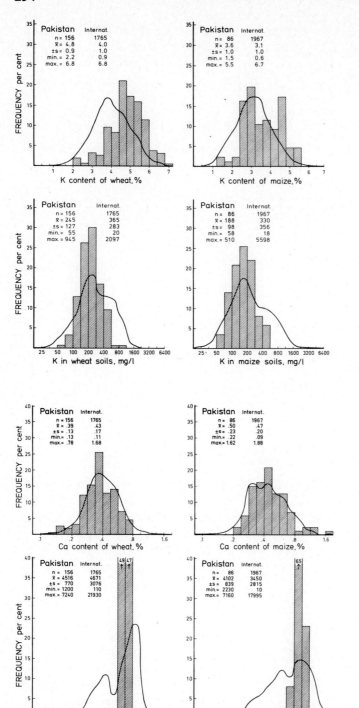

Fig. 293. Potassium, Pakistan.

Fig. 294. Calcium, Pakistan.

than 4.0 percent K had received 123 kg N/ha while plants with less than 3.0 percent K were only fertilized with 44 kg N/ha on the average. Similarly, wheats with 5.5 percent K or more had received 91 kg and wheats with less than 4.0 percent K only 51 kg N/ha on the average.

Fig. 295. Magnesium, Pakistan.

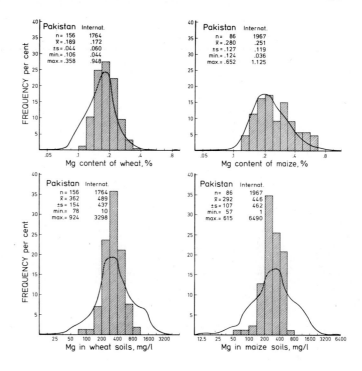

The average **calcium** contents of Pakistani soils and plants correspond closely to the international means (Fig. 9). In wheats and wheat soils the contents are slightly on the low side, while in the case of maize they are on the high side (Fig. 294). Although the exchangeable Ca contents of Pakistani soils vary only little, the variations in the plant Ca are wide.

In general, the **magnesium** situation in Pakistan is quite normal (Figs 10 and 295) and in many respects is reminiscent of that of Ca. The variation of exchangeable Mg contents of soils is rather narrow but wide in the case of plant Mg.

7.4.3 Micronutrients

Boron. In the "international B fields" (Figs 22 and 25) Pakistan is among the countries occupying the highest positions. The correlations between plant B and soil B are very high (Figs 296 and 297). Correction of soil B values for CEC increases the relative share of high B values and moves Pakistan to the right in the "international B fields". About eleven percent of the B values fall in each of the two highest B Zones (IV and V), many of them in extreme positions (Fig. 297). Many of the highest B values are found in Multan but there are also soils rich in B in Sind and elsewhere. Low B values were often recorded in samples from unirrigated sites. If soil B values of 0.3—0.5 mg/l are considered as critical

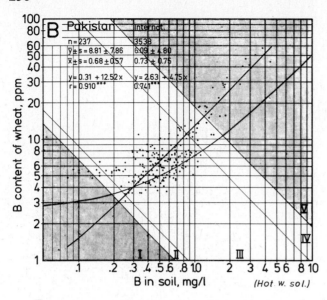

Fig. 296. Regression of B content of pot-grown wheat (y) on hot water soluble soil B (x), Pakistan. For details of summarized international background data, see Chapter 4.

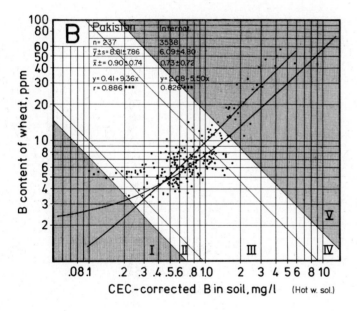

Fig. 297. Regression of B content of pot-grown wheat (y) on CEC-corrected (hot water soluble) soil B (x), Pakistan.

(Section 2.7), a number of Pakistani soils can be suspected of B deficiency.

Copper. Almost all the Cu values measured from the plant and soil samples are within the normal Cu range (Zone III, Figs 298 and 299). The present analytical data indicate neither Cu deficiency nor excess. Only one sample pair (48462) from Multan falls in Zone V but even that is not in a very extreme position.

Iron. Against the international background the Fe situation in Pakistan is relatively normal (Fig. 31). About ten percent of the Fe values are within the two lowest Fe Zones (Fig. 300), mainly in Zone II. Such values as are in Zone I lie towards its upper boundary and, at worst, indicate hidden Fe deficiency. Dry weather during the growing season may

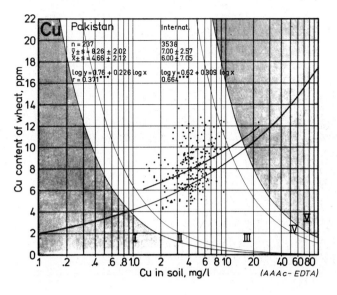

Fig. 298. Regression of Cu content of pot-grown wheat (y) on acid ammonium acetate-EDTA extractable soil Cu (x), Pakistan.

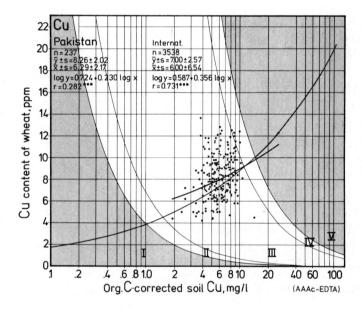

Fig. 299. Regression of Cu content of pot-grown wheat (y) on organic carbon-corrected (AAAc-EDTA extr.) soil Cu (x), Pakistan.

lead to oxidation of ferrous to ferric Fe due to excess soil aeration, and so decrease the availability of Fe to plants and induce Fe deficiency (e.g. De Kock, 1955; Granick, 1958; Mokady, 1961). Most of the lowest Fe values were measured from samples collected in the Peshawar and Lyallpur — Sheikhupura — Lahore areas.

Manganese. With its highly alkaline soils Pakistan is among the countries of the lowest Mn status (Figs 33 and 37). The variations in both plant and soil Mn contents are exceptionally small (Figs 301 and 302), and nearly all Mn values are concentrated in the two lowest Mn Zones or close to the lower boundary of Zone III. There seems to be no clear geographically distinguishable low Mn districts and soils of Mn shortage are scattered

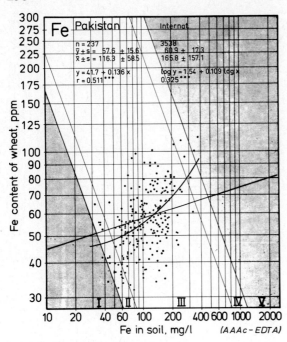

Fig. 300. Regression of Fe contents of pot-grown wheat (y) on acid ammonium acetate-EDTA extractable soil Fe (x), Pakistan.

all over the sampled areas. On the basis of these analytical data it seems likely that Mn deficiency, acute or hidden, is relatively common in Pakistan.

Molybdenum. Unlike Mn, no low Mo contents were found in Pakistani plant samples. The national average for pot-grown wheat Mo content was the highest among the 29 countries studied (Figs 15 and 20). Owing to ammonium oxalate-oxalic acid being an inefficient extractant of Mo from alkaline soils, the national mean soil Mo content is lower than the international average (Fig. 15). This contradiction is resolved by correcting the soil Mo values for soil pH (Fig. 20), whereupon the national mean for soil Mo becomes higher than that of any other country. Almost 40 percent of the Pakistani plant—soil Mo values fall in the two highest Mo Zones, with more than half of these in Zone V (Figs 303 and 304). Geographically there are no distinct high Mo regions. The origins of high Mo values were scattered all over the sampled area, being somewhat less frequent, however, in Rawalpindi and Peshawar than elsewhere.

Zinc. With Turkey, India and Iraq, Pakistan belongs to the group of countries where low Zn values are more common than elsewhere (Figs 41, 43, 305 and 306). The national mean values of plant and soil Zn are similar to those of India but the variations in Pakistani Zn values are smaller. Relatively fewer Zn values therefore fall in the low Zn Zones and in less extreme positions than in the case of India. Nevertheless, the frequency of low Zn values is internationally high and in numerous locations a response to Zn can be expected. Only a few Pakistani samples showed an abundance of Zn. The low Zn values were perhaps more typical for Sind, Multan and Peshawar than for other sampled areas.

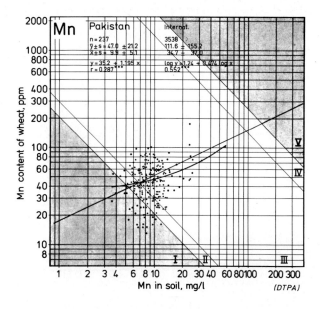

Fig. 301. Regression of Mn content of pot-grown wheat (y) on DTPA extractable soil Mn (x), Pakistan.

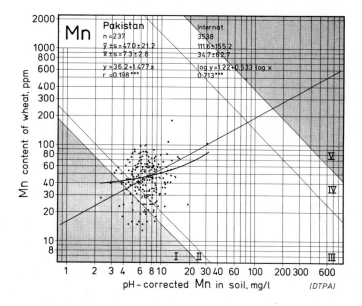

Fig. 302. Regression of Mn content of pot-grown wheat (y) on DTPA extractable soil Mn corrected for pH (x), Pakistan.

7.4.4 Summary

The Pakistani soils are usually coarse to medium textured, low in organic matter and have a low cation exchange capacity. The outstanding feature is their high alkalinity. Very low P and N, relatively low K, medium Mg and Ca and high Na contents are typical of most sampled Pakistani soils.

Out of the six micronutrients included in this study, only the contents of Cu and Fe are usually within the normal international range. The contents of B vary widely, high values being the most common, but in a number of soils shortage of B is apparent. Owing to the high alkalinity of Pakistani soils, the availabilities to plants of Mn and Zn are low but that of Mo is high. Consequently, shortages of Mn and Zn and an excess of Mo are typical of many soils.

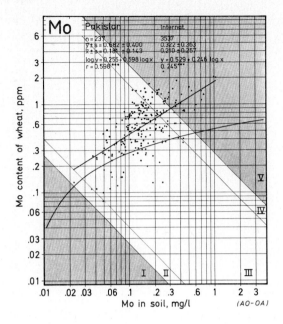

Fig. 303. Regression of Mo content of pot-grown wheat (y) on ammonium oxalate-oxalic acid extractable soil Mo (x), Pakistan.

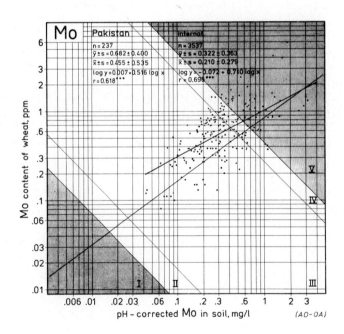

Fig. 304. Regression of Mo content of pot-grown wheat (y) on pH-corrected AO-OA extractable soil Mo (x), Pakistan.

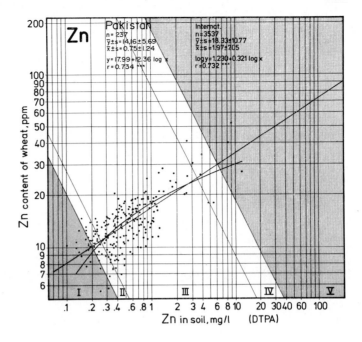

Fig. 305. Regression of Zn content of pot-grown wheat (y) on DTPA extractable soil Zn (x), Pakistan.

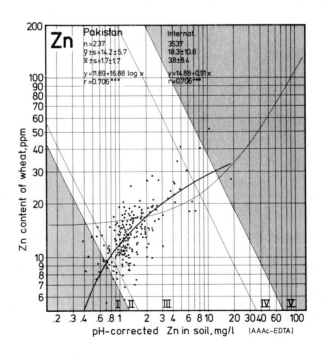

Fig. 306. Regression of Zn content of pot-grown wheat (y) on AAAc-EDTA extractable soil Zn corrected for pH (x), Pakistan.

7.5 Philippines

7.5.1 General

The sampling sites of the 197 maize-soil sample pairs collected from the Republic of the Philippines are well distributed over the most important agricultural areas of the country (Fig. 307). The sampled soils were not classified into FAO soil units but according to the FAO/Unesco Soil Map of the World (1976) the majority of Philippine soils consist of

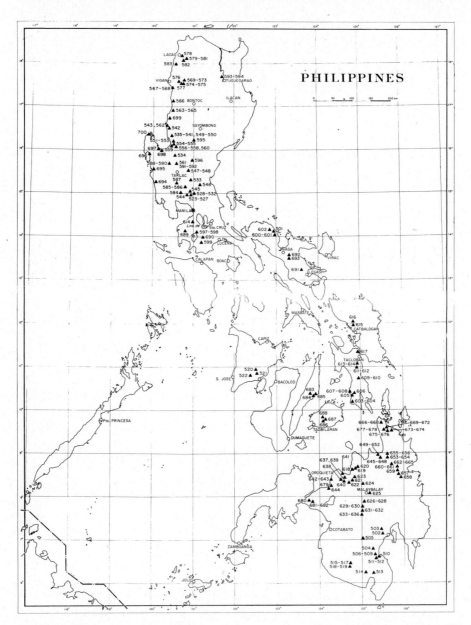

Fig. 307. Sampling sites in the Philippines. The last three numerals of each sample pair number are given.

Nitosols, Acrisols, Cambisols and Andosols. The textural variation of the soils is exceptionally wide (Fig. 308) covering the whole texture scale from very coarse (texture index < 20) to very fine (TI > 90). In the pH of soils the variation is also great and only the most alkaline soils are not represented. The contents of soil organic matter are usually at a medium level but the majority of soils have a relatively high cation exchange capacity. The electrical conductivity and $CaCO_3$ equivalent values and sodium contents of soils are generally low (Appendixes 3 and 4).

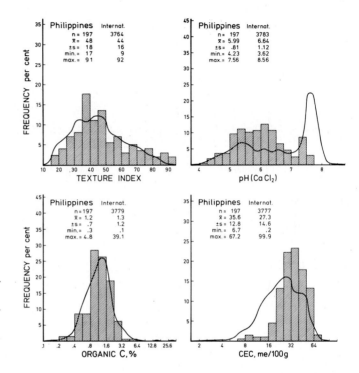

Fig. 308. Frequency distributions of texture, pH, organic carbon content, and cation exchange capacity in soils of Philippines (columns). Curves show the international frequency of the same characters.

7.5.2 Macronutrients

The total **nitrogen** and $NaHCO_3$ extractable **phosphorus** contents of Philippine soils are of the average international level and the N and P contents of maize are slightly lower (Figs 6, 7, 309 and 310). In all cases variations are wide. The sampled crops were fertilized with only small amounts of N and P fertilizers (8 ± 19 kg N/ha and 1 ± 4 kg P/ha).

Wide variation is also typical for the **potassium** contents of Philippine soils as well as for the K content of maize but, in general, the K status is relatively low (Figs 8 and 311). The international minimum soil K value (18 mg K/l) and the minimum maize K content (0.6 %) were both found in samples collected in the Philippines. Only nominal amounts of potassium fertilizers (1 ± 4 kg K/ha) were applied to the sampled maize crops. A good response to potassium is likely at many locations in the Philippine Islands.

With regard to **calcium** and **magnesium** the Philippine maize and maize soils tend to be high on an international scale, Ca contents only slightly so but the national mean Mg contents are among the very highest (Figs 9, 10, 312 and 313). As for many other properties of Philippine soils and plants, extensive variations are typical for Ca and Mg contents.

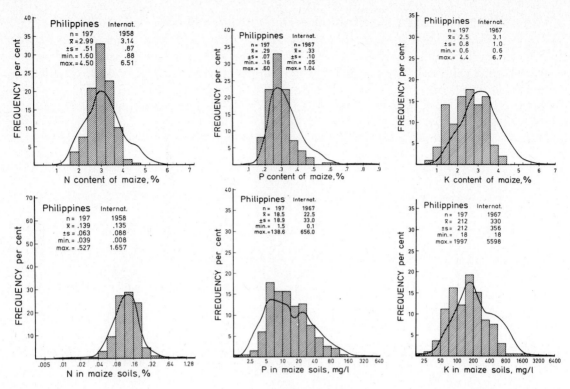

Fig. 309. Nitrogen, Philippines.   Fig. 310. Phosphorus, Philippines.   Fig. 311. Potassium, Philippines.

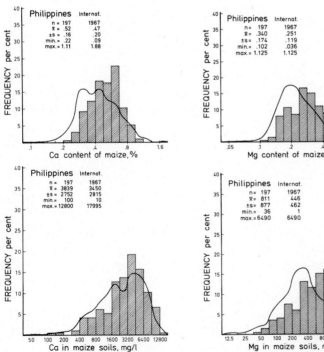

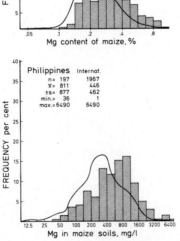

Figs 309–313. Frequency distributions of nitrogen, phosphorus, potassium, calcium and magnesium in original maize samples and respective soils (columns) of Philippines. Curves show the international frequency of the same characters.

Fig. 312. Calcium, Philippines.   Fig. 313. Magnesium, Philippines.

7.5.3 Micronutrients

Boron. Of the six micronutrients included in this study B deficiency appears the most widespread in the Philippines. The national mean B contents of soils and plants are among the lowest recorded in this study (Figs 22 and 25). Every second plant—soil sample pair falls into the two lowest B Zones (Figs 314 and 315). This high frequency is exceeded only by Nepal. The plant—soil B points falling in the normal B Zone (III) occur most frequently toward the lower boundary of the zone. There seems to be no clear geographical distinction between the low and normal B areas, but soils of B shortage seem to be widely distributed over the sampled areas.

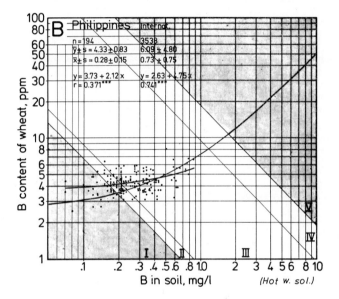

Fig. 314. Regression of B content of pot-grown wheat (y) on hot water soluble soil B (x), Philippines. For details of summarized international background data, see Chapter 4.

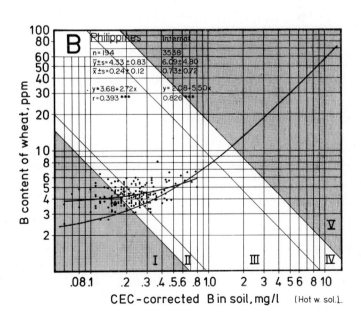

Fig. 315. Regression of B content of pot-grown wheat (y) on CEC-corrected (hot water soluble) soil B (x), Philippines.

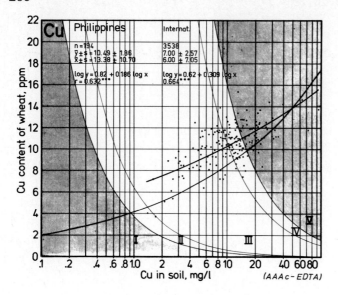

Fig. 316. Regression of Cu content of pot-grown wheat (y) on acid ammonium acetate-EDTA extractable soil Cu (x), Philippines.

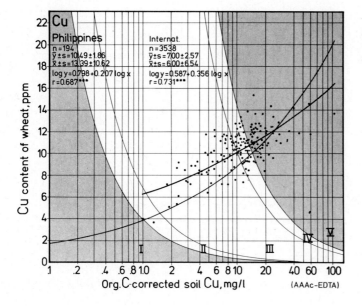

Fig. 317. Regression of Cu content of pot-grown wheat (y) on organic carbon-corrected (AAAc-EDTA extr.) soil Cu (x), Philippines.

Copper. Contrary to the occurrence of B, the soils of the Philippines are exceptionally rich in Cu (Figs 316 and 317). With Brazil and Italy the Philippines occupy the highest positions in the "international Cu fields" (Figs 27 and 29) and high Cu value (Zones IV and V) were found more frequently in the Philippines than in any other country. Only a couple of relatively low Cu values were measured in samples which came from Camarines Norte (Bicol) but there are no indications of Cu shortage anywhere else in the country. The highest Cu values were recorded in Ilocos samples, but high Cu contents were also found in samples from Northern and Eastern Mindanao and occasionally elsewhere.

Fig. 318. Regression of Fe contents of pot-grown wheat (y) on acid ammonium acetate-EDTA extractable soil Fe (x), Philippines.

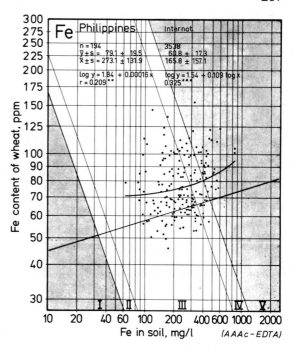

Iron. The Philippine national average Fe content of plants is the highest and that of AAAc-EDTA extractable soil Fe the fifth highest recorded in this study (Fig. 31). Every third sample pair (Fig. 318) falls within the Fe Zones IV and V but none within, nor even close to, the low Fe Zones (I and II). It seems therefore that the possibility of any response to iron fertilization is minimal. Most of the high (Zone V) Fe values originate in Ilocos and Central Luzon.

Manganese. Substantial variations in the plant and soil Mn contents were found in the Philippine material (Figs 319 and 320). In general, high Mn values were more frequent and the national averages are higher than those of most other countries (Figs 33 and 37). Only one sample pair (43688 from Bohol) indicates possible Mn shortage but the relative frequency of high Mn (Zones IV and V) values is almost double the international frequency. Almost all of the highest (Zone V) Mn values were found in samples from Northern and Eastern Mindanao.

Molybdenum. In the "international Mo fields" (Figs 15 and 20) the Philippines are placed centrally, but as in the case of several other soil properties extensive variations in both plant Mo and soil Mo contents are typical (Figs 321 and 322). The correlation between plant Mo and AO-OA extractable soil Mo is substantially improved by pH-correction owing to wide pH variations. The sites for both the high and the low Mo contents are geographically scattered; high contents were associated with soils of relatively high pH and the low contents with soils of low pH. At locations where Mo deficiency might be suspected, raising the soil pH through liming may be a sufficient treatment to increase the availability of Mo to a normal level.

Zinc. Generally high but widely varying Zn contents are typical of Philippine soils and plants (Figs 323 and 324); from several locations abundant Zn contents were measured. In

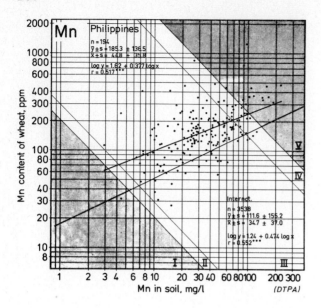

Fig. 319. Regression of Mn content of pot-grown wheat (y) on DTPA extractable soil Mn (x), Philippines.

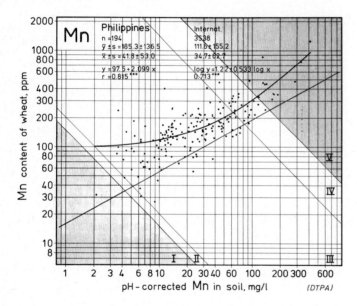

Fig. 320. Regression of Mn content of pot-grown wheat (y) on DTPA extractable soil Mn corrected for pH (x), Philippines.

two sample pairs (43684 from Cebo and 43686 from Bohol) the Zn contents were particularly high. Other high Zn values were distributed in various sites from Southern Mindanao to Ilocos. Unfortunately, the soils of eastern Mindoro where serious Zn deficiency in rice has been reported (FAO, 1980 a) were not sampled for this study.

On the basis of the present analytical data, Zn deficiency is likely to occur in other parts of the Philippines though not as severely nor to such an extensive degree as in many other countries (e.g. Iraq, Turkey, India, Pakistan). See also Section 2.7.

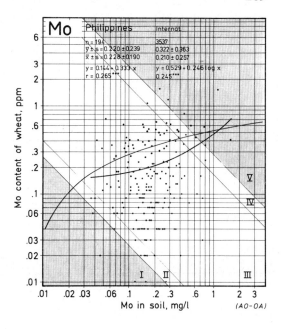

Fig. 321. Regression of Mo content of pot-grown wheat (y) on ammonium oxalate-oxalic acid extractable soil Mo (x), Philippines.

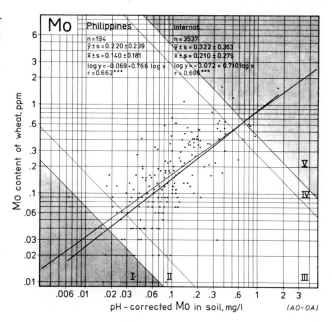

Fig. 322. Regression of Mo content of pot-grown wheat (y) on pH-corrected AO-OA extractable soil Mo (x), Philippines.

7.5.4 Summary

Wide variations in most of the general soil properties as well as in the soil macronutrient contents are typical of the Philippines.

Of the six micronutrients included in this study B deficiency seems to constitute the most widespread micronutrient problem in the country. The level of Mo is effectively at a normal international level while that of Fe, Mn, Zn, and especially of Cu, are internationally at the highest level. As locations where soils are low in Zn, response to Zn application is likely.

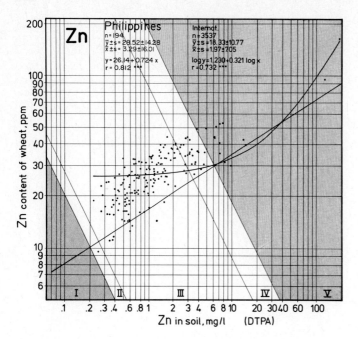

Fig. 323. Regression of Zn content of pot-grown wheat (y) on DTPA extractable soil Zn (x), Philippines.

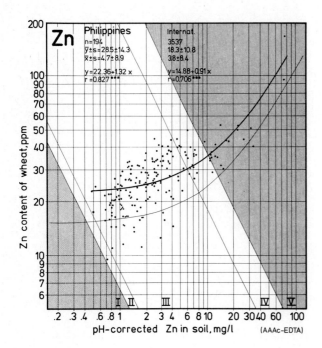

Fig. 324. Regression of Zn content of pot-grown wheat (y) on AAAc-EDTA extractable soil Zn corrected for pH (x), Philippines.

7.6 Sri Lanka

7.6.1 General

Of the 21 soils collected from Sri Lanka, 13 were classified as Luvisols and 8 as Acrisols. The latter soils occur south of the Puttalam-Batticaloa line and most of the Luvisols north of the line (Fig. 325). With three exceptions the soils are coarser textured than the average of soils in this study (Fig. 326). The range of variation in pH is wide but acid soils are

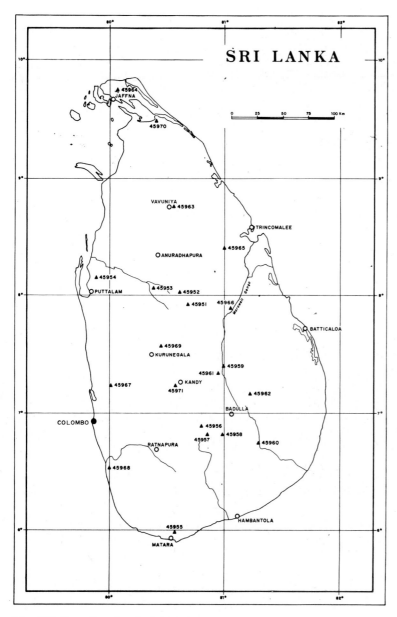

Fig. 325. Sampling sites in Sri Lanka.

more common. A medium organic matter content is typical but the cation exchange capacity, electrical conductivity, $CaCO_3$ equivalent value and sodium content are very low (Fig. 326 and Appendixes 3 and 4).

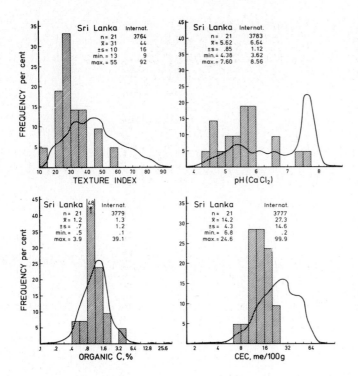

Fig. 326. Frequency distributions of texture, pH, organic carbon content, and cation exchange capacity in soils of Sri Lanka (columns). Curves show the international frequency of the same characters.

7.6.2 Macronutrients

The total **nitrogen** and $NaHCO_3$ extractable **phosphorus** contents of soils and the N and P contents of the sampled maize crops tend to be somewhat on the low side in the "international N and P fields" (Figs 6 and 7). The variation in the N contents of maize is relatively wide (Fig. 327) which may partly be due to varying nitrogen fertilization (24 ± 26 kg N/ha). The relatively low P contents of soils (Fig. 328) are partly compensated for by phosphate applications (10 ± 9 kg P/ha). See also CEC in Fig. 326 and Fig. 14 in Section 2.2.4 and related text.

Potassium applications to sampled maize crops (14 ± 12 kg K/ha, Appendix 5) are high compared to most other developing countries and in spite of the low level of the native soil K the K contents of maize are above the average international level (Figs 329 and 8). To some extent this contradiction may also be due to the relatively higher uptake of K by plants from acid than from neutral or alkaline soils. See Fig. 11 (Section 2.2.4) and related text and Fig. 326.

Calcium and **magnesium** contents of Sri Lanka soils and maize are internationally fairly low (Figs 330, 331, 9 and 10). In all cases the variations are wide. At some sites a response to Mg could be expected. See also texture and CEC in Sri Lanka soils (Fig. 326) and Figs 12 and 14 and related text in Section 2.2.4.

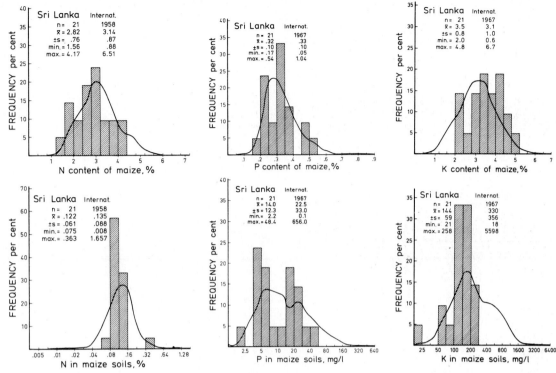

Fig. 327. Nitrogen, Sri Lanka.

Fig. 328. Phosphorus, Sri Lanka.

Fig. 329. Potassium, Sri Lanka.

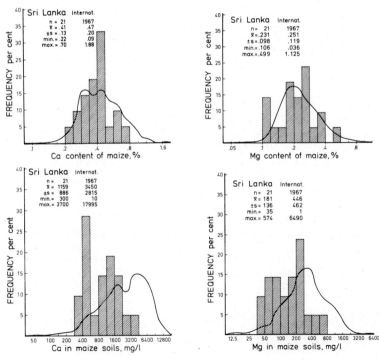

Fig. 330. Calcium, Sri Lanka.

Fig. 331. Magnesium, Sri Lanka.

Figs 327–331. Frequency distributions of nitrogen, phosphorus, potassium, calcium and magnesium in original maize samples and respective soils (columns) of Sri Lanka. Curves show the international frequency of the same characters.

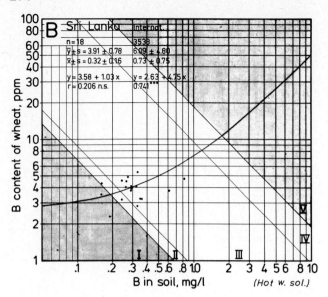

Fig. 332. Regression of B content of pot-grown wheat (y) on hot water soluble soil B (x), Sri Lanka. For details of summarized international background data, see Chapter 4.

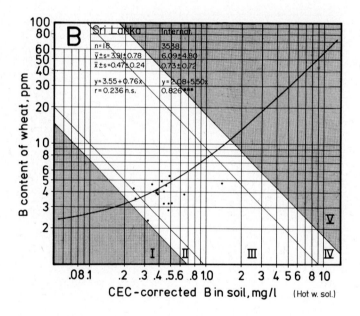

Fig. 333. Regression of B content of pot-grown wheat (y) on CEC-corrected (hot water soluble) soil B (x), Sri Lanka.

7.6.3 Micronutrients

Boron. A low B level is typical of soils and plants of Sri Lanka (Figs 332, 333, 22 and 25). Almost all B points are within the low B range (Zones I and II) or in the lower position of Zone III. No extreme values, however, were recorded. The sites of the lowest B values are in the more northern parts of the country.

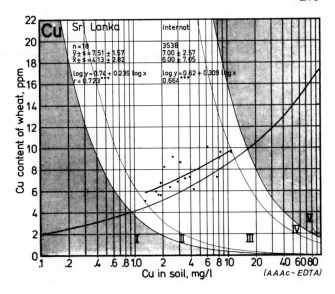

Fig. 334. Regression of Cu content of pot-grown wheat (y) on acid ammonium acetate-EDTA extractable soil Cu (x), Sri Lanka.

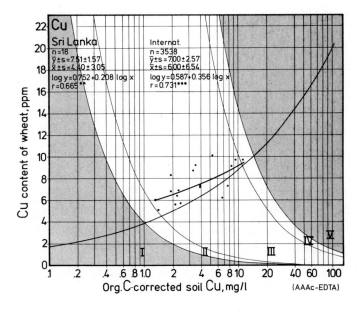

Fig. 335. Regression of Cu content of pot-grown wheat (y) on organic carbon-corrected (AAAc-EDTA extr.) soil Cu (x), Sri Lanka.

Copper and **Iron.** Both Cu and Fe values determined from Sri Lanka soils and plants correspond to the average international level (Figs 27, 29 and 31). No extreme (Zone I or V) values were recorded for either of these elements (Figs 334, 335 and 336).

Manganese. Only Zambia, Brazil and Malawi have national mean Mn contents at levels as high as those of Sri Lanka (Figs 33 and 37). In samples from three locations (45962, -67 and -71) exceptionally high Mn contents were measured (Figs 337 and 338), all having acid

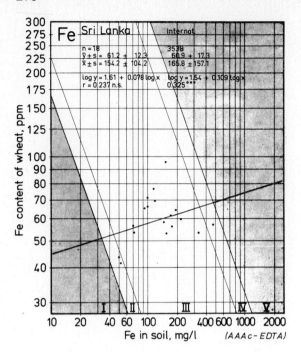

Fig. 336. Regression of Fe contents of pot-grown wheat (y) acid ammonium acetate-EDTA extractable soil Fe (x), Sri Lanka.

soils with pH(CaCl$_2$) 5 or below. The possible Mn toxicity could be corrected by liming.

Molybdenum. The results of plant Mo analyses and AO-OA extractable soil Mo are contradictory and lead to a non-significant negative correlation (r = —0.161 n.s., Fig. 339). Furthermore, the average Mo contents of Sri Lanka plants are about half the international average, but the AO-OA extractable soil Mo contents correspond to the international mean. When the effect of soil pH on AO-OA extraction is taken into account the correlation is improved to 0.650**, and at the same time the mean of soil Mo is reduced to half the international mean (Fig. 340).

In the "international Mo field" (Fig. 20) Sri Lanka is clearly on the low side but even so, only one Mo point falls in Zone II, the others being within the normal range (Fig. 340).

Zinc. Irrespective of the soil extraction method used for Zn, the limited number of samples from Sri Lanka indicates a relatively high Zn level in soils and plants (Figs 41 and 43). However, with the exception of the very high Zn contents measured in one sample pair from Kalutara, the Zn contents are mainly within the "normal" Zn range (Figs 341 and 342).

7.6.4 Summary

Considering the limited soil sample material of Sri Lanka, it may be concluded that typical soils are of relatively coarse texture, varying pH (mainly rather acid), of medium organic matter content and a low cation exchange capacity. The contents of macronutrients (N, P, K, Ca, Mg) are internationally fairly low. Of the six micronutrients, normal values were measured for Cu, Fe, Mo and Zn, but low B and high Mn levels were recorded.

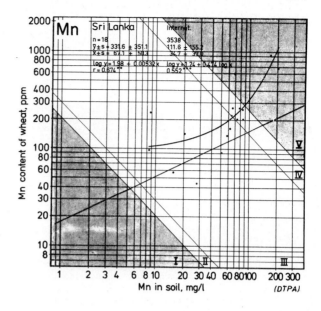

Fig. 337. Regression of Mn content of pot-grown wheat (y) on DTPA extractable soil Mn (x), Sri Lanka.

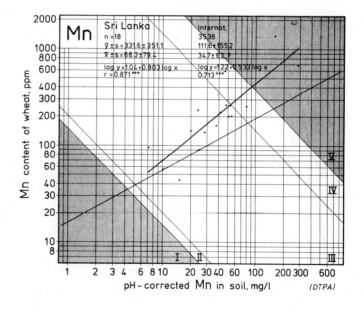

Fig. 338. Regression of Mn content of pot-grown wheat (y) on DTPA extractable soil Mn corrected for pH (x), Sri Lanka.

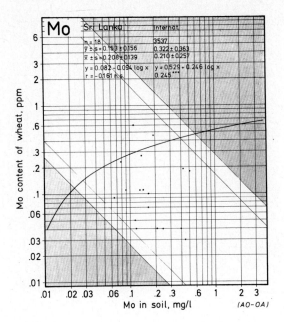

Fig. 339. Regression of Mo content of pot-grown wheat (y) on ammonium oxalate-oxalic acid extractable soil Mo (x), Sri Lanka.

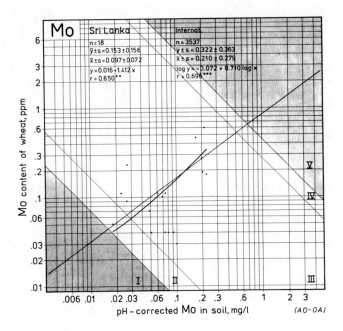

Fig. 340. Regression of Mo content of pot-grown wheat (y) on pH-corrected AO-OA extractable soil Mo (x), Sri Lanka.

Fig. 341. Regression of Zn content of pot-grown wheat (y) on DTPA extractable soil Zn (x), Sri Lanka.

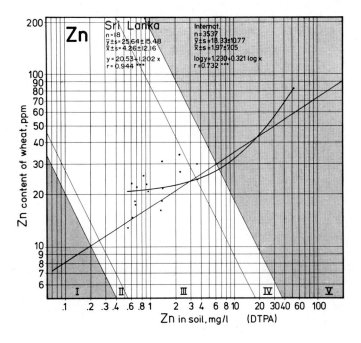

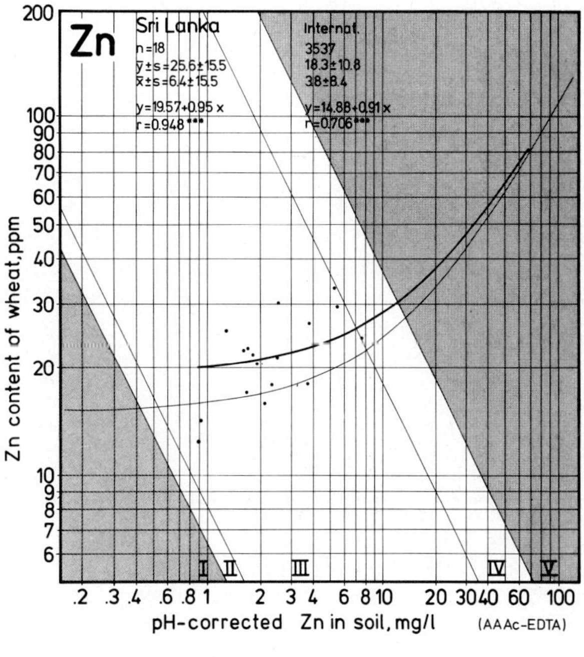

Fig. 342. Regression of Zn content of pot-grown wheat (y) AAAc-EDTA extractable soil Zn corrected for pH (x), Sri Lanka.

7.7 Thailand

7.7.1 General

The sample material received from Thailand consists of maize and respective soil samples only since wheat is not an important crop in this country. According to the geographical distribution of sampling sites, illustrated in Fig. 343, the central parts of the country are

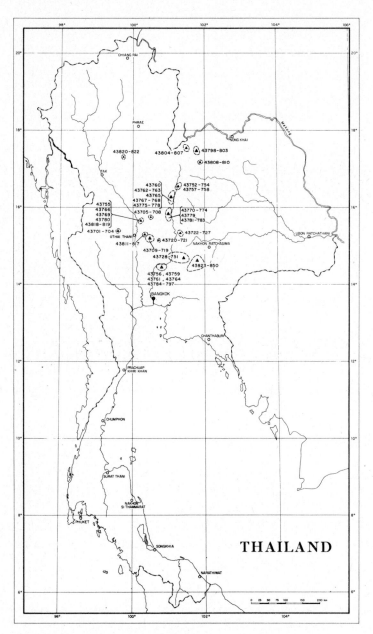

Fig. 343. Sampling sites in Thailand.

those best represented. Of the 150 soils sampled, 33 were classified as Rendzinas, 25 as Acrisols and 21 as Vertisols.

The majority of Thai soils are fine textured, though many medium and a few coarse textured soils are included in the sample material (Fig. 344). The range of variation in soil pH is relatively wide but very extreme pH values are not represented and alkaline or slightly acid soils are most typical. Owing to the relatively high organic matter content and heavy texture, the national average cation exchange capacity of Thai soils is among the highest recorded in this study (Appendixes 3 and 4). The average electrical conductivity and sodium content are somewhat below the respective averages for all the international material. $CaCO_3$ equivalent values for the majority of soils are low but a few quite high values were recorded.

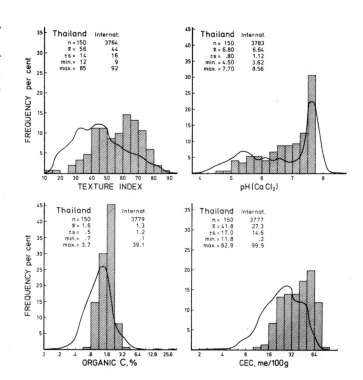

Fig. 344. Frequency distributions of texture, pH, organic carbon content, and cation exchange capacity in soils of Thailand (columns). Curves show the international frequency of the same characters.

7.7.2 Macronutrients

According to the information received from the sample collectors, no chemical fertilizers were applied to the sampled crops. On average, the total **nitrogen** contents of Thai soils and maize are slightly above their respective international averages (Figs 345 and 6).

On average, the **phosphorus** and **potassium** contents of both the Thai soils and maize are a little low by international standards but vary substantially from one sample to another (Figs 7, 8, 346 and 347). At many locations a good response to applications of both P and K should be expected.

The average **calcium** and **magnesium** contents of maize grown in Thailand are low compared to most other countries (Appendix 3, Figs 348 and 349) despite the relatively medium to high exchangeable Mg and high Ca levels found in Thai soils. The variations in values of both these elements are relatively wide. To obtain a general picture of the Ca and

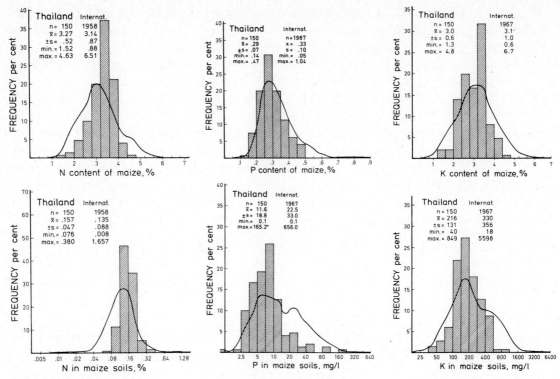

Fig. 345. Nitrogen, Thailand.   *Fig. 346.* Phosphorus, Thailand.   *Fig. 347.* Potassium, Thailand.

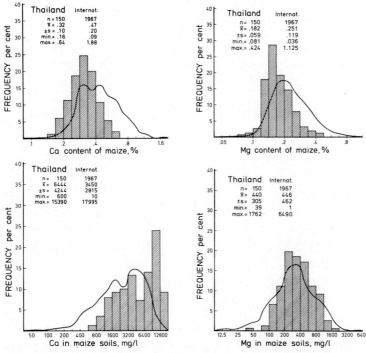

Figs 345–349. Frequency distributions of nitrogen, phosphorus, potassium, calcium and magnesium in original maize samples and respective soils (columns) of Thailand. Curves show the international frequency of the same characters.

Fig. 348. Calcium, Thailand.   *Fig. 349.* Magnesium, Thailand.

Mg status of Thai plants and soils, see also Figs 9 and 10. The heavy texture and high CEC of the Thai soils (Fig. 344) may have contributed to the low uptake of Ca and Mg by plants. See Figs 12 and 14 and related text in Section 2.2.4.

7.7.3 Micronutrients

Boron. In the "international B fields" Thailand features among the low-B countries (Figs 22 and 25). Unlike most other nutrients the B contents of Thai soils and of the respective pot-grown wheats vary within narrow limits. Almost all plant and soil B values are lower than their respective international means. The correction of hot water extractable soil B values for CEC not only improves the plant B—soil B correlation but also moves many of the samples from the "normal" range (Zone III) into the low B Zones (I and II, Figs 350

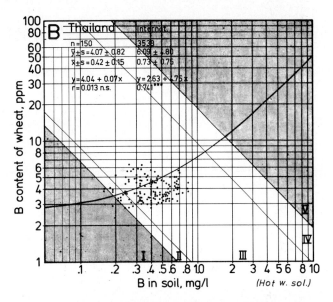

Fig. 350. Regression of B content of pot-grown wheat (y) on hot water soluble soil B (x), Thailand. For details of summarized international background data, see Chapter 4.

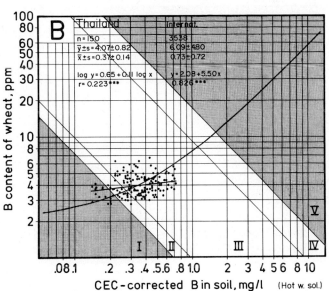

Fig. 351. Regression of B content of pot-grown wheat (y) on CEC-corrected (hot water soluble) soil B (x), Thailand.

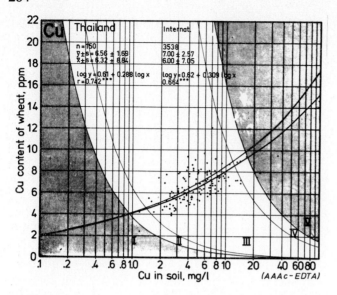

Fig. 352. Regression of Cu content of pot-grown wheat (y) on acid ammonium acetate-EDTA extractable soil Cu (x), Thailand.

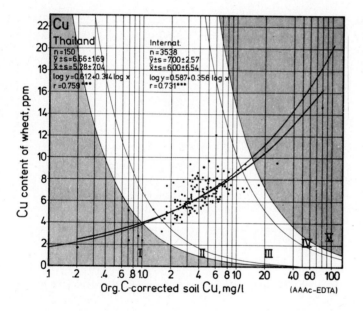

Fig. 353. Regression of Cu content of pot-grown wheat (y) on organic carbon-corrected (AAAc-EDTA extr.) soil Cu (x), Thailand.

and 351). In only a few countries is the relative frequency of low B (Zones I and II) values higher than that in Thailand. On the basis of these analytical results B deficiency, acute or latent, seems to be more likely in Thailand than in most other countries.

Copper. The national average Cu contents of Thai soils and plants correspond closely to the international mean values and Thailand is located at the centres of the "international Cu fields" (Figs 27 and 29). Over 90 percent of the sample pairs are within the normal Cu range (Zone III) but there are a few rather high as well as low values (Figs 352 and 353). According to the present data, Cu deficiency does not seem to be common, but shortage of Cu may be a factor limiting normal crop growth in some localities.

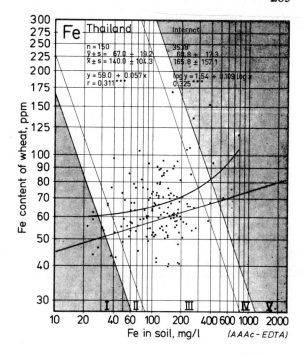

Fig. 354. Regression of Fe contents of pot-grown wheat (y) on acid ammonium acetate-EDTA extractable soil Fe (x), Thailand.

Iron. The analytical data on Fe given in Figs 354 and 31 show that, in general, the Thai soils and plants correspond to the normal international levels and distributions for Fe. Although a few sample pairs are within the two low Zones (I and II) their locations do not indicate any severe Fe deficiency.

Manganese. Thailand stands at the centres of the "international Mn fields" and its national mean values for plant and soil Mn deviate only slightly from the mean values of all the international material (Figs 33 and 37). The ranges of variation for both the plant and soil Mn are comparatively narrow with no low (Zones I and II) values among the Thai samples. Most of the high values are relatively close to the normal range (Figs 355 and 356). Only in one sample pair (43823) were exceptionally high Mn contents measured. These originated from a site in Pakchong with a heavy textured acid soil, TI 81, pH(CaCl$_2$) 4.5. Low pH is also typical of other Thai soils where high Mn contents were recorded.

Molybdenum. The Thai national means for plant and soil Mo correspond well to those of the whole international material (Figs 15 and 20), although some relatively low and high Mo values were found among the samples (Figs 357 and 358). The present analytical data give no indication of any severe Mo problems in Thailand.

Zinc. The Thai national mean values for plant and soil Zn are close to the international means (Figs 41 and 43), as are those of other micronutrients included in this study (B is an exception). In spite of the wide variations in both plant and soil Zn contents, the Thai samples do not occupy very extreme positions in the international scale (Figs 359 and 360). Low Zn values seem to be more typical for Sara Buri and Nakornsawan than the other areas sampled. There, and elsewhere in the country, a good response to Zn fertilization is likely. See also Section 2.7.

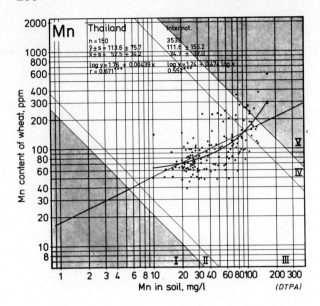

Fig. 355. Regression of Mn content of pot-grown wheat (y) on DTPA extractable soil Mn (x), Thailand.

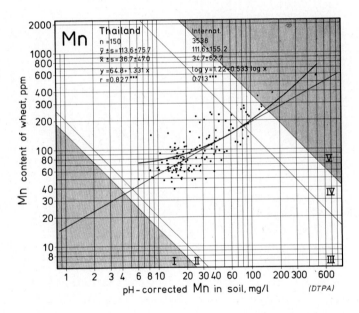

Fig. 356. Regression of Mn content of pot-grown wheat (y) on DTPA extractable soil Mn corrected for pH (x), Thailand.

7.7.4 Summary

Most of the Thai soils are heavy to medium textured, alkaline or slightly acid, relatively high in organic matter and have a high cation exchange capacity. Compared to data obtained from other countries the average Ca content of Thai soils is quite high, N and Mg are at the average international levels and P and K somewhat below. Most of the micronutrient contents (Cu, Fe, Mn, Mo) analysed from the Thai samples represent normal values. At some locations a deficiency of Cu is likely. The most likely micronutrient problem in Thailand is that of B and Zn deficiency.

Fig. 357. Regression of Mo content of pot-grown wheat (y) on ammonium oxalate-oxalic acid extractable soil Mo (x), Thailand.

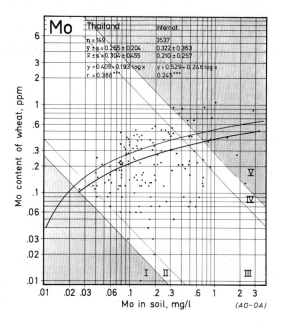

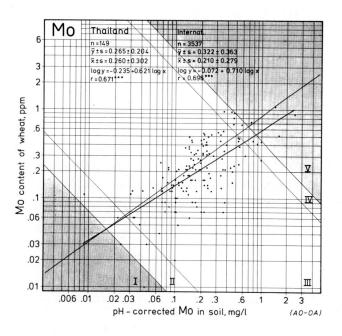

Fig. 358. Regression of Mo content of pot-grown wheat (y) on pH-corrected AO-OA extractable soil Mo (x), Thailand.

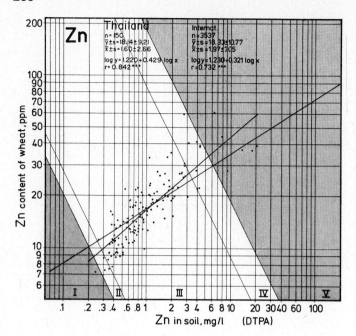

Fig. 359. Regression of Zn content of pot-grown wheat (y) on DTPA extractable soil Zn (x), Thailand.

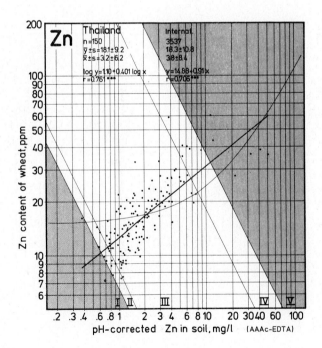

Fig. 360. Regression of Zn content of pot-grown wheat (y) on AAAc-EDTA extractable soil Zn corrected for pH (x), Thailand.

8. Near East

8.1 Egypt

8.1.1 General

The Egyptian sample material consists of 200 plant-soil sample pairs, half of which (100) were collected from wheat fields and half from maize fields. The locations of sampling sites illustrated in Fig. 361 show that the most important agricultural areas of the country,

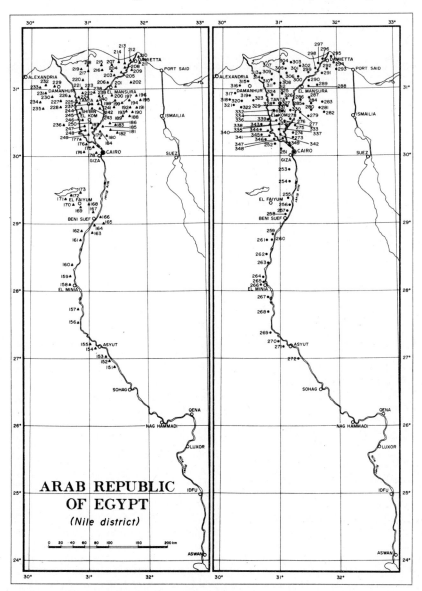

Fig. 361. Sampling sites in the Egypt (maize fields on the left and wheat fields on the right). The last three numerals of each sample pair number (45151—45350) are given.

the Nile valley and delta, are well covered. Almost all (197) the sampled soils were classified as Fluvisols; the remaining three soils as Xerosols.

The range of textural variation of the soils is very wide but the great majority is fine textured (Fig. 362). About 85 percent have a texture index exceeding the international average of soils in this study. Unlike texture, the variation of soil pH(CaCl$_2$) is very limited, from 7.4 to 8.2 only, and the national mean (7.74) is one of the highest recorded in this study. The organic matter contents correspond to the average international level and vary within relatively narrow limits. The national mean cation exchange capacity is one of the highest among the 30 countries studied, and in only a few soils were low CEC values recorded. The values of CaCO$_3$ equivalent are usually at the average international level but those of electrical conductivity are very high and the average sodium contents are higher than in any other country (Appendixes 2—4).

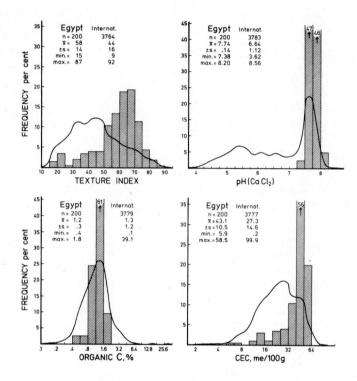

Fig. 362. Frequency distributions of texture, pH, organic carbon content, and cation exchange capacity in soils of Egypt (columns). Curves show the international frequency of the same characters.

8.1.2 Macronutrients

The **nitrogen** contents of original Egyptian plants as well as the total N contents of soils are of average international level (Figs 6 and 363). The variations, especially of soil N, are narrow. Nitrogen fertilizers were applied at high rates to both sampled crops (wheat 129 ± 31 and maize 161 ± 39 kg N/ha).

The **phosphorus** contents of both plants as well as the NaHCO$_3$ extractable P contents of the respective soils tend to be slightly on the low side in the context of the whole international material in this study (Figs 7 and 364). The P values vary relatively more

Figs 363–367. Frequency distributions of nitrogen, phosphorus, potassium, calcium and magnesium in original wheat and maize samples and respective soils (columns) of Egypt. Curves show the international frequency of the same characters.

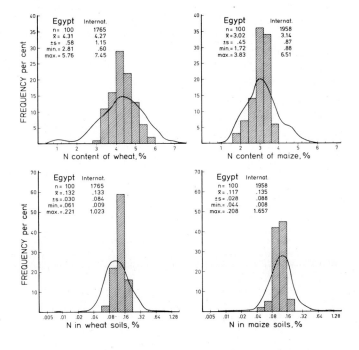

Fig. 363. Nitrogen, Egypt.

Fig. 364. Phosphorus, Egypt.

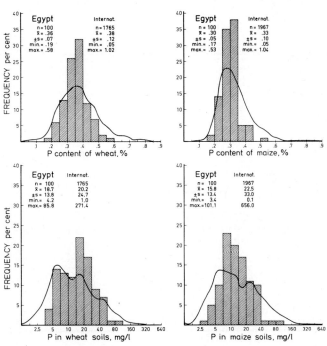

than those of N but no extremes were measured. The dressings of P applied in fertilizers were small (wheat: 9 ± 7 and maize: 2 ± 6 kg P/ha).

The average CH_3COONH_4 exchangeable **potassium** contents of Egyptian wheat and maize soils are about twice as high as the respective international averages (Fig. 365). Only in a few other countries were such high soil K contents recorded (Fig. 8). Although no potassium fertilizers were applied to the sampled wheat and maize crops, their K

292

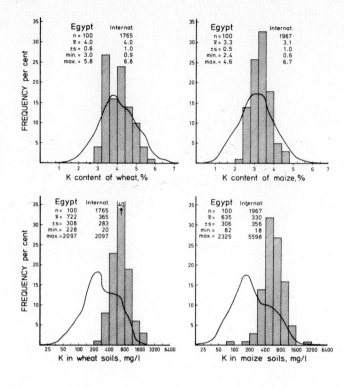

Fig. 365. Potassium, Egypt.

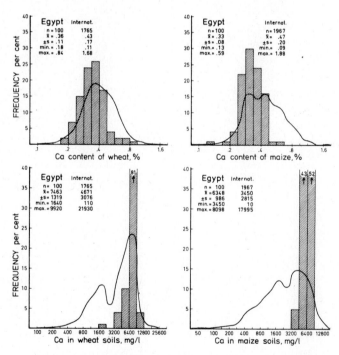

Fig. 366. Calcium, Egypt.

contents correspond well to the average international level.

The soils of Egypt are generally calcareous. Their average exchangeable **calcium** content is higher than that of most other countries but the Ca contents of both plants are slightly lower than the respective international averages (Figs 9 and 366).

Fig. 367. Magnesium, Egypt.

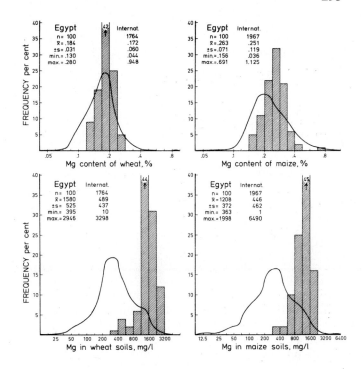

In the case of **magnesium** the situation is similar to Ca. The average contents of plants are only slightly above the respective international mean values although the national averages for exchangeable soil Mg are higher than in any other country (Figs 10 and 367). Obviously, the generally heavy texture and high CEC of Egyptian soils (Fig. 362) have lowered the uptake of Ca and Mg by plants in relation to the exchangeable Ca and Mg of the soils. See Figs 12* and 14 and related text in Section 2.2.4.

8.1.3 Micronutrients

Boron. The B levels of Egyptian soils and pot-grown plants seem quite normal. In the "international B fields" (Figs 22 and 25) Egypt stands near their centres. There were no values indicating severe B deficiency but some were so low as to suggest at least the possibility of hidden B deficiency (Figs 368 and 369). Although many Egyptian samples showed an abundance of B, none of the B values was extraordinarily high. According to

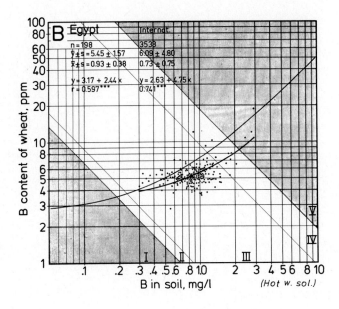

Fig. 368. Regression of B content of pot-grown wheat (y) on hot water soluble soil B (x), Egypt. For details of summarized international background data, see Chapter 4.

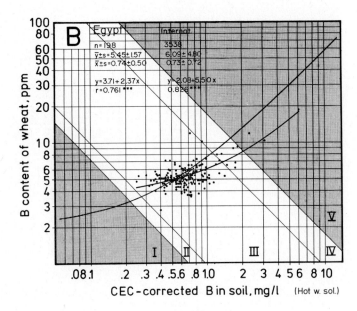

Fig. 369. Regression of B content of pot-grown wheat (y) on CEC-corrected (hot water soluble) soil B (x), Egypt.

Elseewi and Emalky (1979) much of the water soluble B in Egyptian soils comes from Nile water used for irrigation. They also considered the B contents of soils sufficient but incidences of toxicity unlikely.

Copper. Relatively high Cu contents are typical of Egyptian soils and plants (Figs 27, 29, 370 and 371), and about 30 % of the Cu values are within the two highest Cu Zones (IV and V) indicating an abundance but not likely toxicity of Cu. Deficiency of Cu seems unlikely in Egypt.

Iron. According to the present analytical data (Figs 31 and 372) no severe Fe problems are to be expected in Egypt.

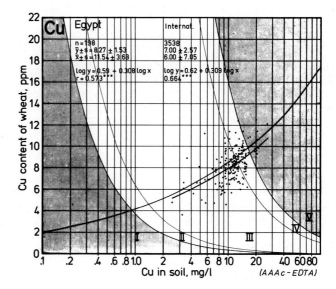

Fig. 370. Regression of Cu content of pot-grown wheat (y) on acid ammonium acetate-EDTA extractable soil Cu (x), Egypt.

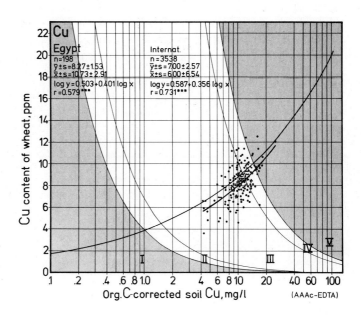

Fig. 371. Regression of Cu content of pot-grown wheat (y) on organic carbon-corrected (AAAc-EDTA extr.) soil Cu (x), Egypt.

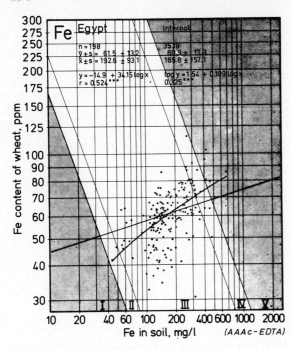

Fig. 372. Regression of Fe contents of pot-grown wheat (y) on acid ammonium acetate-EDTA extractable soil Fe (x), Egypt.

Manganese. Unlike most other nutrients the level of available Mn in Egypt is low (Figs 33, 37, 373 and 374), which is a natural consequence of the high alkalinity of soils. For the same reason the ranges of variation in plant and soil Mn contents are exceptionally narrow. All plant Mn contents, as well as pH-corrected DTPA extractable soil Mn contents, are below the respective international means and about 20 % of the sample pairs fall in the two lowest Mn Zones (I and II). Problems due to the small amount of available Mn may arise at several locations.

Molybdenum. The high level and limited variation of plant and soil Mo contents in the Egyptian sample material are due to the high pH of Egyptian soils. The apparent contradiction between high Mo contents of plants and low AO-OA extractable soil Mo contents (Fig. 375) is eliminated by pH-correction (Fig. 376). In spite of the generally high Mo level only a couple of plant-soil sample pairs fall in the highest Mo Zone (V). On the basis of these data Mo problems in Egypt are unlikely.

Zinc. Whichever extraction method is used for determining soil Zn contents, the Egyptian Zn levels are very similar. The national mean values for plant and soil Zn are internationally somewhat low (Figs 41 and 43), the variation ranges are relatively narrow and, therefore, only few Zn values fall outside the normal Zn range (Zone III; Figs 377 and 378). Many of the plant Zn contents, however, are unusually low and indicate lower Zn availability than do the soil analyses. At these sites response to Zn may be obtained. See also Section 2.7 where the critical limits for Zn are discussed.

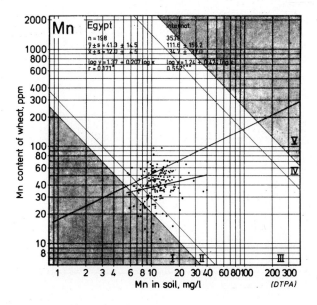

Fig. 373. Regression of Mn content of pot-grown wheat (y) on DTPA extractable soil Mn (x), Egypt.

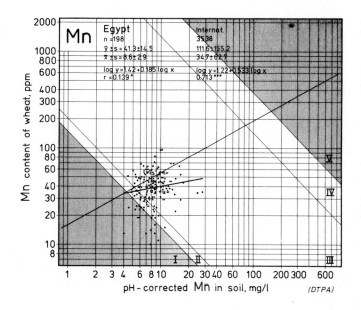

Fig. 374. Regression of Mn content of pot-grown wheat (y) on DTPA extractable soil Mn corrected for pH (x), Egypt.

8.1.4 Summary

Typical Egyptian soils are fine textured, highly alkaline, have a medium organic matter content and high cation exchange capacity. The macronutrient content of soils is generally good. The national average of P is slightly below the international P mean, but K, Ca, and Mg contents of Egyptian soils are higher than in most other countries.

Micronutrient problems concerning Cu, Fe and Mo are unlikely but occasionally responses to B, Mn and Zn can be expected.

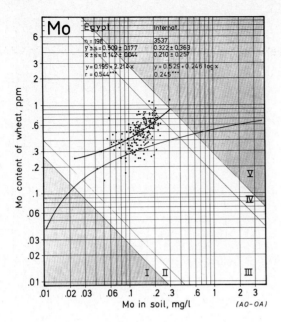

Fig. 375. Regression of Mo content of pot-grown wheat (y) on ammonium oxalate-oxalic acid extractable soil Mo (x), Egypt.

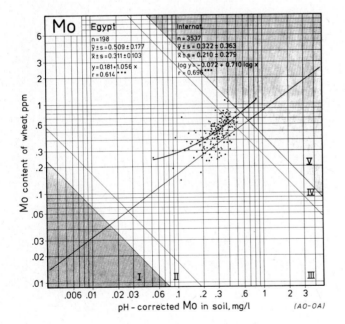

Fig. 376. Regression of Mo content of pot-grown wheat (y) on pH-corrected AO-OA extractable soil Mo (x), Egypt.

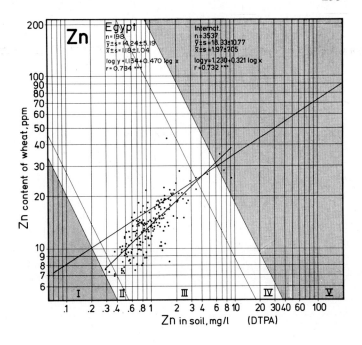

Fig. 377. Regression of Zn content of pot-grown wheat (y) on DTPA extractable soil Zn (x), Egypt.

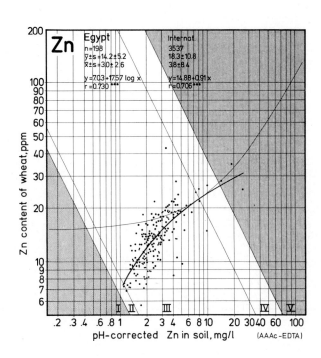

Fig. 378. Regression of Zn content of pot-grown wheat (y) on AAAc-EDTA extractable soil Zn corrected for pH (x), Egypt.

8.2 Iraq

8.2.1 General

The sample material from Iraq consists of 150 plant—soil sample pairs, of which 119 were collected from wheat fields and 31 from maize fields. The sampling sites are illustrated in Fig. 379 and cover the agriculturally important areas of the country. The Western Desert and Jazirah regions are excluded. One-third of the sampled crops had been grown under rainfed conditions in the northern part of the country where the annual precipitation usually exceeds 400 mm, two-thirds were irrigated and came mainly from the Mesopotamian plain.

The Iraqi soils were classified as follows: Halosols (all Solonchaks, 45 soils), Xerosols (35), Fluvisols (31), Vertisols (30) and Lithosols (8). Most soils are fine to

Fig. 379. Sampling sites in Iraq. The last three numerals of each sample pair number are given (dots = wheat fields, triangles = maize fields).

medium textured and all are alkaline; the national average pH is one of the highest (Fig. 380). The organic matter contents are low and the cation exchange capacity is at the average international level. The national mean value for electrical conductivity is the highest and those for $CaCO_3$ equivalent and sodium content the second highest recorded for countries participating in this study (Appendix 4).

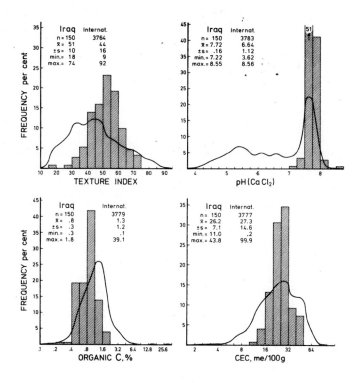

Fig. 380. Frequency distributions of texture, pH, organic carbon content, and cation exchange capacity in soils of Iraq (columns). Curves show the international frequency of the same characters.

8.2.2 Macronutrients

The total **nitrogen** contents of Iraqi soils are quite low (Figs 6 and 381). Low N contents are typical of Iraqi wheats but for maize the national mean equals the international mean. This may be due to nitrogen fertilization; the sampled wheats had only received 5 ± 9 kg N/ha while the maize crops were fertilized with 27 ± 24 kg N/ha.

The national average **phosphorus** contents of both plants and soils are among the lowest of this study (Figs 7 and 382). Wheat crops were fertilized with only nominal amounts of phosphates (2 ± 4 kg P/ha) but maize received somewhat more (10 ± 10 kg P/ha). A good response to phosphorus fertilizers could be expected at most locations examined, provided there are no other soil factors limiting the crop growth.

The **potassium** contents of Iraqi soils as well as the K contents of both indicator plants seem to correspond to the average international level (Figs 8 and 383). Potassium fertilizers were applied only occasionally (wheat 1 ± 3 and maize 0 ± 1 kg K/ha).

Both the exchangeable **calcium** and **magnesium** contents of Iraqi soils are high but the Ca and Mg contents of plants correspond more or less to the average international level (Figs 9, 10, 384 and 385).

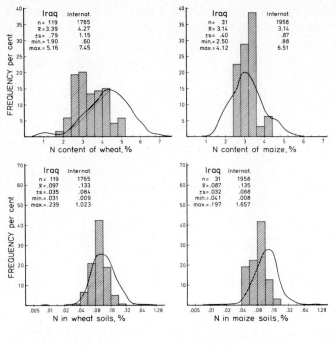

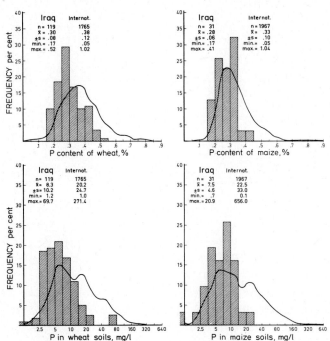

Figs 381–385. Frequency distributions of nitrogen, phosphorus, potassium, calcium and magnesium in original wheat and maize samples and respective soils (columns) of Iraq. Curves show the international frequency of the same characters.

Fig. 381. Nitrogen, Iraq.

Fig. 382. Phosphorus, Iraq.

Of the 119 wheat samples, 88 were classified as HYV ('Mexipak') and 30 as local varieties. This last group consisted of varieties 'Sabir Biek' (11 samples), unclassified, marked 'Local' (9), 'Harama' (2), 'Moseele' (2), 'Bakra Toe' (1), 'Rashboal' (1) and four samples called 'Italian'. The contents of all macronutrients in the HYV wheat were somewhat higher than of the local varieties. However, this comparison is of limited value because in most cases the HYV had either grown on soils richer in macronutrients or had

Fig. 383. Potassium, Iraq.

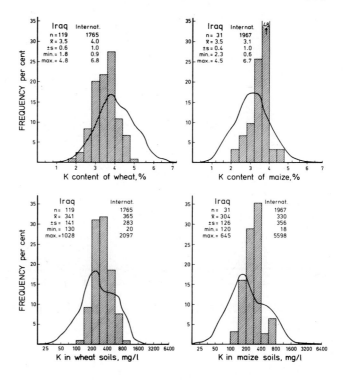

Fig. 384. Calcium, Iraq.

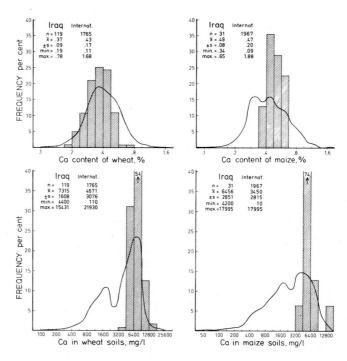

received more macronutrient fertilizers than the local varieties. Furthermore, the majority of local varieties were grown under rainfed conditions and the majority of HYV under irrigated conditions, rendering the comparison between the variety groups still less justified. See Section 8.2.4.

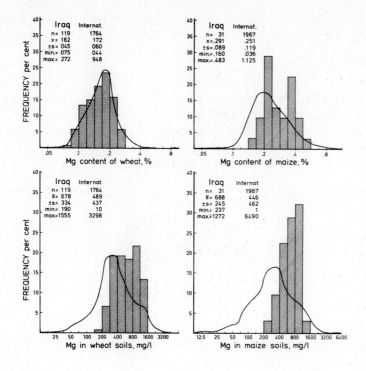

Fig. 385. Magnesium, Iraq.

8.2.3 Micronutrients

Boron. The national mean B contents of Iraqi soils and pot-grown wheat were considerably higher than those for any other country (Figs 22, 25, 386 and 387). A third of the plant—soil sample pairs falls in the highest B Zone (V) and half in Zones IV and V. The ranges of variation of both plant and soil B contents are exceptionally wide with some values falling in the two low-B Zones. The latter do not indicate any serious B deficiency but it is apparent that a response to B could be obtained at several locations in Northern Iraq. (See also Section 2.7). B toxicity could become a problem at several locations in Iraq if crops sensitive to excess B are grown. This would be most likely in the Mesopotamian

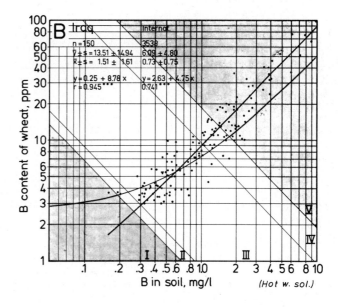

Fig. 386. Regression of B content of pot-grown wheat (y) on hot water soluble soil B (x), Iraq. For details of summarized international background data, see Chapter 4.

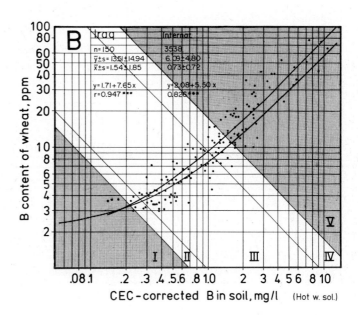

Fig. 387. Regression of B content of pot-grown wheat (y) on CEC-corrected (hot water soluble) soil B (x), Iraq.

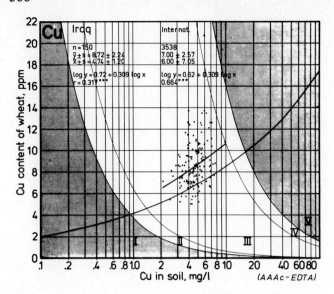

Fig. 388. Regression of Cu content of pot-grown wheat (y) on acid ammonium acetate-EDTA extractable soil Cu (x), Iraq.

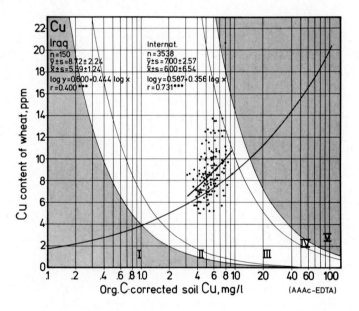

Fig. 389. Regression of Cu content of pot-grown wheat (y) on organic carbon-corrected (AAAc-EDTA extr.) soil Cu (x), Iraq.

plains where the soils are very rich in B and where the irrigation waters apparently contain B. See Section 8.2.4.

Copper. The locations of Iraq are relatively close to the centres of the "international Cu fields" (Figs 27 and 29). The variation of soil Cu contents is exceptionally narrow and with a single exception[1] the samples are within the normal Cu range (Zone III, Figs 388 and 389). Response to Cu in Iraq is unlikely at the sites examined.

---

[1] The extremely high Cu content of pot-grown wheat in this sample (47541) may be due to contamination. Neither the soil Cu value nor the Cu content of the respective original wheat indicated excess Cu. The Fe and Mn contents of this sample were also exceptionally high.

Fig. 390. Regression of Fe contents of pot-grown wheat (y) on acid ammonium acetate-EDTA extractable soil Fe (x), Iraq.

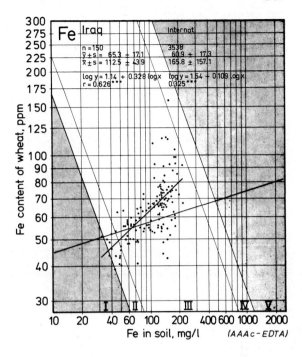

Iron. According to the data presented in Figs 31 and 390, problems with Fe are unlikely in Iraq unless the soil oxidation-reduction conditions are such as to limit its availability. These conditions are more likely to develop in the northern parts of the country where crops are grown under rainfed conditions, and where increased oxidation potential may lead to the oxidation of Fe from ferrous to ferric forms so decreasing the Fe availability to plants. See also Section 8.2.4.

Manganese. Because of the generally high and only slightly varying pH of Iraqi soils the Mn contents of both plants and soils are low and their ranges of variation exceptionally narrow (Figs 33, 37, 391 and 392). Although a few samples fall in the low-Mn Zones (I and II) these do not indicate any severe degree of Mn deficiency. However, a response to Mn could be obtained at several locations especially at sites where soil moisture conditions are unfavourable for Mn uptake by plants. See also Section 8.2.4.

Molybdenum. Unlike Mn the contents of Mo vary widely in spite of a uniformly high soil pH (Figs 393 and 394). In general, high values dominate and in the "international Mo fields" Iraq is among the high-Mo countries (Figs 15 and 20). Although the national plant Mo—soil Mo correlation is not greatly affected by pH correction, the location of the whole Iraqi Mo data on the regression graphs is more consistent with those of other countries when the AO-OA extractable soil Mo values are corrected for soil pH. See also Section 8.2.4.

Zinc. Irrespective of the extraction method used for determining soil Zn, the positions of Iraq are the lowest in the "international Zn fields" (Figs 41 and 43) and both soil methods give good correlations with the results of plant Zn analyses (Figs 395 and 396). Still, the DTPA extractable soil Zn values are relatively lower than those extracted with AAAc-EDTA. In other words, more than half of the plant Zn—DTPA Zn points are

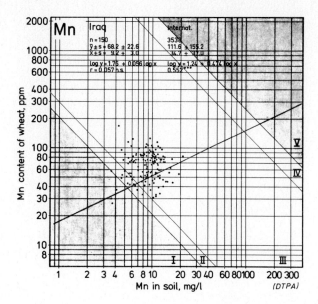

Fig. 391. Regression of Mn content of pot-grown wheat (y) on DTPA extractable soil Mn (x), Iraq.

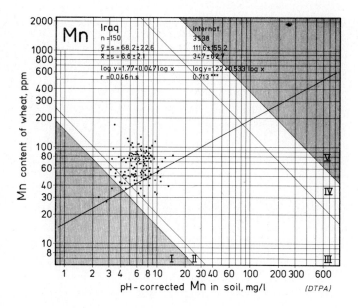

Fig. 392. Regression of Mn content of pot-grown wheat (y) on DTPA extractable soil Mn corrected for pH (x), Iraq.

within the two lowest Zn Zones while in case of AAAc-EDTA there is a strong concentration of samples in the lower side of Zone III. As pointed out in Section 2.3.7.2 (Table 18 and related text), the DTPA extractable soil Zn is generally better correlated with plant Zn in areas with alkaline soils and pH-corrected AAAc-EDTA Zn in areas where acid soils predominate. Therefore, it would seem that in the case of Iraq the DTPA extraction method gives a better estimate of the soil Zn status (Fig. 395) than does the AAAc-EDTA method. See also Section 2.7. According to these analytical data Zn deficiency seems more common in Iraq than in any other country studied. The lowest values were measured from samples originating from widely scattered areas, and no distinct low-Zn areas could be distinguished. See also Section 8.2.4.

Fig. 393. Regression of Mo content of pot-grown wheat (y) on ammonium oxalate-oxalic acid extractable soil Mo (x), Iraq.

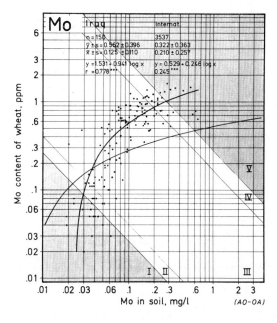

Fig. 394. Regression of Mo content of pot-grown wheat (y) on pH-corrected AO-OA extractable soil Mo (x), Iraq.

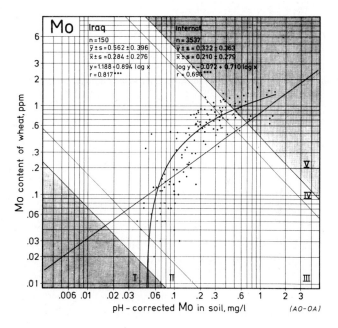

8.2.4 Nutrient status of soils and plants of Iraq with special reference to irrigation

Successive civilizations have lived in the Mesopotamian riverplains for approximately 6000 years. They founded their existence on irrigated agriculture, and consequently, irrigation practices have greatly influenced the soil formation. At present almost the whole plain is covered by a layer of irrigation deposits of usually several metres in thickness (Delver, 1960). About two-thirds of the samples in this study originate from the Mesopotamian plains and represent irrigated agriculture while one-third were collected from Northern Iraq where the crops were grown under rainfed conditions. The analytical

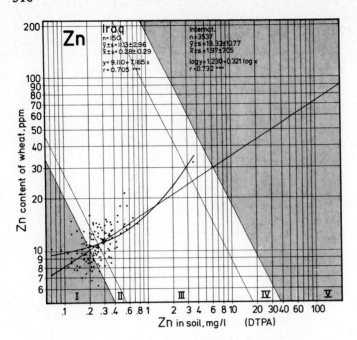

Fig. 395. Regression of Zn content of pot-grown wheat (y) on DTPA extractable soil Zn (x), Iraq.

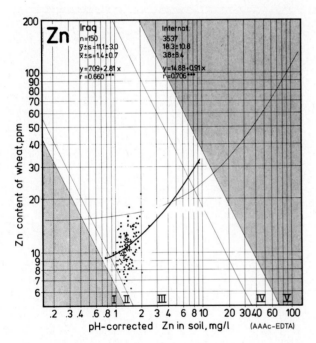

Fig. 396. Regression of Zn content of pot-grown wheat (y) on AAAc-EDTA extractable soil Zn corrected for pH (x), Iraq.

data on the soils and wheat[1] of these two groups are given for comparison in Table 27.

When comparing the data it should be borne in mind that although the irrigation methods practised since ancient times and their consequences may be the most likely cause for the differences between the soil and plant properties given in Table 27, the samples under comparison are from different geographical areas and this may account for

---

[1] Maize crops and respective soils are omitted from the comparison because the original maize crops, with one exception, were all irrigated.

Table 27. Comparison of analytical data from non-irrigated and irrigated soils of Iraq and respective plants[1]. Statistical differences (t-values) between the groups are given at 5*, 1** and 0.1*** percent significance levels.

Characteristics	Non-irrigated (n = 44)	Irrigated (n = 75)	(t)
pH(CaCl$_2$)	7.55 ± 0.08	7.77 ± 0.11	11.98***
Texture index	54 ± 11	50 ± 10	2.19*
Organic C, %	0.97 ± 0.33	0.77 ± 0.27	3.43**
CEC, me/100 g	30.9 ± 8.3	24.5 ± 5.6	4.54***
El. conductivity, $\frac{10^{-4} S}{cm}$	1.7 ± 0.9	12.2 ± 10.9	8.27***
CaCO$_3$ equivalent, %	23.8 ± 8.3	26.0 ± 4.9	1.55 n.s.
Na in soil, mg/l	21 ± 64	510 ± 465	8.97***
N in orig. wheat, %	2.91 ± 0.59	3.67 ± 0.76	6.06***
N in soil, %	0.104 ± 0.034	0.093 ± 0.036	1.76 n.s.
P in orig. wheat, %	0.269 ± 0.062	0.317 ± 0.087	3.59***
P in soil, mg/l	6.1 ± 3.7	9.6 ± 12.4	2.25*
K in orig. wheat, %	3.20 ± 0.52	3.65 ± 0.60	4.34***
K in soil, mg/l	430 ± 117	288 ± 127	6.19***
Ca in orig. wheat, %	0.383 ± 0.092	0.370 ± 0.095	0.77 n.s.
Ca in soil, mg/l	7929 ± 1447	6955 ± 1597	3.41**
Mg in orig. wheat, %	0.124 ± 0.025	0.185 ± 0.038	12.70***
Mg in soil, mg/l	375 ± 107	856 ± 260	12.95***
B in orig. wheat, ppm	3.64 ± 2.84	18.32 ± 16.00	7.74***
B in pot wheat, ppm	4.23 ± 1.34	17.07 ± 13.83	7.98***
B in soil, mg/l	0.45 ± 0.37	1.89 ± 1.68	7.10***
Cu in orig. wheat, ppm	7.31 ± 2.81	10.95 ± 15.05	2.03*
Cu in pot wheat, ppm	7.02 ± 1.02	10.18 ± 2.14	10.85***
Cu in soil, mg/l	4.97 ± 0.85	5.86 ± 1.26	4.49***
Fe in pot wheat, ppm	53.3 ± 7.0	74.7 ± 18.2	9.09***
Fe in soil, mg/l	64.0 ± 17.7	137.3 ± 32.7	15.84***
Mn in orig. wheat, ppm	57.4 ± 17.8	99.7 ± 42.5	7.57***
Mn in pot wheat, ppm	69.5 ± 17.7	70.3 ± 25.8	0.21 n.s.
Mn in soil, mg/l	6.8 ± 1.5	6.0 ± 1.8	2.09*
Mo in orig. wheat, ppm	0.28 ± 0.23	1.72 ± 0.85	13.85***
Mo in pot wheat, ppm	0.12 ± 0.08	0.72 ± 0.26	18.59***
Mo in soil, mg/l	0.096 ± 0.049	0.370 ± 0.280	8.28***
Zn in orig. wheat, ppm	21.2 ± 4.6	21.1 ± 6.5	0.07 n.s.
Zn in pot wheat, ppm	10.8 ± 1.8	12.0 ± 3.5	2.39*
Zn in soil, mg/l	0.28 ± 0.11	0.28 ± 0.40	0.00 n.s.

[1] For extraction and analytical methods, see Section 1.3.
    B: hot water extraction + CEC correction
    Cu: AAAc-EDTA + org. C correction
    Fe: AAAc-EDTA
    Mn: DTPA + pH correction
    Mo: AO-OA + pH correction
    Zn: DTPA

the differences. Since the effects of irrigation depend on the quality of the irrigation water in relation to soil properties, and as no analytical data of the water used at different sampling sites are available, too far-reaching conclusions on these effects cannot be justified. However, the very large differences in the contents of such highly soluble elements as Na and B would suggest that much of the extractable Na and B in the irrigated soils is due to accumulation of these elements brought to the soils by irrigation water. The same may be true for Mo and Fe; e.g. Fireman and Kraus (1965) found Mo in irrigation waters. Irrigation seems to have practically no effect on Zn levels, and for Cu the quantitative differences are relatively small. There are only small and statistically non-significant differences between the Mn contents of wheat grown in pots of formerly irrigated and rainfed soils, though the original irrigated wheats contained significantly more Mn than did the rainfed wheats. It would seem that in irrigated soils the moisture (redox) conditions are more favourable to Mn absorption by plants than in the drier field conditions in the rainfed areas. This difference does not appear in the Mn contents of pot-grown wheat since all pots were maintained under similar soil moisture conditions.

El-Dujaili and Ismail (1971) measured 2.6 me Ca and 2.2 me Mg per litre of Tigris River water at Aziziya. If these concentrations are typical of the water used for irrigating the present crops, it it possible that the irrigation water leached Ca from the soil which is high in Ca and contributed Mg to the soil which is low in Mg. For N, P and K the comparison may be obscured to some degree by fertilizer application, even though the amounts applied were generally small. According to the field information data, the average N, P and K applications to irrigated wheat were 8, 3 and 1 kg/ha and to non-irrigated wheat 2, 1 and 0 kg/ha, respectively. On the debit side the higher yields obtained from irrigated fields would remove more nutrients from the irrigated than from the non-irrigated soils. This applies especially to K which crops extract from soils in considerable quantities. It is apparent that the accumulation of elements such as Na and Mg in the irrigated soils contributes to the higher electrical conductivity and pH of these soils.

The above results indicate in a general way some of the differences between the irrigated and non-irrigated soils. Further studies, including detailed data on the quality of irrigation water and ground water at various locations, are needed to obtain a more comprehensive picture of the behaviour of different nutrients and other elements and of their leaching-accumulation characteristics.

8.2.5 Summary

Most of the Iraqi soils sampled for this study are medium to fine textured, highly alkaline, low in organic matter and have a medium cation exchange capacity. Very high electrical conductivity, $CaCO_3$ equivalent, and sodium content are typical, especially for the soils of the Mesopotamian plains.

In general, the P and N contents of soils are low, the K contents correspond to the average international level and the Ca and Mg contents are high. The most likely problems to arise with micronutrients would be due to deficiency of Zn but those due to both shortage and excess of B can also be expected. High Mo and low Mn are typical for Iraqi soils but most of the Cu and Fe values are within the "normal" range. There are distinct differences concerning most nutrients and other soil properties between the northern parts of the country, where the crops are grown under rainfed conditions, and the Mesopotamian plain where irrigated agriculture has been practised since ancient times.

8.3 Lebanon

8.3.1 General

The geographical sources of the Lebanese maize-soil sample material is shown in Fig. 397. The soils are medium to fine textured and alkaline with relatively low organic matter content and high cation exchange capacity (Fig. 398). The high $CaCO_3$ equivalents and electrical conductivities and relatively low sodium contents are typical of most Lebanese soils sampled (Appendixes 3 and 4).

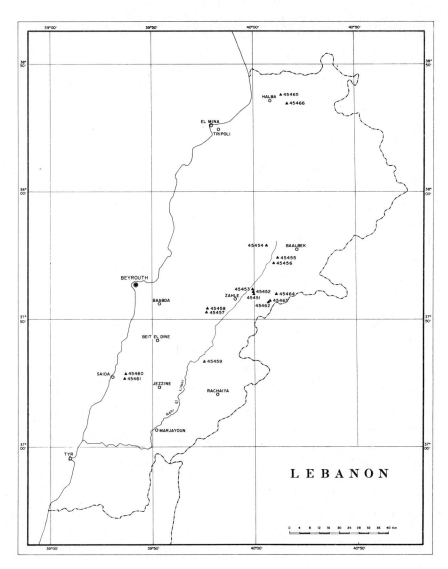

Fig. 397. Sampling sites in Lebanon.

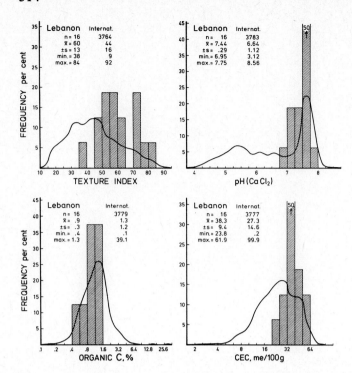

Fig. 398. Frequency distributions of texture, pH, organic carbon content, and cation exchange capacity in soils of Lebanon (columns). Curves show the international frequency of the same characters.

8.3.2 Macronutrients

The total **nitrogen** contents of soils are low. However, owing to the relatively high rates of applied nitrogen fertilizers (53 ± 99 kg N/ha), the mean N contents of maize are slightly above the international average (Figs 6 and 399). The wide variation in N contents of maize may be attributable to varying fertilizer rates.

The **phosphorus** contents of soils and maize are relatively high (Figs 7 and 400). The **potassium** contents also slightly exceed the respective international means (Figs 8 and 401). The sampled crops had received relatively heavy dressings of phosphates (33 ± 54 kg P/ha) but no potassium.

The national mean for the exchangeable **calcium** content of Lebanese soils is the second highest (after Syria) in this study but the average Ca content of maize exceeds only slightly the respective international mean (Fig. 402). The mean exchangeable **magnesium** content of Lebanese soils exceeds the international mean but the mean Mg content of maize is lower on the international scale (Fig. 403). These results may be due to the relatively low uptake of Ca and Mg by plants from soils of heavy texture and high CEC which characterize most Lebanese soils. See Figs 12 and 14 and related text in Section 2.2.4.

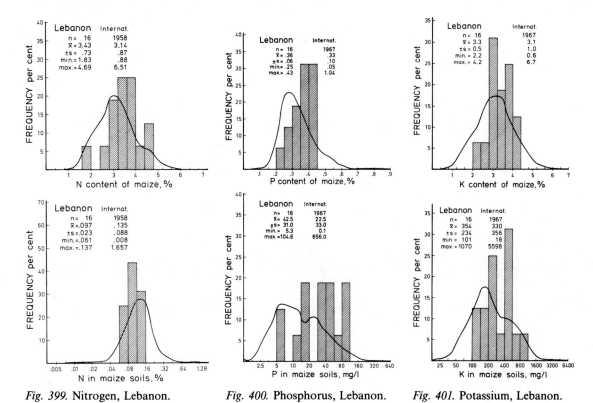

Fig. 399. Nitrogen, Lebanon.   Fig. 400. Phosphorus, Lebanon.   Fig. 401. Potassium, Lebanon.

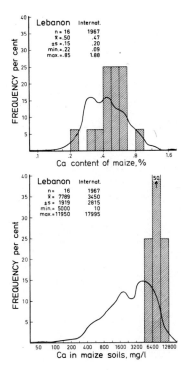

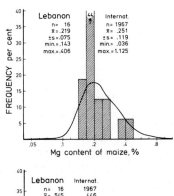

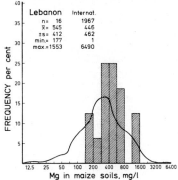

Fig. 402. Calcium, Lebanon.   Fig. 403. Magnesium, Lebanon.

Figs 399–403. Frequency distributions of nitrogen, phosphorus, potassium, calcium and magnesium in original maize samples and respective soils (columns) of Lebanon. Curves show the international frequency of the same characters.

8.3.3 Micronutrients

Judging from the limited number of samples the micronutrient levels in Lebanon seem to be quite satisfactory. In the "international micronutrient fields" the position of Lebanon is relatively close to their centres, often slightly lower (Section 2.3: Figs 15, 20, 22, 25, 27, 29, 31, 33, 37, 41 and 43). Since the ranges of variation are usually narrow the points

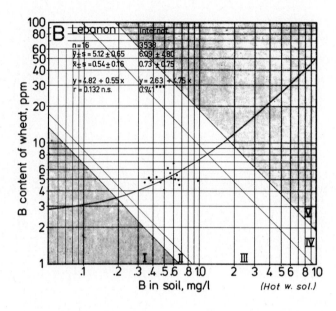

Fig. 404. Regression of B content of pot-grown wheat (y) on hot water soluble soil B (x), Lebanon. For details of summarized international background data, see Chapter 4.

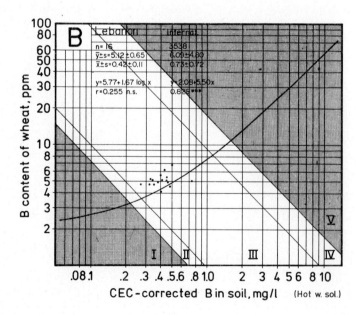

Fig. 405. Regression of B content of pot-grown wheat (y) on CEC-corrected (hot water soluble) soil B (x), Lebanon.

indicating micronutrient values for single plant—soil sample pairs are usually within the "normal" range (Zone III in Figures 404 to 414). Even the few Fe, Mn and Mo values outside Zone III are not in any extreme positions. However, at some locations response to B and Zn may be obtained (see Section 2.7). On the basis of these limited analytical data, micronutrient problems in Lebanon seem less likely than in most other countries involved in this study.

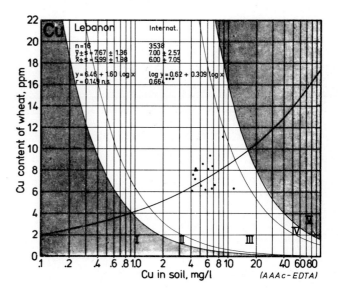

Fig. 406. Regression of Cu content of pot-grown wheat (y) on acid ammonium acetate-EDTA extractable soil Cu (x), Lebanon.

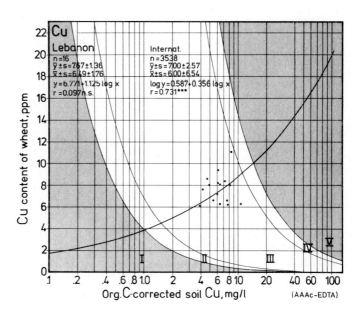

Fig. 407. Regression of Cu content of pot-grown wheat (y) on organic carbon-corrected (AAAc-EDTA extr.) soil Cu (x), Lebanon.

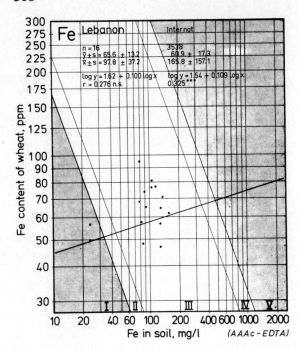

Fig. 408. Regression of Fe contents of pot-grown wheat (y) on acid ammonium acetate-EDTA extractable soil Fe (x), Lebanon.

8.3.4 Summary

With exception of relatively low total N contents the macronutrient status of sampled Lebanese soils is generally good. Fewer problems concerning micronutrients could be expected than in most other countries.

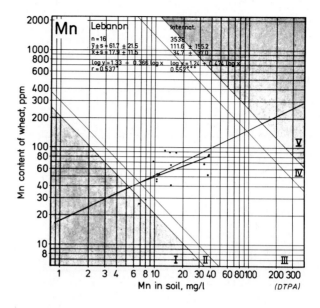

Fig. 409. Regression of Mn content of pot-grown wheat (y) on DTPA extractable soil Mn (x), Lebanon.

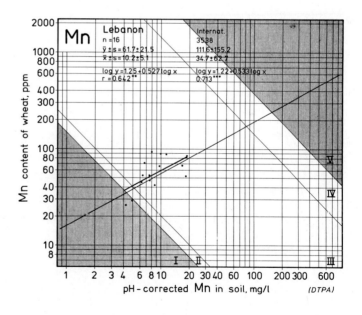

Fig. 410. Regression of Mn content of pot-grown wheat (y) on DTPA extractable soil Mn corrected for pH (x), Lebanon.

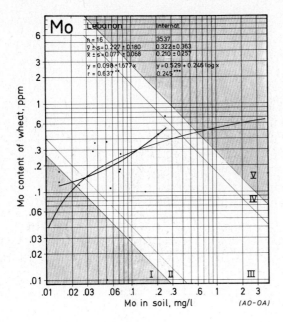

Fig. 411. Regression of Mo content of pot-grown wheat (y) on ammonium oxalate-oxalic acid extractable soil Mo (x), Lebanon.

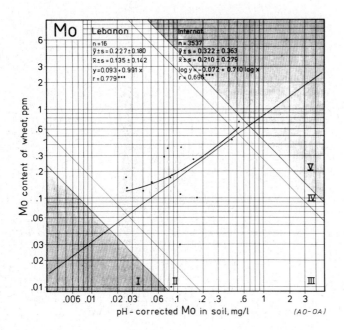

Fig. 412. Regression of Mo content of pot-grown wheat (y) on pH-corrected AO-OA extractable soil Mo (x), Lebanon.

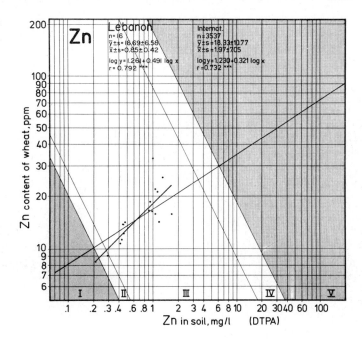

Fig. 413. Regression of Zn content of pot-grown wheat (y) on DTPA extractable soil Zn (x), Lebanon.

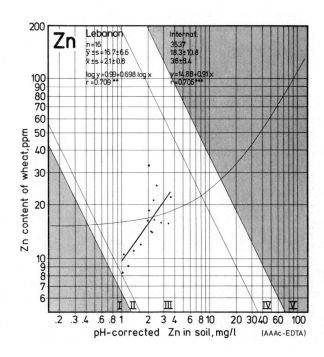

Fig. 414. Regression of Zn content of pot-grown wheat (y) on AAAc-EDTA extractable soil Zn corrected for pH (x), Lebanon.

8.4 Syria

8.4.1 General

The original Syrian sample material consisting of 20 wheat-soil and 18 maize-soil sample pairs was collected from seven out of 11 Syrian Provinces: Latakia (11), Aleppo (7), Hama (7), Deir-ez-Zor (7), Derá (2), Homs (2) and Raqqa (2). The geographical distribution of sampling sites is illustrated in Fig. 415. Of the sampled soils, 15 were classified as Fluvisols and 12 as Luvisols.

The Syrian soils vary widely in texture but the majority are fine textured (Fig. 416). Very high pH, medium organic matter content and a high cation exchange capacity are typical. The national average for $CaCO_3$ equivalent is one of the highest recorded in this study. Electrical conductivity values are relatively high but the sodium contents correspond to the average international level (Appendixes 2—4).

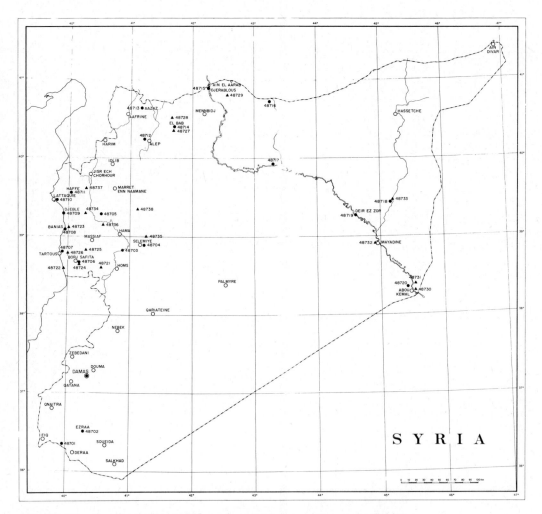

Fig. 415. Sampling sites in Syria (dots = wheat fields, triangles = maize fields).

8.4.2 Macronutrients

The total **nitrogen** contents of the sampled Syrian soils as well as the N contents of the original wheat and maize are at the average international level or slightly below it (Figs 6 and 417). The sampled crops, especially maize, were fertilized with high, but widely varying dressings of nitrogen (wheat: 52 ± 59 and maize 117 ± 87 kg N/ha) which may have extended the range of variation for the N contents of plants.

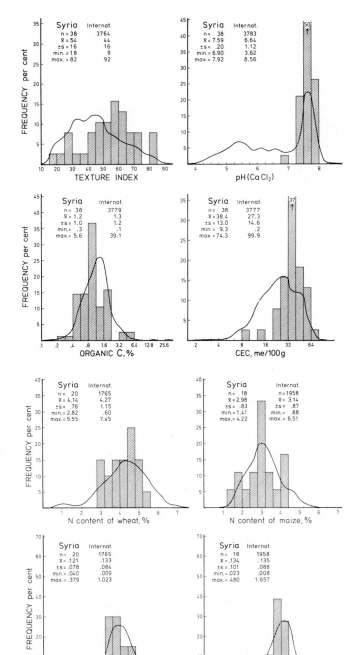

Fig. 416. Frequency distributions of texture, pH, organic carbon content, and cation exchange capacity in soils of Syria (columns). Curves show the international frequency of the same characters.

Figs 417–421. Frequency distributions of nitrogen, phosphorus, potassium, calcium and magnesium in original wheat and maize samples and respective soils (columns) of Syria. Curves show the international frequency of the same characters.

Fig. 417. Nitrogen, Syria.

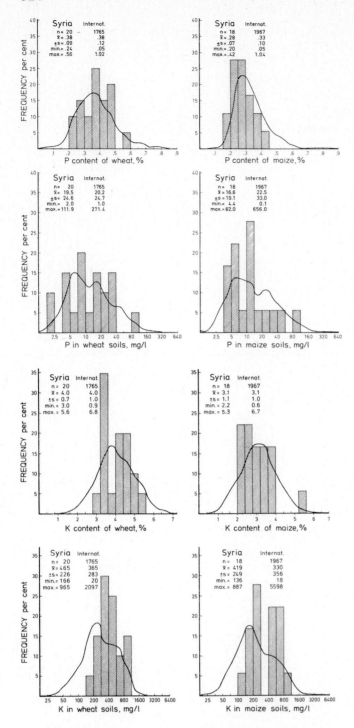

Fig. 418. Phosphorus, Syria.

Fig. 419. Potassium, Syria.

The NaHCO$_3$ extractable **phosphorus** contents of Syrian soils as well as the P contents of plants are, on average, at the normal international level while varying substantially from one sample to another (Figs 7 and 418). Only moderate amounts of phosphates were applied to the sampled crops (wheat: 7 ± 11 and maize 7 ± 12 kg P/ha).

The mean exchangeable **potassium** contents of soils and K contents of plants are at the average international level but the ranges of variation are quite wide (Figs 8 and 419). No

Fig. 420. Calcium, Syria.

Fig. 421. Magnesium, Syria.

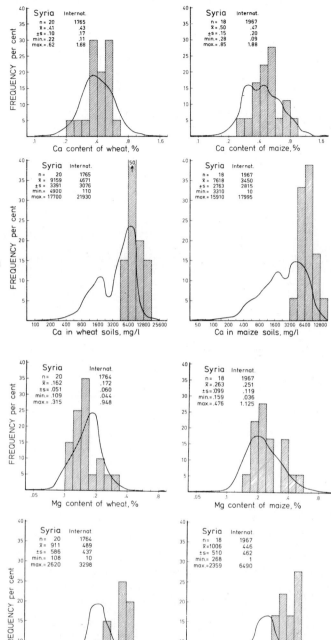

potassium fertilizers were applied. The national mean values for exchangeable **calcium** and **magnesium** in Syrian soils are among the very highest recorded in this study, but the average Ca and Mg contents of both original indicator plants correspond to their respective international mean values (Figs 9, 10, 420 and 421). See also texture and CEC of Syrian soils (Fig. 416) and the effect of these soil characteristics on the relationship between Ca and Mg in soils and plants (Figs 12 and 14, Section 2.2.4).

8.4.3 Micronutrients

Boron. On average the Syrian soils and plants contain somewhat more B than usually found in the soils and plants of this material (Figs 22 and 25). Although the B content of soils and plants varies considerably from one sample to another (Figs 422 and 423) and several high (Zones IV and V) plant—soil B values were measured, these are still much lower than those found in neighbouring Iraq. The differences in the B contents of soils and

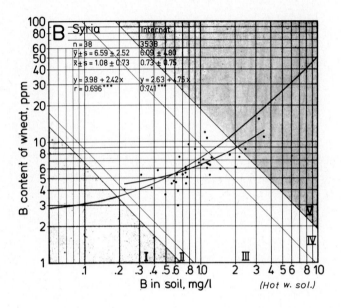

Fig. 422. Regression of B content of pot-grown wheat (y) on hot water soluble soil B (x), Syria. For details of summarized international background data, see Chapter 4.

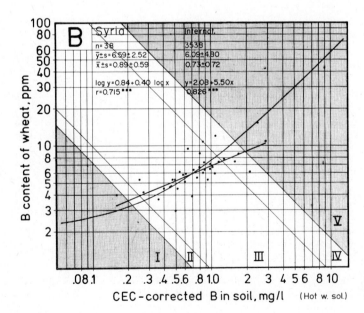

Fig. 423. Regression of B content of pot-grown wheat (y) on CEC-corrected (hot water soluble) soil B (x), Syria.

plants between the irrigated and non-irrigated sites are smaller than those found in Iraq. The present limited data from Syria do not indicate any serious problems due to shortage of B but at some locations a response to applied B may be obtained (see also Section 2.7).

Copper. Syria's position in the "international Cu fields" (Figs 27 and 29) is central. Since only a couple of sample pairs fall outside the normal Cu range (Figs 424 and 425) no problems concerning this micronutrient are to be expected in the areas sampled in Syria.

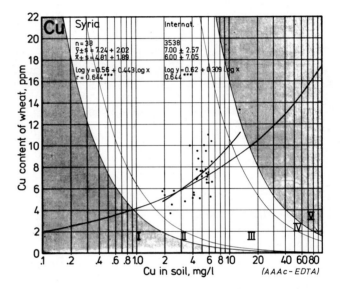

Fig. 424. Regression of Cu content of pot-grown wheat (y) on acid ammonium acetate-EDTA extractable soil Cu (x), Syria.

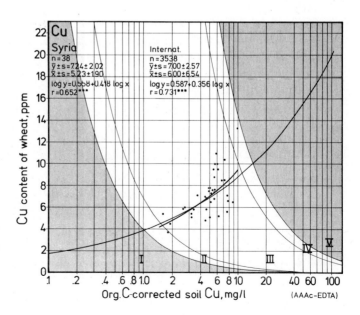

Fig. 425. Regression of Cu content of pot-grown wheat (y) on organic carbon-corrected (AAAc-EDTA extr.) soil Cu (x), Syria.

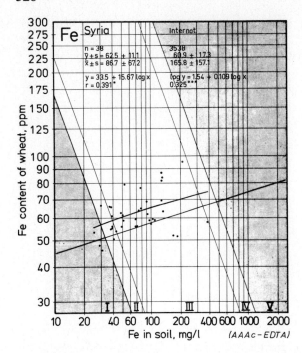

Fig. 426. Regression of Fe content of pot-grown wheat (y) on acid ammonium acetate-EDTA extractable soil Fe (x), Syria.

Iron. The Fe values measured from pot-grown wheat samples grown on Syrian soils vary from 46 to 95 ppm and approach the international mean closely (Figs 31 and 426). The AAAc-EDTA extractable soil Fe values are below the international mean. On the basis of these limited data Fe deficiency should not be expected to be common in Syria. However, it has been reported in the deciduous orchards of the Damascus Ghouta and in certain areas along the eastern fringe of the Anti-Lebanon mountains (Loizides, 1967).

Manganese. Low Mn availability is typical for countries with alkaline soils and Syria is no exception (Figs 33 and 37). Almost all the Syrian plant and soil Mn values are lower than the international means and about 25 percent of the sample pairs fall in the two lowest Mn Zones (I and II, Figs 427 and 428). Low Mn availability may be a factor limiting the normal growth of crops in Syria although extremely low Mn contents were measured in only one sample pair.

Molybdenum. Despite the high pH of Syrian soils the average Mo content of pot-grown wheat grown on these soils only slightly exceeds the international mean (Figs 15, 20, 429 and 430). The low soil Mo values (mean 0.119 mg/l) given in Fig. 429 (due to low extractability of Mo by AO-OA in alkaline soils) is doubled (mean 0.235 mg/l) by correction for pH (Fig. 430) and the apparent contradiction between the analytical results of plant and soil analyses is resolved. The corrected data indicate no Mo deficiency in Syria, rather, both the plant and soil Mo values are seen to be normal or somewhat high.

Zinc. Both the methods used for extracting Zn from soils give similar estimates of Zn levels in Syria (Figs 41, 43, 431 and 432). High Zn contents were recorded for a few locations only, the majority of values are quite low, a few falling in the two lowest Zn Zones. Although none of the values indicates a very severe Zn deficiency it is possible that a favourable response to Zn fertilization would be obtained at several locations.

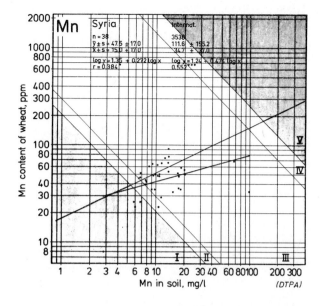

Fig. 427. Regression of Mn content of pot-grown wheat (y) on DTPA extractable soil Mn (x), Syria.

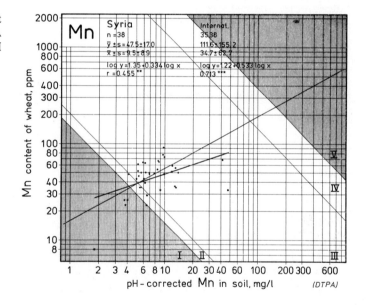

Fig. 428. Regression of Mn content of pot-grown wheat (y) on DTPA extractable soil Mn corrected for pH (x), Syria.

8.4.4 Summary

Most of the Syrian soils are medium to fine textured and have medium organic matter contents. The other essential features are their high alkalinity, high cation exchange capacity, high $CaCO_3$ equivalent and relatively high electrical conductivity. High Ca and Mg and medium N, P and K contents are typical of most sampled Syrian soils.

Most of the micronutrient values measured from Syrian sample material are within the "normal" international range. Due to the high soil alkalinity the availabilities of Mn and Zn to plants are low, and the most likely micronutrient problems in Syria are shortages of these elements. At some locations a response to B fertilization may be obtained.

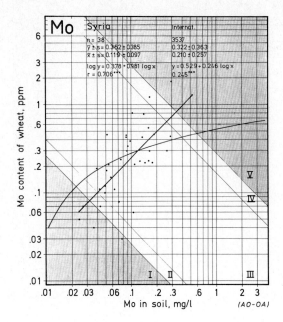

Fig. 429. Regression of Mo content of pot-grown wheat (y) on ammonium oxalate-oxalic acid extractable soil Mo (x), Syria.

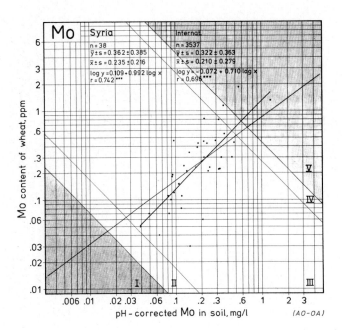

Fig. 430. Regression of Mo content of pot-grown wheat (y) on pH-corrected AO-OA extractable soil Mo (x), Syria.

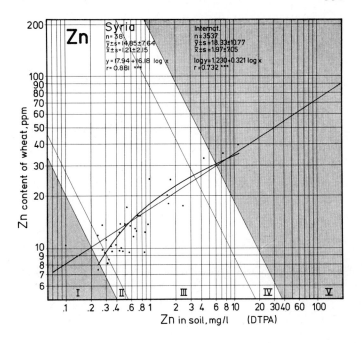

Fig. 431. Regression of Zn content of pot-grown wheat (y) on DTPA extractable soil Zn (x), Syria.

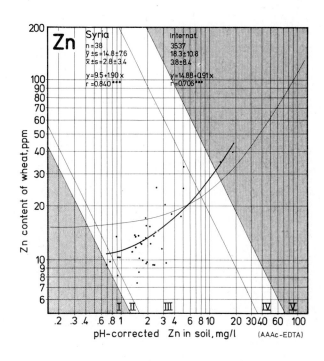

Fig. 432. Regression of Zn content of pot-grown wheat (y) on AAAc-EDTA extractable soil Zn corrected for pH (x), Syria.

8.5 Turkey

8.5.1 General

The sampling sites of the 250 wheat—soil and 50 maize—soil sample pairs collected from Turkey, are well distributed over the agriculturally important areas of the country (Fig. 433). Five out of the seven natural regions (Black Sea, Marmara, Aegean, Central

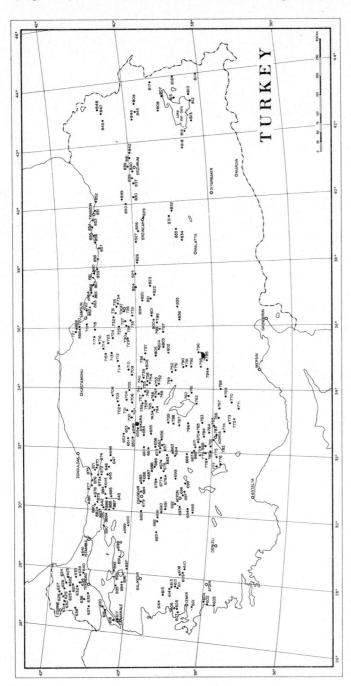

Fig. 433. Sampling sites in Turkey. The last three numerals of each sample pair number are given (dots = wheat fields, triangles = maize fields).

Anatolia and Eastern Anatolia) are represented and these constitute about four-fifths of Turkey's arable land area. Of the total Turkish plant—soil sample material, 87 percent had been collected from non-irrigated and 13 percent from irrigated sites. About two-thirds of the sampled soils were classified by FAO/Unesco soil units: Fluvisols (81 soils), Luvisols (62), Kastanozems (17), Vertisols (16), Lithosols (12), Rhegosols (2), and Xerosols (2).

The sampled soils vary extensively in texture from TI 17 to 85 (Fig. 434). Highly alkaline soils, $pH(CaCl_2) > 7.5$, predominate but some neutral and a few relatively acid soils are included. The organic matter contents of soils are usually rather low but their cation exchange capacity tends to be high compared to the international distributions of these properties in the whole material of this study. Relatively low electrical conductivity values and sodium contents are typical of most Turkish soils but the values for $CaCO_3$ equivalent are usually quite high (Appendixes 2—4).

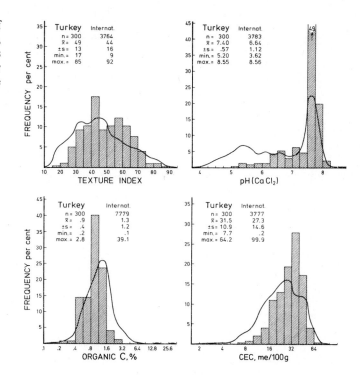

Fig. 434. Frequency distributions of texture, pH, organic carbon content, and cation exchange capacity in soils of Turkey (columns). Curves show the international frequency of the same characters.

8.5.2 Macronutrients

The total **nitrogen** contents of Turkish soils are low compared to most other countries (Figs 6 and 435). The N contents of the original Turkish wheat are also internationally rather low but those of maize tend to be high. In both cases the N contents vary substantially. The sampled maize crops were fertilized with nitrogen (61 ± 50 kg N/ha) at double the rate applied to the wheats (34 ± 30 kg N/ha). The relatively higher N contents of maize as well as the wide variations in the N contents of both crops may partly be due to fertilization.

The $NaHCO_3$ extractable **phosphorus** contents of Turkish soils, especially of wheat grown soils, are internationally somewhat low (Figs 7 and 436). Although moderate

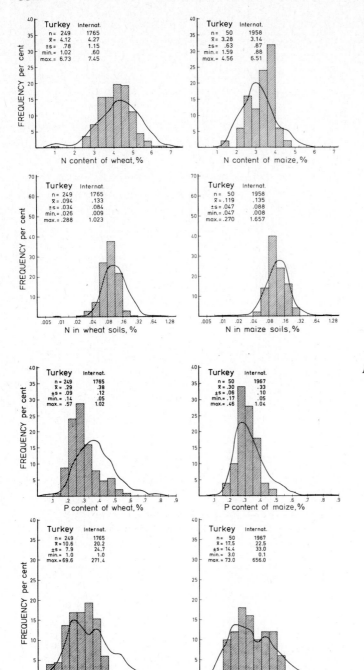

Figs 435–439. Frequency distributions of nitrogen, phosphorus, potassium, calcium and magnesium in original wheat and maize samples and respective soils (columns) of Turkey. Curves show the international frequency of the same characters.

Fig. 435. Nitrogen, Turkey.

Fig. 436. Phosphorus, Turkey.

dressings of phosphates were applied (wheat: 18 ± 13 and maize: 6 ± 11 kg P/ha) the average P contents of both plants remain below the respective international means.

The level of exchangeable **potassium** in Turkish wheat soils is almost double that of maize soils (Fig. 437) but, on average, the K contents of Turkish soils correspond closely to the respective international mean for soil K (Fig. 8). Practically no potassium fertilizers were applied, and the national mean K contents of both indicator crops remain rather

Fig. 437. Potassium, Turkey.

Fig. 438. Calcium, Turkey.

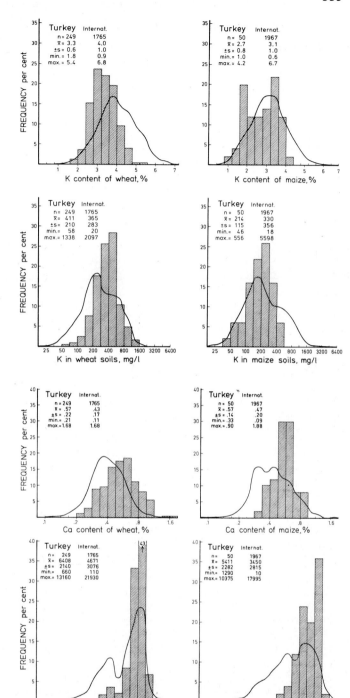

low internationally. Apart from the lack of potassium fertilization, the low K content of wheats may be due to the effect of soil pH which was considerably higher in wheat soils than in maize soils (means 7.52 and 6.79, respectively, Appendixes 2 and 3). For the effect of pH on the plant K—soil K relationships see Fig. 11 and related text in Section 2.2.4.

On average, the **calcium** contents of Turkish soils and plants are high and the average **magnesium** contents correspond closely to the respective international means (Figs 9, 10, 438 and 439).

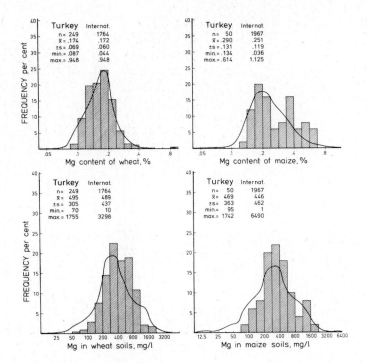

Fig. 439. Magnesium, Turkey.

8.5.3 Micronutrients

Boron. Wide variations in B contents are typical of Turkish soils and plants (Figs 440 and 441) but high B values dominate and in the "international B fields" Turkey clearly stands high (Figs 22 and 25). In spite of the variation, 84 percent of the plant—soil B values are within the "normal" B range (Zone III, Fig. 441) and only two percent of these fall in the two low B Zones, one percent in each. The lowest B values occur in the Black Sea, the Aegean and the Marmara regions. The majority of the high (Zones V and IV) B values are found in Central Anatolia. The highest B values were measured in sample pair No 46673 which came from Ankara Province in an irrigated field with highly calcareous soil, pH(CaCl$_2$) 7.9. Although the effect of irrigation is not as marked as, for example, in Iraq, it is apparent that in many cases the practice of irrigation is partly responsible for the high B values found in Turkish plants and soils. The relative frequency of high

Fig. 440. Regression of B content of pot-grown wheat (y) on hot water soluble soil B (x), Turkey. For details of summarized international background data, see Chapter 4.

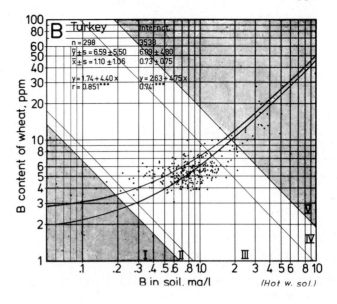

Fig. 441. Regression of B content of pot-grown wheat (y) on CEC-corrected (hot water soluble) soil B (x), Turkey.

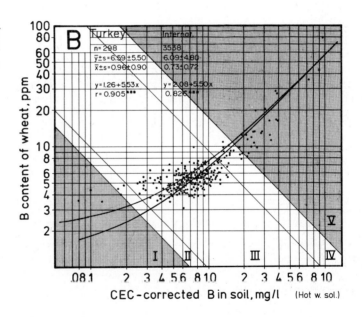

(Zone V) B values is about three times as high for irrigated as for non-irrigated sites, while none of the sample pairs from irrigated sites fall in the two lowest B Zones (I and II). See also Section 8.2.4.

Judging from these analytical data, B deficiency occurs in Turkey but is not widespread. The analysed B values are usually "normal", but at several locations attention should be paid to possible disorders due to an excess of available B caused either by naturally high B contents of soils or induced by high-B irrigation waters. See also Section 2.7.

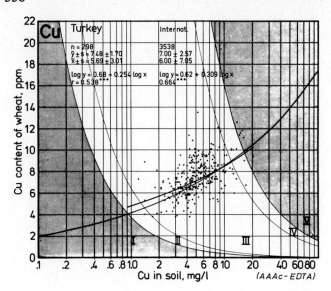

Fig. 442. Regression of Cu content of pot-grown wheat (y) on acid ammonium acetate-EDTA extractable soil Cu (x), Turkey.

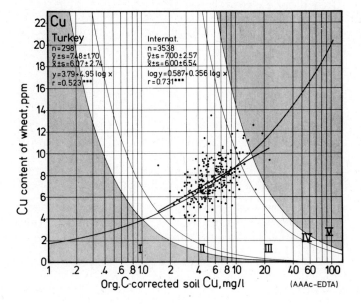

Fig. 443. Regression of Cu content of pot-grown wheat (y) on organic carbon-corrected (AAAc-EDTA extr.) soil Cu (x), Turkey.

Copper. The Turkish national averages of plant and soil Cu contents correspond closely to the international respective mean values in this study (Figs 27, 29, 442 and 443). Relatively small variations in Turkish plant and soil Cu values are typical, and no extremely low or high Cu contents were recorded. The present analytical data suggest that problems due to shortage or excess of Cu are unlikely in Turkey.

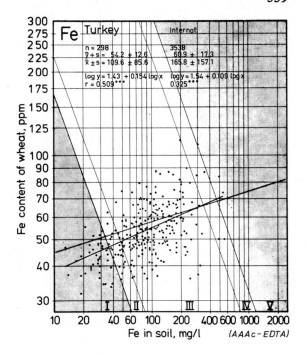

Fig. 444. Regression of Fe contents of pot-grown wheat (y) on acid ammonium acetate-EDTA extractable soil Fe (x), Turkey.

Iron. Compared to most other countries the Fe content of Turkish soils and plants is low (Figs 31 and 444). More than 20 percent of the Fe values fall in the two low-Fe Zones (I and II), and some of these show distinctly low plant and soil Fe contents. It seems likely that in many places crops sensitive to Fe deficiency would respond to iron fertilization. Almost 90 percent of the low (Zone I) Fe values occur in Central Anatolia and half of these are in the Province of Konya.

Manganese. Turkey's placing in the "international Mn fields" tends to be low (Figs 33 and 37). However, in spite of the generally low Mn content, only about two percent of the Turkish samples fall in the lowest Mn Zone (I), and none of these indicates any severe shortage of available Mn (Figs 445 and 446).

Molybdenum. The two Mo graphs (Figs 447 and 448) give dissimilar pictures of the Mo status in Turkey. This is because AO-OA is unable to extract Mo from alkaline soils in quantities related to its availability to plants (see Section 2.3.2.2), so distorting the relationship between soil and plant Mo. After pH correction the soil Mo data are in better conformity with the plant data and only a few sample pairs indicate shortage of available Mo. Several relatively high, but not extreme, Mo values were recorded from the samples. No distinct geographical division between low and high Mo areas could be drawn but, for example, of the 62 pot-grown wheats with Mo content < 0.1 ppm only two were grown on soils originating in irrigated sites. See also Section 8.2.4.

Zinc. The standing of Zn in Turkey is one of the very lowest recorded in this study (Figs 41 and 43). Irrespective of the method used for extracting Zn from the soils, the great majority of extractable Zn contents of soils and plants are below the respective international means. About 35 percent of the sample pairs fall in the two lowest Zn Zones (Figs 449 and 450) and only a few in the high Zn zones. Low (Zones I and II) Zn values were recorded most frequently in samples which came from Central and Eastern Anatolia where almost every sample pair gave such low figures. About 20 percent of samples from the Black Sea, Marmara and Aegean Regions fell in the low Zn zones. It appears, therefore, that Zn deficiency occurs commonly in Turkey.

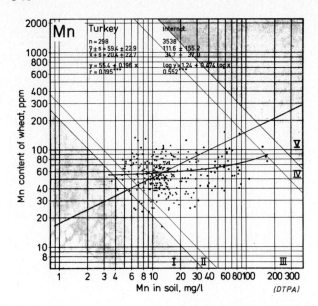

Fig. 445. Regression of Mn content of pot-grown wheat (y) on DTPA extractable soil Mn (x), Turkey.

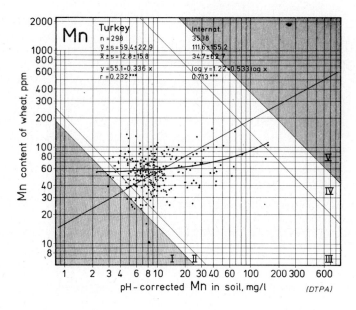

Fig. 446. Regression of Mn content of pot-grown wheat (y) on DTPA extractable soil Mn corrected for pH (x), Turkey.

8.5.4 Summary

The Turkish soils vary greatly in texture. Most of the soils are alkaline and low in organic matter but have a relatively high cation exchange capacity. Compared to other countries participating in this study the N and P contents of the soils are usually low, while those of K and Mg are at a medium level and Ca contents are high.

The most evident micronutrient disorders in Turkey are those due to deficiency of Zn. The level of Fe is also low as is that of Mn. Low B values were recorded occasionally. In general, the B and Mo values are "normal" but some of them are very high. Problems due to shortage or excess of Cu are unlikely.

Fig. 447. Regression of Mo content of pot-grown wheat (y) on ammonium oxalate-oxalic acid extractable soil Mo (x), Turkey.

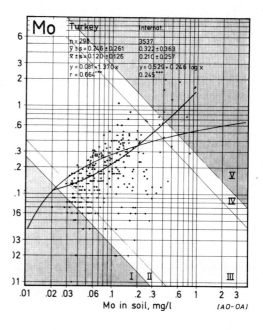

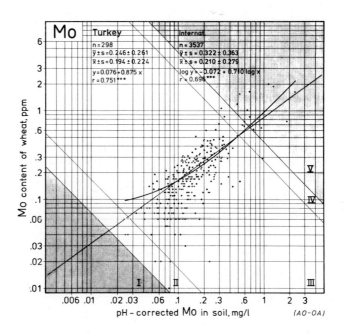

Fig. 448. Regression of Mo content of pot-grown wheat (y) on pH-corrected AO-OA extractable soil Mo (x), Turkey.

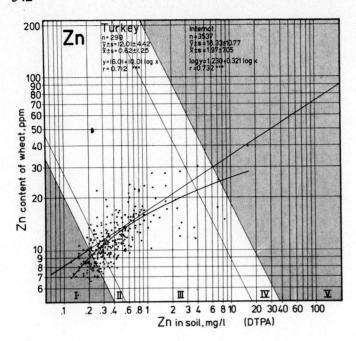

Fig. 449. Regression of Zn content of pot-grown wheat (y) on DTPA extractable soil Zn (x), Turkey.

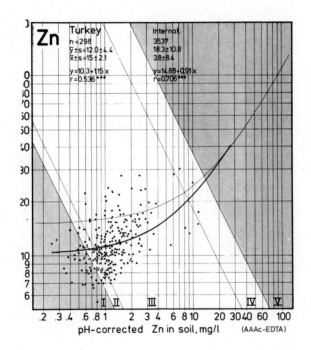

Fig. 450. Regression of Zn content of pot-grown wheat (y) on AAAc-EDTA extractable soil Zn corrected for pH (x), Turkey.

9. Africa

9.1 Ethiopia

9.1.1 General

Soil and original wheat and maize samples received for this study from Ethiopia came from Shoa (40 sample pairs), Sidamo (32), Arussi (27), Wollega (16), Harar (5), Gamu-Gofa (4) and Kaffa (3). These provinces represent about half on the arable land of the country. The approximate sampling sites are given in Fig. 451.

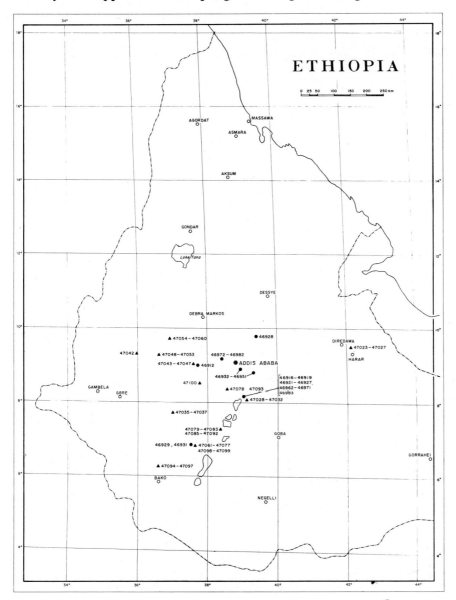

Fig. 451. Sampling sites in Ethiopia (dots = wheat fields, triangles = maize fields).

The range of textural variation of Ethiopian soils is very wide, although heavy textured soils predominate (Fig. 452). The national mean texture index (62) is exceeded only by that of Brazil (Appendix 4). The pH values vary widely, from 4.1 to 7.7, but most of the soils show moderate to strong acidity. Due to the relatively high organic matter contents and heavy textures the cation exchange capacities of Ethiopian soils are very high (see also Fig. 3, h and m). The electrical conductivity and $CaCO_3$ equivalent values and sodium contents are generally low (Appendixes 2—4).

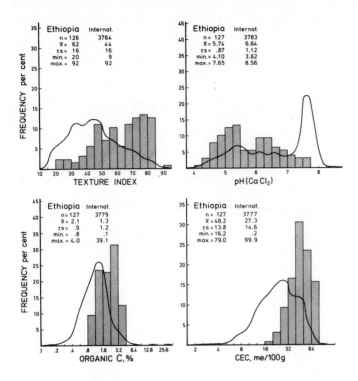

Fig. 452. Frequency distributions of texture, pH, organic carbon content, and cation exchange capacity in soils of Ethiopia (columns). Curves show the international frequency of the same characters.

9.1.2 Macronutrients

In assessing the macronutrient contents of the original wheat crops it must be borne in mind that the Ethiopian wheats were sampled at rather late stages of growth (mainly between 54 and 62 days after planting). See Section 1.2.2. The analytical results of Ethiopian wheats therefore show lower macronutrient contents than they would have done if sampled about three weeks earlier as was done in the case of maize. This may also explain some of the contradictions between the results of soil and original wheat analyses.

The mean total **nitrogen** contents of Ethiopian soils are somewhat higher than the respective averages for the whole international data of this study (Figs 453 and 6). The average N contents of maize also exceeds the international average but that of wheat is lower. Relatively light dressings of nitrogen fertilizer were applied to the sampled wheat and maize crops, 11 ± 10 and 7 ± 11 kg/N ha, respectively (Appendix 5).

From an international point of view the **phosphorus** contents of both soils and plants are somewhat low but the **potassium** contents are relatively high (Figs 454, 455, 7 and 8). Perhaps arising out of previous experience of fertilization, the sampled wheat and maize crops were given moderate amounts of phosphorus (10 ± 10 and 5 ± 7 kg P/ha, re-

Figs 453–457. Frequency distributions of nitrogen, phosphorus, potassium, calcium and magnesium in original wheat and maize samples and respective soils (columns) of Ethiopia. Curves show the international frequency of the same characters.

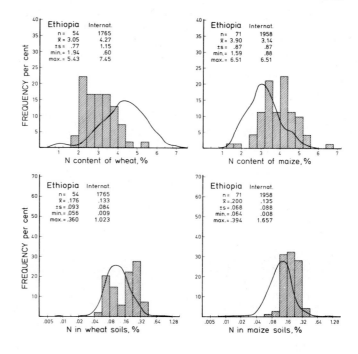

Fig. 453. Nitrogen, Ethiopia.

Fig. 454. Phosphorus, Ethiopia.

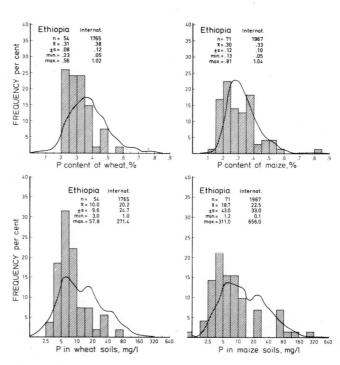

spectively), but no potassium (Appendix 5). For example, Gertsch (1972), when reporting the results of several hundred fertilizer trials, stated that a response to K was found only occasionally in Ethiopia and to a lesser extent than to P and N.

The **calcium** and **magnesium** contents of maize and respective soils are lightly low on the international scale (Figs 456 and 457) and in spite of the relatively high Ca and Mg

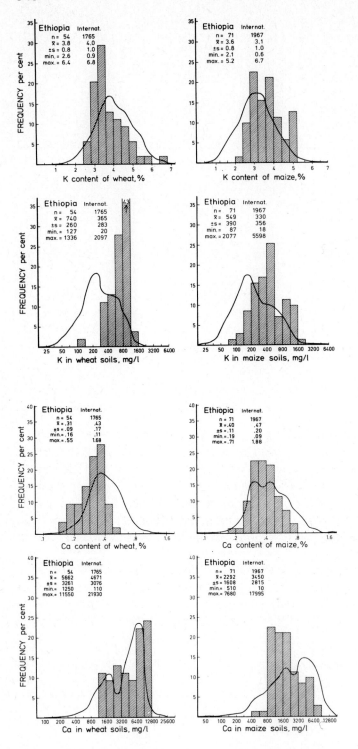

Fig. 455. Potassium, Ethiopia.

Fig. 456. Calcium, Ethiopia.

contents of wheat soils the contents of these elements are low in wheats. This contradiction may be due to the advanced maturity of the wheats when sampled, as pointed out earlier. Other factors which also may be responsible for the strong contradiction between low Ca and Mg contents of Ethiopian wheats and high exchangeable Ca and Mg contents

Fig. 457. Magnesium, Ethiopia.

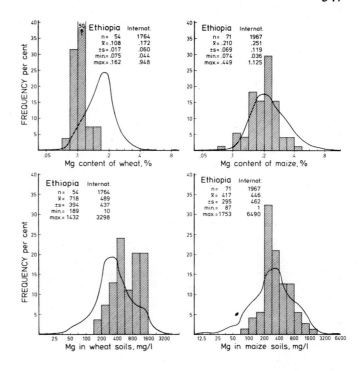

of wheat soils may be soil texture and CEC. The relative availability of Ca and Mg to plants in heavy textured soils with high CEC is low. See Figs 12 and 14 and related text in Section 2.2.4. In the case of Ethiopian wheat soils the average texture index was 68 and CEC 56.7 me/100 g which were the highest national means for wheat soils recorded in this study (Appendix 2). These two soil factors affect less the uptake of Ca and Mg by maize since maize soils are somewhat coarser textured (aver. TI 56) and have a lower CEC (aver. 41.5 me/100 g). For a general assessment of Ca and Mg status in Ethiopia against the international background, see Figs 9 and 10.

The majority of the Ethiopian plant samples consisted of local varieties (43 local and 11 HYV wheats; 53 local and 18 HYV maize samples). Differences in Ca contents between the two variety groups were obscure but the HYV of both crops show higher average Mg contents than the local varieties in spite of the change-around in the exchangeable Mg contents of the respective soils:

	HYV (n = 11)	Local (43)		HYV (18)	Local (53)
Wheat, Mg %	0.116	0.106	Maize	0.243	0.199
Soil, mg Mg/l	393	809	Soil	360	436

The above differences between the two variety groups are statistically significant at the 5 percent level in Mg contents of both plants, at the 0.1 percent level in wheat soils but non-significant in maize soils. See also Section 9.1.4.

9.1.3 Micronutrients

Boron. The average B contents of Ethiopian soils and pot-grown wheat are lower than the respective international averages (Figs 458 and 459) and consequently, Ethiopia stands somewhat low in the "international B fields" (Figs 22 and 25, Section 2.3.3). The ranges of variation in B contents compared to many other countries are relatively narrow, and in spite of the generally low B level no extremely low values were recorded. According to these data no severe deficiency of B exists in Ethiopia but at several locations a response to B should be expected.

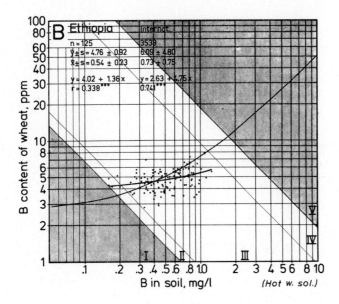

Fig. 458. Regression of B content of pot-grown wheat (y) on hot water soluble soil B (x), Ethiopia. For details of summarized international background data, see Chapter 4.

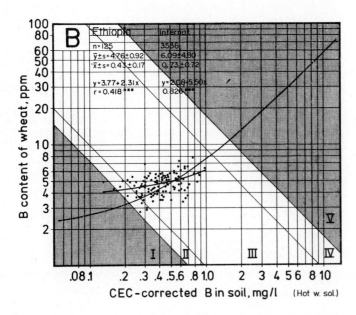

Fig. 459. Regression of B content of pot-grown wheat (y) on CEC-corrected (hot water soluble) soil B (x), Ethiopia.

Copper. In the "international Cu fields" (Figs 27 and 29, Section 2.3.4) Ethiopia is among the low-Cu countries. Contrary to B, the Cu contents vary greatly and very low plant and soil Cu contents (including the minima of the whole international material) were measured (Figs 460 and 461). Of the whole Ethiopian sample material, 21 percent of the sample pairs fall in the lowest Cu Zone (I) and 6 percent in Zone II (Fig. 461). Most of the low (Zone I) Cu values occur in Sidamo Province, where two-thirds of the sample pairs show very low contents, and most of the remainder occur in Zone II. Although a few low Cu values were also recorded in samples from the neighbouring Provinces of Shoa, Gamu-Gofa and Arussi, the Cu levels of soils in these and other provinces seem to be

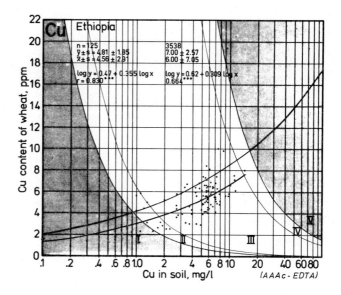

Fig. 460. Regression of Cu content of pot-grown wheat (y) on acid ammonium acetate-EDTA extractable soil Cu (x), Ethiopia.

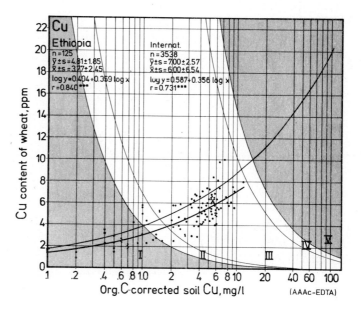

Fig. 461. Regression of Cu content of pot-grown wheat (y) on organic carbon-corrected (AAAc-EDTA extr.) soil Cu (x), Ethiopia.

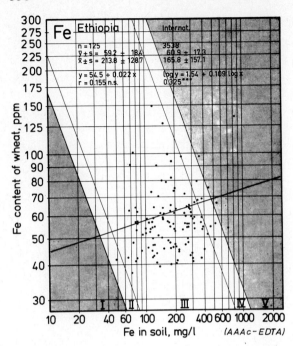

Fig. 462. Regression of Fe contents of pot-grown wheat (y) on acid ammonium acetate-EDTA extractable soil Fe (x), Ethiopia.

within the normal range. No values showing a possible excess of Cu were recorded. These analytical data point to the presence of Cu deficiency in Ethiopia, especially in the Sidamo area, and a response to Cu fertilization in several locations is very probable.

Iron. On average, the Fe contents of wheat grown in pots on Ethiopian soils correspond closely to those of the international material as a whole, but the extractable Fe contents of soils are somewhat higher (Figs 462 and 31). Normal (Zone III) values dominate and in spite of a few samples within Zones II, IV and V, these are insufficient as an indication of Fe problems in Ethiopia.

Manganese. Ethiopia stands clearly on the high side in the "international Mn fields" (Figs 33 and 37) and in spite of a relatively wide variation of Mn contents (Figs 463 and 464) no low (Zone I or II) values were recorded. Instead, a substantial percentage of Ethiopian samples are found in the high Mn Zones, IV and V. Most of the highest (Zone V) Mn values were measured from samples which came from Sidamo Province but occasionally some were obtained from the samples from other provinces. All sites where excess Mn was recorded had acid soils, the pH(CaCl$_2$) was usually below 5.0, and therefore, by raising the soil pH through liming the availability of Mn could be reduced and the possibility of its toxic effects eliminated.

Molybdenum. In Ethiopia the problems associated with Mo differ from those of most other countries because typical Ethiopian soils are exceptionally heavy textured and usually acid. In such soils the availability of Mo to plants is low although soil analyses may show relatively high Mo values (Fig. 16). This explains why many Ethiopian low plant Mo contents are combined with moderately high AO-OA extractable soil Mo values (Fig. 465). This contradiction is partly overcome when the soil Mo values are corrected for pH (Fig. 466). Since the correction for texture (see Section 2.3.2.2) is not applied to the Ethiopian Mo data, the effect of texture must be taken into account when interpreting the

Fig. 463. Regression of Mn content of pot-grown wheat (y) on DTPA extractable soil Mn (x), Ethiopia.

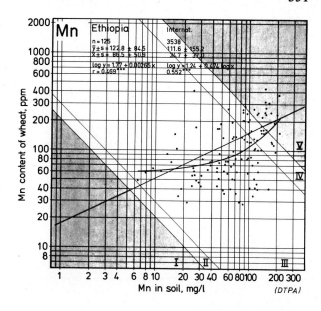

Fig. 464. Regression of Mn content of pot-grown wheat (y) on DTPA extractable soil Mn corrected for pH (x), Ethiopia.

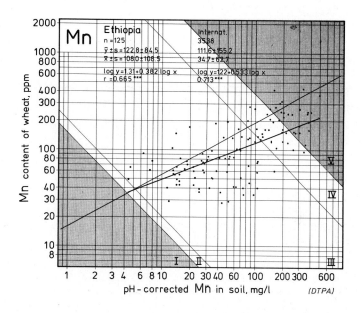

Mo results sample by sample. See also Figs 15, 20, 21 and Table 8 in Section 2.3.2.

Low Mo values are rather more common in Sidamo than in other provinces but are also found in Shoa, Wollega and Arussi, usually in sites with soils of low pH and of heavy texture. Many of the highest Mo values were analysed in Sidamo samples from sites where the soils were less acid or alkaline. The highest plant Mo content (9.79 ppm) and the highest AO-OA extractable soil Mo content (uncorrected 3.6 and pH-corrected 4.8 mg/l) in the whole international material were both obtained from a Sidamo sample pair (47079). In this case Mo toxicity should be suspected. On several sites low Mo values are combined with high Mn values. The application of lime would increase the availability

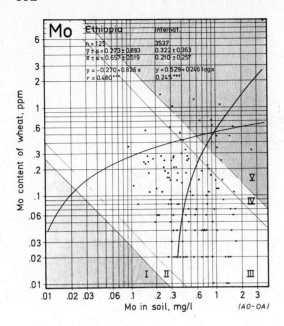

Fig. 465. Regression of Mo content of pot-grown wheat (y) on ammonium oxalate-oxalic acid extractable soil Mo (x), Ethiopia.

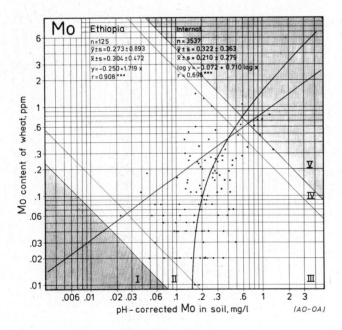

Fig. 466. Regression of Mo content of pot-grown wheat (y) on pH-corrected AO-OA extractable soil Mo (x), Ethiopia.

of the former and decrease that of the latter, and would be the recommended treatment.

Zinc. The Ethiopian national mean values for plant and soil Zn are high compared to those of most other countries (Figs 41 and 43). The variation, however, is wide and both high and relatively low Zn contents are found. Independent of the soil extraction method the plant—soil correlations are good (Figs 467 and 468). According to both methods every third sample pair falls in the high Zn Zones (IV and V), a few of them showing very

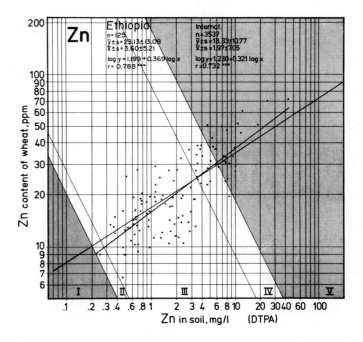

Fig. 467. Regression of Zn content of pot-grown wheat (y) on DTPA extractable soil Zn (x), Ethiopia.

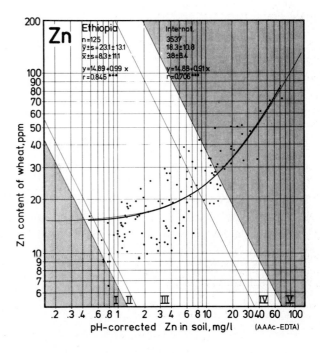

Fig. 468. Regression of Zn content of pot-grown wheat (y) on AAAc-EDTA extractable soil Zn corrected for pH (x), Ethiopia.

high Zn contents. When Zn contents are low, the DTPA method reveals a slightly better Zn content of soils than does the AAAc-EDTA extraction. Neither of them indicates severe Zn deficiency but at many locations a hidden shortage of Zn is likely. The lowest Zn contents mainly come from Harar and Wollega and the majority of high Zn values from Sidamo.

9.1.4 Micronutrient contents of original plants with special reference to varieties

The most obvious varietal differences in the micronutrient contents of original indicator plants were those between the Mn and Zn contents of the HYV and local maize plants. The HYV maize (varieties H 512, H 611, H 613, H 632, Composite II and KCB, n = 18) showed higher average Mn and Zn contents that the local varieties (n = 53) although they were grown on soils poorer in these elements.

	Mn		Zn	
	HYV	Local	HYV	Local
Maize, ppm	90	75	52	43
Soil, mg/l[1ature]	98	132	2.7	5.6

[1]) pH-corr. DTPA extr. Mn, DTPA extr. Zn

When one wheat cultivar ('Apu') was grown on the above soils in pots, the average Mn and Zn contents were as follows:

	Mn		Zn	
Wheat ('Apu'), ppm	154	161	22	29
Soil, mg/l[1]	98	132	2.7	5.6

[1]) pH-corr. DTPA extr. Mn, DTPA extr. Zn

The latter Mn and Zn contents are more compatible with the results of soil analyses than those obtained by analyzing the original plants of different varieties.

The above differences between the two groups did not quite reach statistical significance at the 5 percent level, except in the case of soil Zn. However, since the differences in the original maize Mn and Zn contents were diametrically opposed to these in the respective soils, it would seem possible that there are some genetic differences between the two variety groups and that the HYV maize plants are the more efficient in absorbing the above micronutrients from the soil.

This above example hopes to draw attention to one of the difficulties investigators face when interpreting results of plant analyses obtained from materials consisting of several varieties.

9.1.5 Summary

Typical Ethiopian soils included in this study are fine to medium textured, and show moderate to strong acidity. They are relatively high in organic matter and have a high cation exchange capacity. Compared to the international general means, the mean N and especially K contents of Ethiopian soils are high, P contents are somewhat low, and Ca and Mg contents are at the international average.

The micronutrient content of Ethiopian soils and plants varies considerably depending on the element. Of the six micronutrients studied, the one most likely to be deficient is Cu, but responses to B and Zn can also be expected at several locations. The contents of Fe are usually at the normal international level but many high Mn values were recorded. The Mo and Zn contents typically vary widely and range between relatively low and high.

9.2 Ghana

9.2.1 General

The Ghana sample material consists of 93 maize and related soil samples and was collected from geographically limited but agriculturally quite important areas, namely the Central and the Ashanti regions. All the sampled crops were grown under rainfed conditions. Of the sampled soils 46 were classified as Nitosols, 39 Ferralsols, 5 Gleysols, 2 Lithosols and 1 Vertisol.

In Ghana coarse textured soils predominate (Fig. 469) and only a few have a medium texture. The national average texture index (TI = 30) is one of the lowest among the 30 countries investigated (Appendixes 3 and 4). Strong to moderate acidity, low cation exchange capacity, low electrical conductivity, low $CaCO_3$ equivalent and low sodium content are typical for Ghana soils.

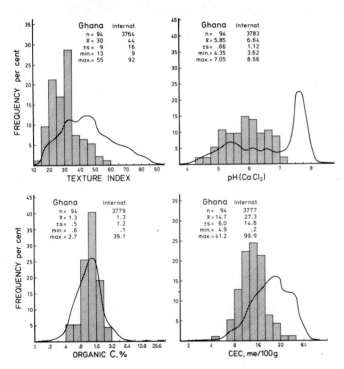

Fig. 469. Frequency distributions of texture, pH, organic carbon content, and cation exchange capacity in soils of Ghana (columns). Curves show the international frequency of the same characters.

9.2.2 Macronutrients

The total **nitrogen** contents of soils are at the average international level (Figs 6 and 470). In spite of relatively high rates of nitrogen fertilizers (48 ± 40 kg N/ha) applied to the sampled crops, the average N content of maize remains below the international mean.

The $NaHCO_3$ extractable **phosphorus** and CH_3COONH_4 exchangeable **potassium** contents of Ghana soils are very low compared to most other countries (Figs 7, 8, 471 and 472). The applied dressings of phosphate and potassium fertilizers (11 ± 11 kg P/ha and 14 ± 12 kg K/ha), although not high, were about double the average P and K applications

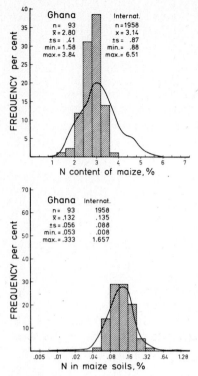
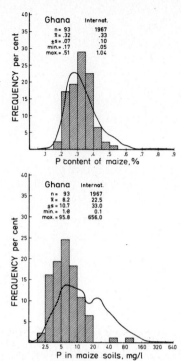
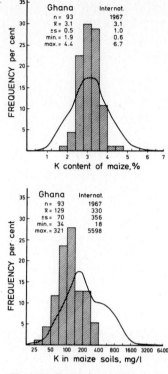

Fig. 470. Nitrogen, Ghana.   *Fig. 471.* Phosphorus, Ghana.   *Fig. 472.* Potassium, Ghana.

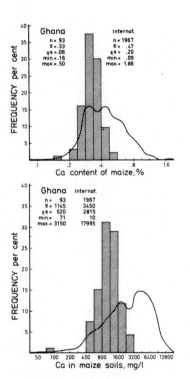
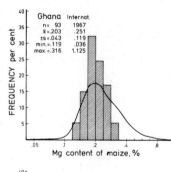
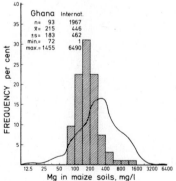

Figs 470–474. Frequency distributions of nitrogen, phosphorus, potassium, calcium and magnesium in original maize samples and respective soils (columns) of Ghana. Curves show the international frequency of the same characters.

Fig. 473. Calcium, Ghana.   *Fig. 474.* Magnesium, Ghana.

to maize crops in this study. These may have contributed to raise the P and K contents of the Ghana maize to the respective average international levels. Also the uptake by plants of K from acid soils and that of P from soils of low CEC is high in relation to the contents of these nutrients in soils. See Figs 14 and 11 and related texts. The exchangeable **calcium** and **magnesium** contents of Ghana soils and the Ca and Mg contents of maize are lower than in most other countries (Figs 9, 10, 473 and 474).

9.2.3 Micronutrients

Boron. The B contents of Ghana soils, and of wheat grown in pots on these soils, vary within quite narrow limits (Figs 475 and 476). The national B averages correspond

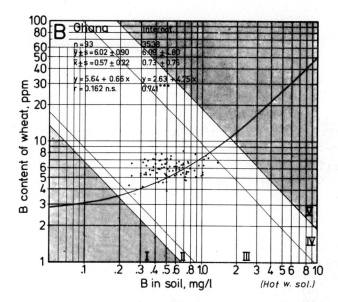

Fig. 475. Regression of B content of pot-grown wheat (y) on hot water soluble soil B (x), Ghana. For details of summarized international background data, see Chapter 4.

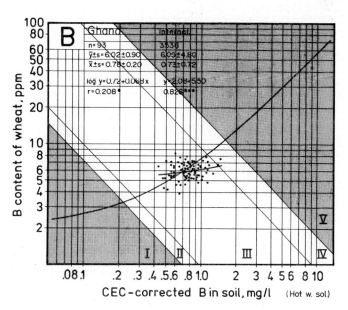

Fig. 476. Regression of B content of pot-grown wheat (y) on CEC-corrected (hot water soluble) soil B (x), Ghana.

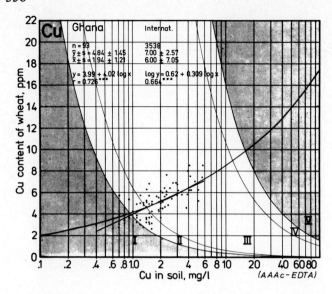

Fig. 477. Regression of Cu content of pot-grown wheat (y) on acid ammonium acetate-EDTA extractable soil Cu (x), Ghana.

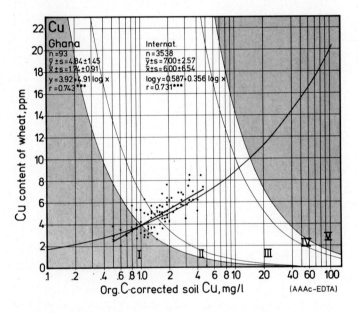

Fig. 478. Regression of Cu content of pot-grown wheat (y) on organic carbon-corrected (AAAc-EDTA extr.) soil Cu (x), Ghana.

closely to respective international mean values and in the "international B fields" Ghana stands at the centre (Figs 22 and 25). Since practically all B values are "normal" and fall in B Zone III, these limited analytical data do not indicate any severe B disorders in Ghana.

Copper. Most plant and soil Cu contents measured from Ghana sample material are lower than the respective mean values in the whole international data of this study, and Ghana occupies one of the lowest positions in the "international Cu fields" (Figs 477, 478, 27 and 29). Only about half of the plant-soil Cu values are "normal" so falling in Cu Zone III, and the other half is distributed evenly in the two low Cu Zones. In general, Cu deficiency, acute or hidden, seems to be more likely in Ghana than in most other countries.

Fig. 479. Regression of Fe contents of pot-grown wheat (y) on acid ammonium acetate-EDTA extractable soil Fe (x), Ghana.

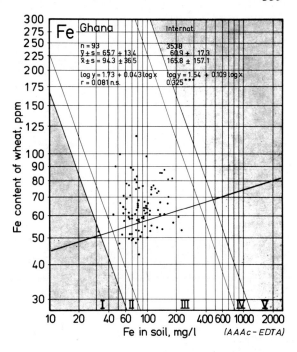

Iron. The Fe data presented in Figs 31 and 479 do not indicate any disorders due to this micronutrient in Ghana.

Manganese. The variation range of the plant Mn contents is relatively wide (from 45 to 601 ppm) and clearly due to the varying pH in Ghana soils, but the values for DTPA extractable Mn vary only from 14 to 138 mg/l (Figs 469 and 480). Correction of DTPA Mn values for pH more than doubles the variation range and improves the plant Mn—soil Mn correlation (Fig. 481).

Ghana stands above many other countries in the "international Mn fields" (Figs 33 and 37). Nine percent of the plant-soil sample pairs fall in the highest Mn Zone (V) and 12 percent in Zone IV, all representing acid soils, pH(CaCl$_2$) < 6.0, and most of these come from Ashanti. No low Mn values were recorded from the sample material of Ghana. See also Mo in the next paragraph.

Molybdenum. Owing to the generally low pH of Ghana soils, the availability of Mo to plants is low and the national average for plant Mo content is less than one-third of the international mean and is one of the lowest among the countries investigated (Figs 482, 483, 15 and 20). The correlation between plant Mo and AO-OA extractable soil Mo contents is non-significant but becomes highly significant when the effect of pH is taken into account by pH-correction.

More than one-third of the Ghana plant—soil Mo points fall in the two lowest Mo Zones (I and II). The low Mo values were most frequently found in samples coming from the acid soils of the Ashanti region and were often combined with high Mn values. Since low pH seems to be the main reason for the low availability of Mo, as well as for the high availability of Mn, the first logical approach to correct both possible disorders would be

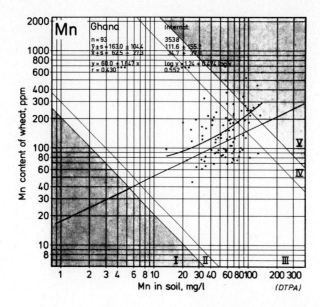

Fig. 480. Regression of Mn content of pot-grown wheat (y) on DTPA extractable soil Mn (x), Ghana.

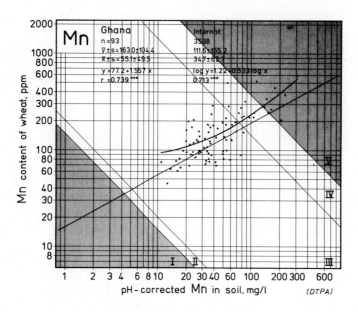

Fig. 481. Regression of Mn content of pot-grown wheat (y) on DTPA extractable soil Mn corrected for pH (x), Ghana.

to raise the soil pH through liming.

Zinc. The mean Zn contents of Ghana soils and plants are somewhat on the low side in the "international Zn fields" (Figs 41 and 43). Comparison of the results obtained by the two extraction methods (Figs 484 and 485) shows, however, that DTPA extraction yields a generally somewhat higher soil Zn level than does the AAAc-EDTA extraction with pH correction. In both cases the correlations with the results of plant analyses are very good.

The lowest plant—soil Zn values, almost without exception, were measured from samples which came from the Ashanti region. The present analytical data do not indicate very severe Zn deficiency but response to Zn may be obtained at several locations. See also Section 2.7.

Fig. 482. Regression of Mo content of pot-grown wheat (y) on ammonium oxalate-oxalic acid extractable soil Mo (x), Ghana.

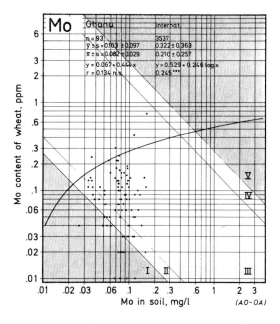

Fig. 483. Regression of Mo content of pot-grown wheat (y) on pH-corrected AO-OA extractable soil Mo (x), Ghana.

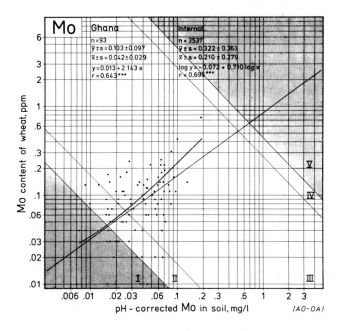

9.2.4 Summary

Coarse texture, low pH, medium organic matter content and low CEC are typical of most Ghana soils. With the exception of total N contents which are medium, the soils are generally poor in macronutrients. The most likely disorders due to micronutrients are those of Cu and Zn deficiency. Shortage of available Mo and excessively available Mn may cause problems with acid soils. Occasional responses to B are possible but disorders due to Fe are unlikely.

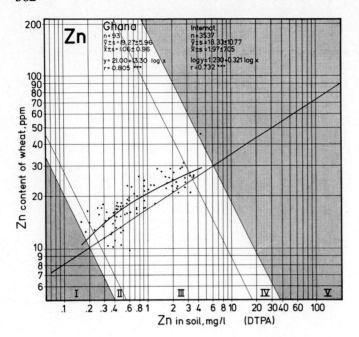

Fig. 484. Regression of Zn content of pot-grown wheat (y) on DTPA extractable soil Zn (x), Ghana.

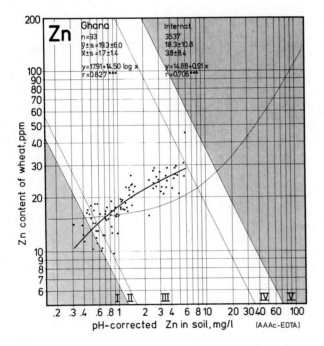

Fig. 485. Regression of Zn content of pot-grown wheat (y) on AAAc-EDTA extractable soil Zn corrected for pH (x), Ghana.

9.3 Malawi

9.3.1 General

The sample material collected from Malawi consists of 100 maize-soil sample pairs. The sampling sites, shown in Fig. 486, are well distributed over the country.

The majority of soils are coarse textured and acid, have a medium organic matter content and a low cation exchange capacity (Fig. 487). Low $CaCO_3$ equivalent, low sodium content and medium electrical conductivity are typical of most Malawi soils (Appendixes 3 and 4).

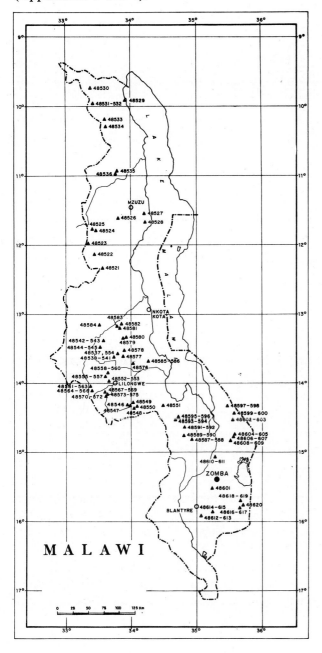

Fig. 486. Sampling sites in Malawi.

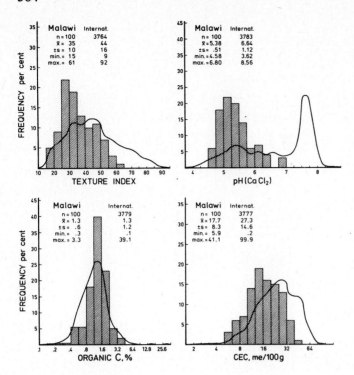

Fig. 487. Frequency distributions of texture, pH, organic carbon content, and cation exchange capacity in soils of Malawi (columns). Curves show the international frequency of the same characters.

9.3.2 Macronutrients

The average **nitrogen** contents of soils is slightly lower than the mean for the whole international soil material in this study (Figs 488 and 6). The N contents of maize remain at a relatively low level in spite of the large nitrogen dressings applied to the sampled maize crops (81 ± 63 kg N/ha). The wide variation in plant N contents is obviously due to nitrogen fertilization. All 15 maize plants with a N content of 3.6 percent or higher (upper graph in Fig. 488) had been fertilized with more than 100 kg N/ha. On the other hand, the 15 plants with the lowest N contents ($< 1.7 \%$) had received no nitrogen fertilizer.

With a few exceptions the **phosphorus** status of Malawi soils is good (Figs 7 and 489). The P contents of maize are usually at the normal international level but the national mean is higher because of a few plants with exceptionally high P contents. The latter plants (P $\% > 0.65$) were grown on soils rich in P (> 74 mg P/l) and (with one exception) received large amounts of phosphate (> 20 kg P/ha). On average, the sampled Malawi maize crops had received 17 ± 14 kg P/ha.

The average **potassium** contents of soils and plants are slightly below the respective international means (Figs 8 and 490). Varying quantities of potassium (11 ± 26 kg K/ha) were applied to the sampled maize crops. None of the plants with a K content of less than 2.0 percent had been fertilized with potassium. The plants with the highest K contents ($> 4.0 \%$) had either received large amounts of potassium (> 45 kg K/ha) or their K uptake was accelerated by large dressings of nitrogen fertilizers (Rinne *et al.*, 1974 a).

The exchangeable **calcium** and **magnesium** contents of Malawi soils and the Mg contents of maize are somewhat low in an international context but the Ca contents of maize correspond to the international average (Figs 9, 10, 491 and 492).

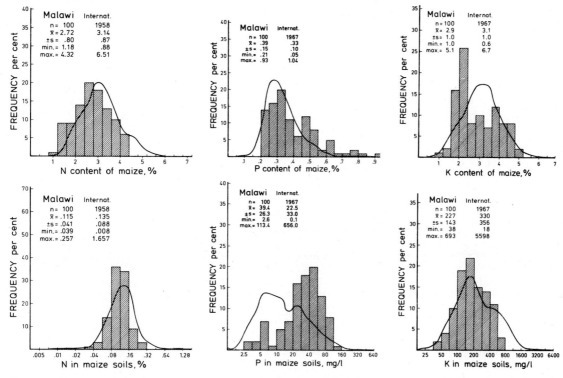

Fig. 488. Nitrogen, Malawi.   *Fig. 489.* Phosphorus, Malawi.   *Fig. 490.* Potassium, Malawi.

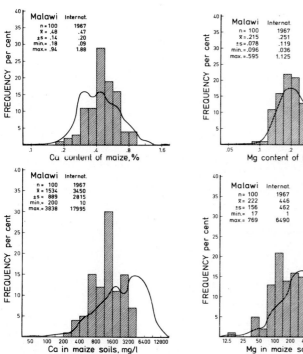

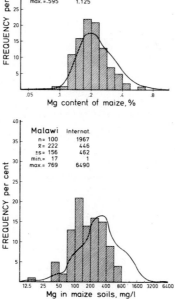

Figs 488–492. Frequency distributions of nitrogen, phosphorus, potassium, calcium and magnesium in original maize samples and respective soils (columns) of Malawi. Curves show the international frequency of the same characters.

Fig. 491. Calcium, Malawi.   *Fig. 492.* Magnesium, Malawi.

9.3.3 Micronutrients

Boron. Almost all B values determined from Malawi samples are lower than the respective international mean B values and the country stands very low in the "international B fields" (Figs 493, 494, 22 and 25). The relative frequency of low (Zones I and II) B levels is higher than in most other countries and consequently, deficiency of B can be suspected at several locations especially for crops with high B requirements. Low B values are not typical of any distinct geographical area but seem to be rather more frequent in the Southern Region of Malawi than elsewhere in the country.

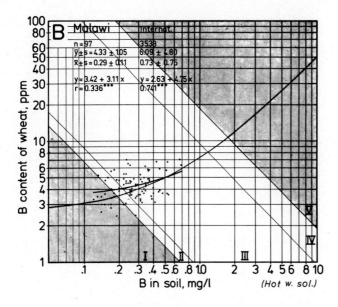

Fig. 493. Regression of B content of pot-grown wheat (y) on hot water soluble soil B (x), Malawi. For details of summarized international background data, see Chapter 4.

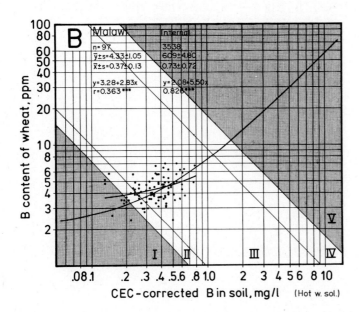

Fig. 494. Regression of B content of pot-grown wheat (y) on CEC-corrected (hot water soluble) soil B (x), Malawi.

Copper. The level of Cu in Malawi soils and plants seems to be somewhat better than that of B (Figs 495 and 496). Nevertheless, low Cu values predominate and in the "international Cu fields" Malawi's positions are on the low side (Figs 27 and 29). The relative frequency of low (Zone I and II) Cu values slightly exceeds the international frequency and at some locations a response to Cu is likely. Most of the low Cu values were measured from samples which came from sites of coarse textured soils and were relatively more frequent in the Southern and Northern Regions than in the Central Region.

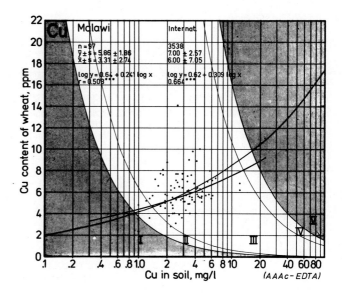

Fig. 495. Regression of Cu content of pot-grown wheat (y) on acid ammonium acetate-EDTA extractable soil Cu (x), Malawi.

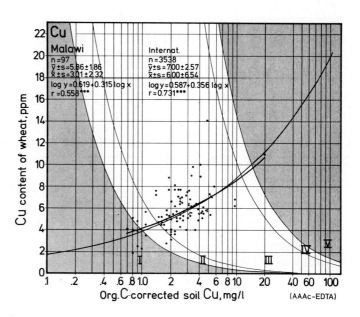

Fig. 496. Regression of Cu content of pot-grown wheat (y) on organic carbon-corrected (AAAc-EDTA extr.) soil Cu (x), Malawi.

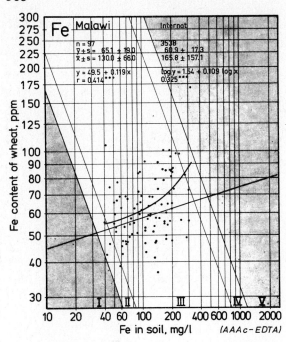

Fig. 497. Regression of Fe contents of pot-grown wheat (y) on acid ammonium acetate-EDTA extractable soil Fe (x), Malawi.

Iron. The Fe contents of Malawi plants and soils are usually within the "normal" Fe range (Zone III) or relatively close to it (Figs 497 and 31).

Manganese. The Mn contents of Malawi soils and plants are on average two to three times as high as those in the other countries, and up to ten times as high as those in countries of the lowest Mn contents (Figs 33 and 37). More than half the sample pairs are within the two highest Mn Zones (Figs 498 and 499). The highest Mn values were typical for sites on acid soils and were recorded most frequently for samples which came from the Northern Region, where 10 out of 16 sample pairs fall in the Mn Zone V.

Molybdenum. Unlike Mn, the average Mo content of wheat grown on Malawi soils is low being less than one-third of the international mean plant Mo content (Figs 500, 501, 15 and 20). This is due to the low availability of Mo to plants in acid soils (Fig. 16, b). Contrariwise, the extractability of Mo to AO-OA from acid soils is high (Fig. 16, a) and, therefore, the soil Mo values given in Fig. 500 and 15 are almost of the average international level. This contradiction is largely eliminated through pH correction. In consequence, the internal correlation between Malawi plant Mo and soil Mo values improves from being non-significant ($r = 0.062$) to highly significant ($r = 0.487***$) and Malawi's position in the "international Mo field" (Fig. 15) moves from its earlier remoter position to a point closer to the international regression line (Fig. 20). Almost every third Malawi plant—soil Mo value is found in the lowest Mo Zones (Fig. 501). These are often combined with high Mn values. As in the case of the high Mn values the low Mo values are more typical for samples from the Northern Region than from other Regions.

Zinc. Irrespective of the extraction method used to extract Zn from Malawi soils, the plant Zn—soil Zn correlation is high and the data obtained on Zn levels by the two methods are much alike (Figs 502, 503, 41 and 43). "Normal" Zn values are typical and even the few sample pairs falling outside Zn Zone III are not in any extreme positions. However, many Zn values are low enough to indicate the possibility of Zn deficiency at several of the sampled sites.

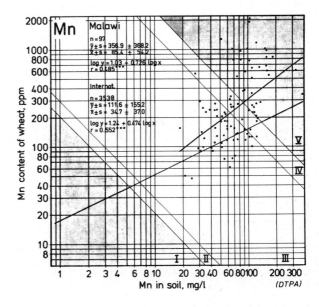

Fig. 498. Regression of Mn content of pot-grown wheat (y) on DTPA extractable soil Mn (x), Malawi.

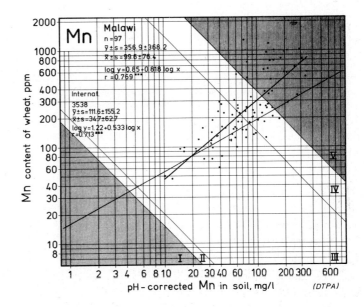

Fig. 499. Regression of Mn content of pot-grown wheat (y) on DTPA extractable soil Mn corrected for pH (x), Malawi.

9.3.4 Summary

The soils of Malawi are mostly coarse textured with low pH, low cation exchange capacity and medium organic matter content. Relatively high P, low to medium N and K, and low Ca and Mg contents are typical of most Malawi soils. Of the six micronutrients included in this study, the analytical data for Fe and Zn show usually "normal" values, but at some locations shortage of Zn likely. The contents of B and Cu are low, those of Mo lower than in most other countries, while the Mn contents are among the highest.

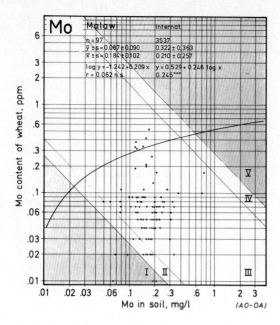

Fig. 500. Regression of Mo content of pot-grown wheat (y) on ammonium oxalate-oxalic acid extractable soil Mo (x), Malawi.

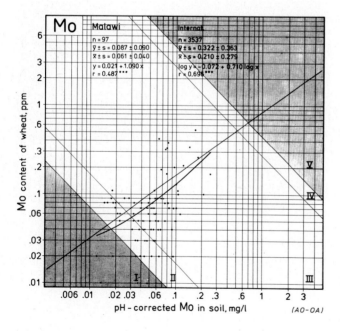

Fig. 501. Regression of Mo content of pot-grown wheat (y) on pH-corrected AO-OA extractable soil Mo (x), Malawi.

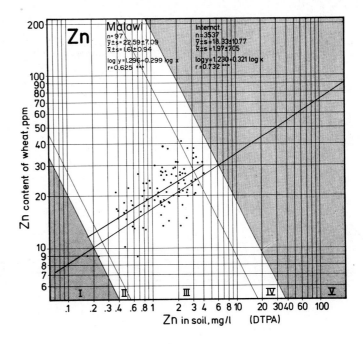

Fig. 502. Regression of Zn content of pot-grown wheat (y) on DTPA extractable soil Zn (x), Malawi.

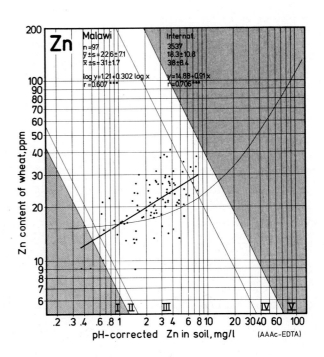

Fig. 503. Regression of Zn content of pot-grown wheat (y) on AAAc-EDTA extractable soil Zn corrected for pH (x), Malawi.

9.4 Nigeria

9.4.1 General

The Nigerian sample material consisting of 103 maize-soil, 42 wheat-soil sample pairs plus 31 soil samples was collected from the sites shown in Fig. 504. About half the samples, including all wheat-soil sample pairs, originate from the Northern States and half from the Western, Mid-Western and Eastern States. Of the 91 soils classified into FAO/Unesco soil units, Ferralsols (31 soils) and Acrisols (27) were the most common. The other soils were classified as Luvisols (8), Fluvisols (6), Nitosols (5), Regosols (5), Cambisols (4), Vertisols (2), Arenosol (1), Gleysol (1) and Lithosol (1).

The great majority of the sampled Nigerian soils are very coarse textured but a few medium and fine textured are included (Fig. 505). The soil pH varies widely covering the range from pH 4 to 8 quite evenly. Most of the soils are low in organic matter and have a low cation exchange capacity. With a few exceptions the values for electrical conductivity, $CaCO_3$ equivalent and sodium content are low (Appendixes 2—4).

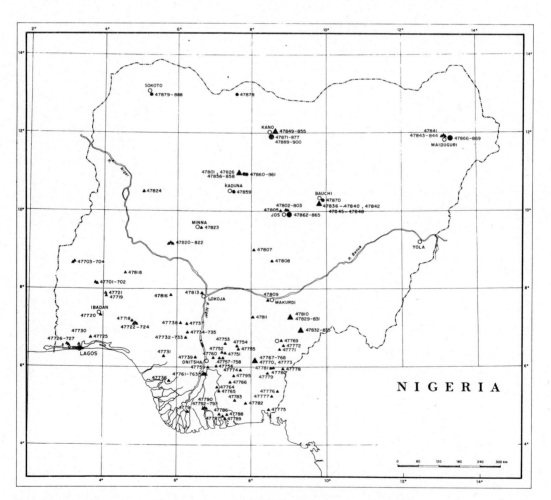

Fig. 504. Sampling sites in Nigeria (dots = wheat fields, triangles = maize fields).

Fig. 505. Frequency distributions of texture, pH, organic carbon content, and cation exchange capacity in soils of Nigeria (columns). Curves show the international frequency of the same characters.

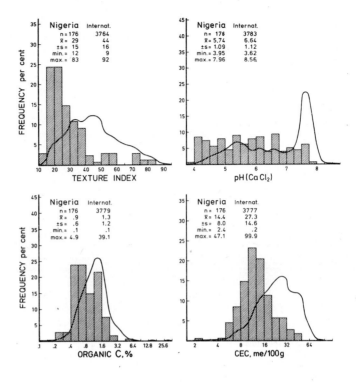

Figs. 506–510. Frequency distributions of nitrogen, phosphorus, potassium, calcium and magnesium in original wheat and maize samples and respective soils (columns) of Nigeria. Curves show the international frequency of the same characters.

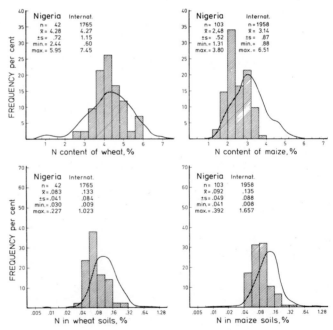

Fig. 506. Nitrogen, Nigeria.

9.4.2 Macronutrients

The total **nitrogen** contents of Nigerian soils are very low compared to those of most other countries (Figs 506 and 6). Because of the relatively high rates of nitrogen fertilizers applied to the sampled wheat crops (85 ± 27 kg N/ha), the national mean N content of

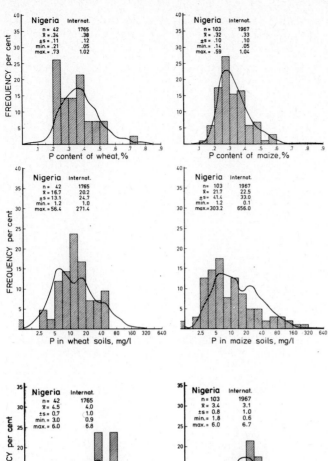

Fig. 507. Phosphorus, Nigeria.

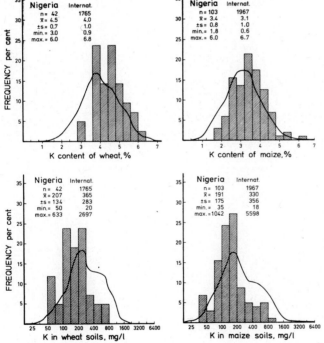

Fig. 508. Potassium, Nigeria.

Nigerian wheats (4.28 %) corresponds to the international mean (4.27 %). Much less nitrogen was applied to maize (13 ± 23 kg N/ha) and the N contents of maize remain at a low level. This national mean is the second lowest after Tanzania (Appendix 3).

The average $NaHCO_3$ extractable **phosphorus** content of maize soils and P content of

Fig. 509. Calcium, Nigeria.

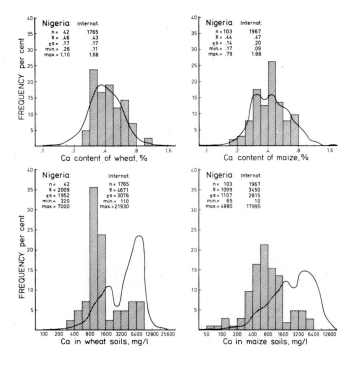

Fig. 510. Magnesium, Nigeria.

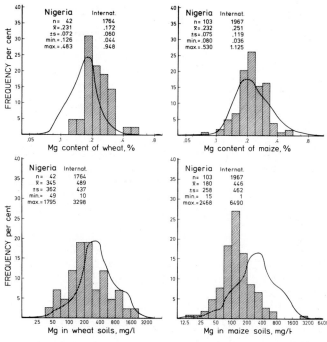

maize correspond to the international averages but those of wheat soils and wheat are somewhat low (Figs 507 and 7). In all cases the variations are wide. The Nigerian crops received somewhat less phosphate fertilizers than the average for crops in this study (Appendix 5).

The average exchangeable **potassium** content of Nigerian soils is slightly on the low side internationally but the mean K contents of both original indicator plants are somewhat higher than the respective international averages (Figs 8 and 508), although only small to moderate applications of potassium fertilizers were used (Appendix 5). This contradiction may partly be due to the general acidity of Nigerian soils. The plants seem to be able to absorb more K from acid soils in relation to exchangeable soil K than from alkaline soils. See Fig. 11 and related text.

In spite of the relatively low exchangeable **calcium** and **magnesium** contents of Nigerian soils, the Ca and Mg contents of the indicator crops are at the average international level (Figs 9 and 10), and in the case of wheat even slightly above it (Figs 509 and 510). As pointed out in Section 2.2.4 (Figs 12 and 14 and related text) these elements are relatively readily available to plants in coarse textured soils of low CEC, and these characteristics are typical of most Nigerian soils.

9.4.3 Macronutrient contents of original plants with special reference to varieties

Since the original Nigerian plant sample material included only a few HYV maize samples, comparisons between HYV and local varieties were not possible. In the case of wheat, 24 samples represented high yielding varieties and 18 were classified as local varieties. The HYV group consisted of cultivars 'Indus' (16 samples), 'Siete Cerros' (7), and 'Inia' (1). The average N, P, K, Ca and Mg contents of each variety group, their respective soils and amounts of applied fertilizers are given in Table 28.

Table 28. Comparison of macronutrient contents of high yielding and local varieties of original Nigerian wheats, respective soils and fertilizer applications. Differences between the mean contents of the two groups followed by the same index letter are not statistically significant. Letters a—b indicate significant differences at 10 and a—c at 5 percent level.

Nutrient	Average Macronutrient Content				Average N, P and K application kg/ha	
	Original wheat %		Wheat soils N %, others mg/l			
	HYV (n = 24)	Local (n = 18)	HYV (n = 24)	Local (n = 18)	(n = 24)	(n = 18)
Nitrogen	4.54[a]	4.00[c]	0.080[a]	0.081[a]	87[a]	82[a]
Phosphorus	0.371[a]	0.311[b]	12.4[a]	20.3[c]	18[a]	15[a]
Potassium	4.74[a]	4.33[b]	218[a]	185[a]	4[a]	3[a]
Calcium	0.495[a]	0.411[b]	2345[a]	1506[a]	—	—
Magnesium	0.232[a]	0.230[a]	343[a]	333[a]	—	—

The HYV wheats were sampled 40 days and the local wheats 35 days from planting, on the average. Therefore, the effect of physiological age is more likely to increase the differences in macronutrient contents of wheat between the two variety groups than to decrease these (see Section 1.2.2).

The data in Table 28 indicate that the two variety groups may differ genetically in their ability to absorb P and N. In the cases of plant K and Ca the differences are more likely to be due to similar differences in the soil K and Ca. With regard to Mg and micronutrients no clear differences were found between the two variety groups.

9.4.4 Micronutrients

Boron. With relatively few exceptions the B contents of Nigerian soils and wheats grown in pots of these soils are lower than the respective international averages (Figs 511 and 512), and Nigeria is one of the countries occupying the lowest positions in the "international B fields" (Figs 22 and 25). A quarter of the Nigerian samples falls in the two lowest B zones (I and II) with some of them in quite extreme positions (Fig. 512). According to these analytical data B deficiency is likely at several locations in Nigeria, and response to B, especially by crops with a high B requirement such as root crops, legumes and some fruits and vegetables could be expected. With only a few exceptions the lowest (Zones I and II)

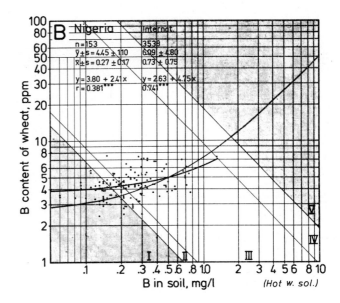

Fig. 511. Regression of B content of pot-grown wheat (y) on hot water soluble soil B (x), Nigeria. For details of summarized international background data, see Chapter 4.

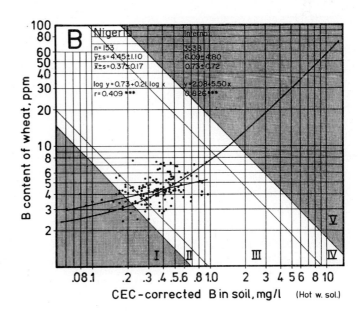

Fig. 512. Regression of B content of pot-grown wheat (y) on CEC-corrected (hot water soluble) soil B (x), Nigeria.

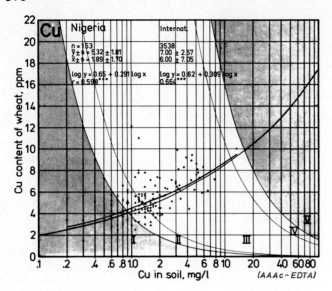

Fig. 513. Regression of Cu content of pot-grown wheat (y) on acid ammonium acetate-EDTA extractable soil Cu (x), Nigeria.

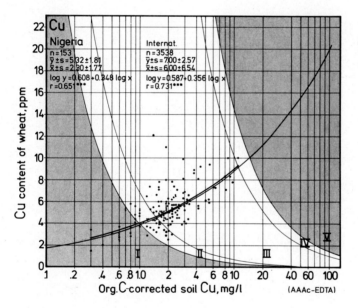

Fig. 514. Regression of Cu content of pot-grown wheat (y) on organic carbon-corrected (AAAc-EDTA extr.) soil Cu (x), Nigeria.

B values originate from the Northern States. A response to B by cotton has been reported e.g. in Niger State (FAO, 1979).

Copper. Only about two-thirds of the Nigerian Cu values are within the "normal" Cu range (Zone III) and one-third falls in the low Cu zones (Figs 513 and 514). Nigeria's position in the "international Cu fields" is well down (Figs 27 and 29). The great majority of the low (Zones I and II) Cu values originate from the Eastern States where such low values were recorded for more than half the sample pairs. In this part of the country Cu deficiency seems much more likely than elsewhere in Nigeria where low Cu values are relatively rare.

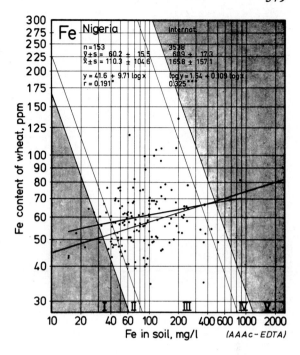

Fig. 515. Regression of Fe contents of pot-grown wheat (y) on acid ammonium acetate-EDTA extractable soil Fe (x), Nigeria.

Iron. In the "international Fe field" (Fig. 31) Nigeria stands slightly low. Although several quite low Fe values were measured from the Nigerian sample material, shortage of Fe is unlikely to be among the primary micronutrient problems of the country (Fig. 515).

Manganese. The national mean Mn content of Nigerian plants is almost double the international mean but the respective DTPA extractable soil Mn values about equal it (Fig. 516). Correction of soil Mn values for pH raises the national mean from 35.3 to 55.1 mg/l, triples the standard deviation and improves the plant Mn—soil Mn correlation (r) from 0.395*** to 0.729*** (Fig. 517). See also Figs 33 and 37.

About a quarter of the sample pairs are within the high Mn Zones (IV and V) but only a few fall in the low Zones (I and II) indicating that problems due to excess Mn are more likely than those due to Mn shortage. Two-thirds of the highest (Zone V) values were measured from samples from the Eastern States.

Molybdenum. The non-significant correlation between plant Mo and AO-OA extractable soil Mo (Fig. 518) becomes highly significant when the soil Mo values are corrected for pH (Fig. 519). The ranges of variation of both plant Mo and soil Mo are wide, but the majority of values are low and Nigeria's position in the "international Mo field" is relatively low (Fig. 20). Only a few Mo points are in the high zones and the relative frequency of low (Zones I and II) Mo values is exceeded by only a few other countries. About a fifth of the samples fall in Zone I and a third in Zones I and II (Fig. 519) indicating that problems due to shortage of Mo are likely in Nigeria, especially in the case of legumes which have a specific need for Mo. A response by groundnut to Mo has been reported e.g. in Niger State (FAO, 1979). Two-thirds of the Mo values in Zone I originate from the Eastern States at sites with a soil pH(CaCl$_2$) of 5.0 or below and these are often associated with high Mn values.

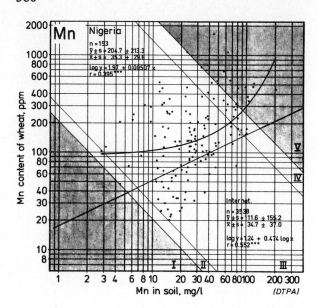

Fig. 516. Regression of Mn content of pot-grown wheat (y) on DTPA extractable soil Mn (x), Nigeria.

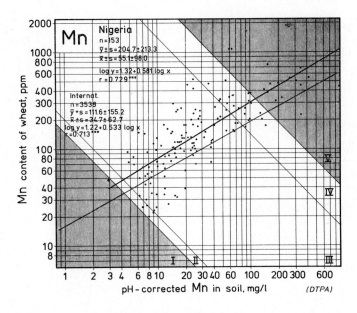

Fig. 517. Regression of Mn content of pot-grown wheat (y) on DTPA extractable soil Mn corrected for pH (x), Nigeria.

Zinc. Irrespective of the method used for determining extractable soil Zn contents, the patterns of Zn levels in Nigeria are similar (Figs 41, 43, 520 and 521). In the "international Zn fields" Nigeria stands clearly high. However, the Zn values vary considerably from one site to another and both very high and relatively low values were determined. The highest Zn values were measured in samples from the Eastern and Northern States, but low Zn contents were more typical for sites in the Northern States than elsewhere in Nigeria. The above analytical data do not indicate very severe Zn deficiency but at many locations a response to Zn fertilization is likely (see also Section 2.7). Also Osiname (1976) reported increases in yield of between 240 and 1920 kg/ha for maize after applying Zn at rates varying from 1 to 8 kg/ha at three locations in Nigeria.

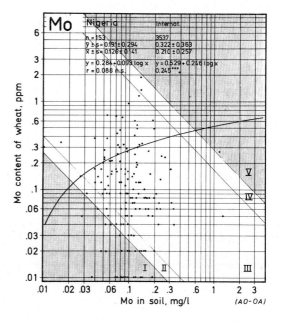

Fig. 518. Regression of Mo content of pot-grown wheat (y) on ammonium oxalate-oxalic acid extractable soil Mo (x), Nigeria.

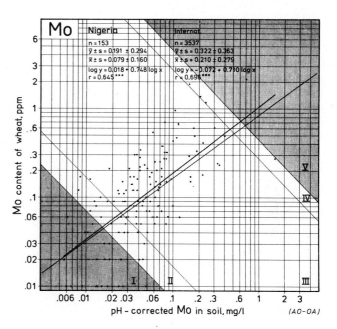

Fig. 519. Regression of Mo content of pot-grown wheat (y) on pH-corrected AO-OA extractable soil Mo (x), Nigeria.

9.4.5 Summary

Most Nigerian soils studied are coarse in texture, low in organic matter and have a low cation exchange capacity. The pH varies widely. The macronutrient status of soils is relatively low.

The micronutrient status of Nigerian soils varies considerably from one element to another. Micronutrient problems due to shortage of B, Cu, Mo and Zn and excess of Mn are likely.

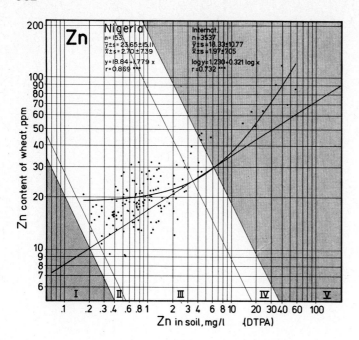

Fig. 520. Regression of Zn content of pot-grown wheat (y) on DTPA extractable soil Zn (x), Nigeria.

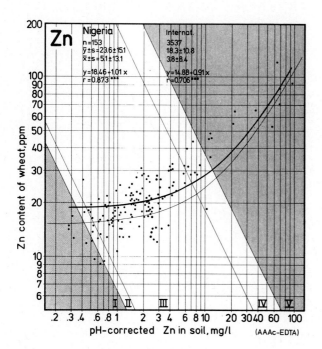

Fig. 521. Regression of Zn content of pot-grown wheat (y) on AAAc-EDTA extractable soil Zn corrected for pH (x), Nigeria.

9.5 Sierra Leone

9.5.1 General

The geographical distribution of the sampling sites for the 50 maize-soil sample pairs collected from Sierra Leone is shown in Fig. 522. All the sampled crops were grown under rainfed conditions in areas where the annual precipitation varies from 2000 to 4000 mm.

The majority of the soils are coarse textured and with a few exceptions very acid (Fig. 523). The national mean pH is lower than that for any other country. In spite of a generally high organic matter content the cation exchange capacity of soils remains at a

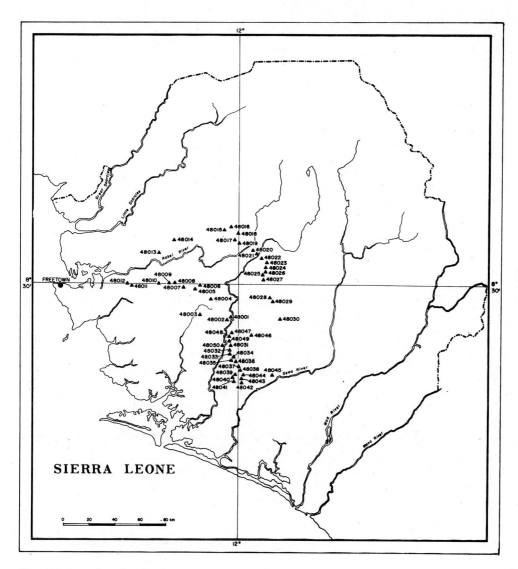

Fig. 522. Sampling sites in Sierra Leone.

relatively low level due to their coarse texture (see Fig. 3, h and m). Low electrical conductivity, CaCO$_3$ equivalent, and sodium content are typical of Sierra Leone soils (Appendixes 3 and 4).

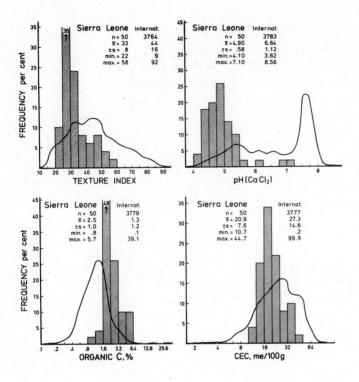

Fig. 523. Frequency distributions of texture, pH, organic carbon content, and cation exchange capacity in soils of Sierra Leone (columns). Curves show the international frequency of the same characters.

9.5.2 Macronutrients

The total **nitrogen** contents of the soils are relatively high due to their high organic matter contents (Fig. 524). The average use of nitrogen fertilizers was minimal (5 ± 22 kg N/ha), so explaining the low N contents of maize. In fact, nitrogen had only been applied to three sampled crops (samples No 48045, -48, and 50; 20, 100 and 112 kg N/ha, respectively). Consequently, the N contents of these maize samples were high (3.32, 3.76 and 4.14 percent, respectively) compared to the other maize samples from Sierra Leone.

Most of the NaHCO$_3$ extractable **phosphorus** contents of Sierra Leone soils are lower than the average in this study (Figs 525 and 7). The low soil P content is reflected in the P contents of maize which are lower than in most other countries. Only three of the 50 sampled maize crops had received phosphate fertilizer.

The average exchangeable **potassium** content of soil is lower but the K content of maize is higher than those of most other countries (Figs 526 and 8). This contradiction may partly be due to the higher K uptake from acid than from neutral or alkaline soils. See Fig. 11 and related text.

The national means for exchangeable **calcium** and **magnesium** contents of Sierra Leone soils and those for the Ca and Mg contents of maize are among the very lowest of this study (Figs 527, 528, 9, 10 and Appendix 3).

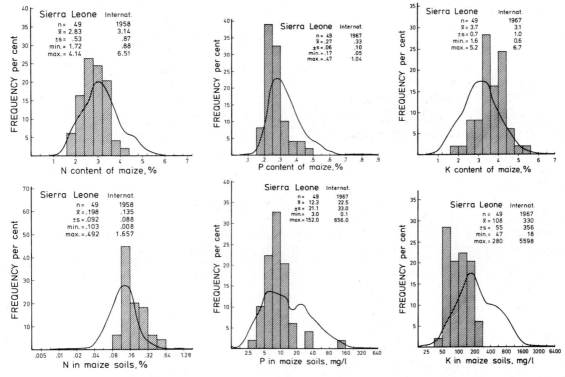

Fig. 524. Nitrogen, Sierra Leone.   *Fig. 525.* Phosphorus, Sierra Leone.   *Fig. 526.* Potassium, Sierra Leone.

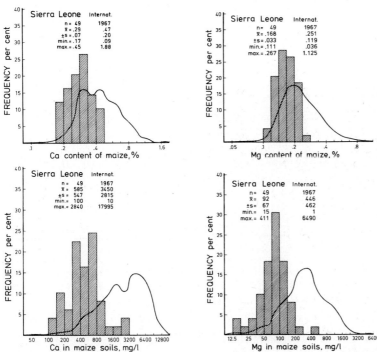

Figs 524–528. Frequency distributions of nitrogen, phosphorus, potassium, calcium and magnesium in original maize samples and respective soils (columns) of Sierra Leone. Curves show the international frequency of the same characters.

Fig. 527. Calcium, Sierra Leone.   *Fig. 528.* Magnesium, Sierra Leone.

9.5.3 Micronutrients

Boron. Almost all Sierra Leone soils and wheat grown in pots on these soils have B contents lower than the respective international averages (Figs 529 and 530) and the country occupies relatively low positions in the "international B fields" (Figs 22 and 25). However, the B contents vary very little from one sample to another and only a few B values fall in the low B Zones (I and II). According to these analytical data, deficiency of B occurs at several of the sampled Sierra Leone sites and a response to B is likely, especially where crops with high B requirements are grown. See also Section 2.7.

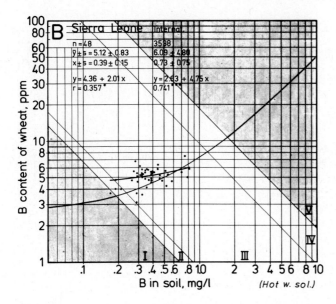

Fig. 529. Regression of B content of pot-grown wheat (y) on hot water soluble soil B (x), Sierra Leone. For details of summarized international background data, see Chapter 4.

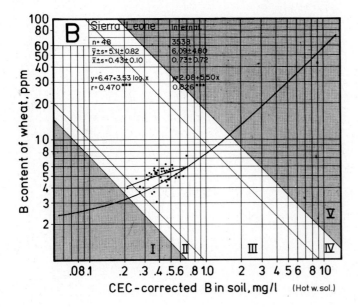

Fig. 530. Regression of B content of pot-grown wheat (y) on CEC-corrected (hot water soluble) soil B (x), Sierra Leone.

Copper. The average Cu contents of Sierra Leone soils and plants are lower than those in any other country participating in this study (Figs 27, 29, 531 and 532). Only 10 percent of the sample material show "normal" (Zone III) Cu contents and 69 percent fall in the lowest Cu Zone (Fig. 532). It seems that the typical characteristics of Sierra Leone soils: low pH, coarse texture, high organic matter content and low cation exchange capacity, contribute to the low Cu levels of the soils and plants in this country (See Figs 523, 28 and 30). Cu deficiency has been reported in coffee and cacao groves (Haque and Godfrey-Sam-Aggrey, 1980).

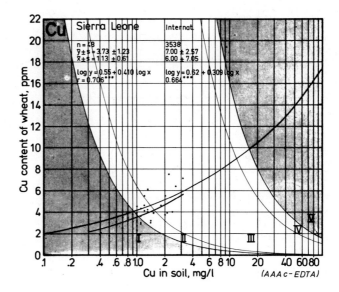

Fig. 531. Regression of Cu content of pot-grown wheat (y) on acid ammonium acetate-EDTA extractable soil Cu (x), Sierra Leone.

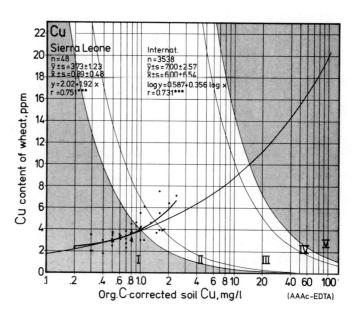

Fig. 532. Regression of Cu content of pot-grown wheat (y) on organic carbon-corrected (AAAc-EDTA extr.) soil Cu (x), Sierra Leone.

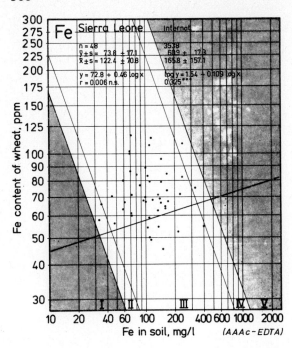

Fig. 533. Regression of Fe contents of pot-grown wheat (y) on acid ammonium acetate-EDTA extractable soil Fe (x), Sierra Leone.

Iron. The Fe situation in Sierra Leone seems quite "normal", with almost all the Fe sample pairs falling within the Fe Zone III (Fig. 533), and consequently, problems due to Fe are unlikely.

Manganese. Owing to very low pH (< 5) of most Sierra Leone soils, there is high availability of Mn to plants but only moderate extractability by DTPA (Fig. 34). This is the reason for the apparent contradiction between high plant and low soil Mn values given in Fig. 534. This contradiction is partly eliminated by pH correction which doubles the average soil Mn values and improves the plant Mn—soil Mn correlation from a non-significant to a highly significant level (Fig. 535). Because the variation of Mn values in the Sierra Leone sample material is relatively narrow, and only "normal" values are included, Mn problems seem unlikely.

Molybdenum. The absorption of Mo by plants is lowest from soils of low pH, but coarse texture, high organic matter content, low CEC, low electrical conductivity and low $CaCO_3$ equivalent are factors which also contribute toward low Mo availability (Fig. 16). Many of these factors affect the extractability of Mo by AO-OA in different ways but when the AO-OA soil Mo values are corrected for pH the effects of these factors on soil Mo are very similar to their effects on plant Mo. That the above mentioned soil properties are typical of Sierra Leone soils (Fig. 523 and Appendix 4) explains why Sierra Leone occupies the lowest positions in the "international Mo fields" (Figs 15 and 20). The correlation between plant Mo and uncorrected AO-OA soil Mo is not significant but is highly significant for pH-corrected soil Mo values (Figs 536 and 537). More than 80 percent of the Sierra Leone Mo sample pairs fall in the lowest Mo Zone (I) and over 90 percent in the two lowest Zones (Fig. 537). Only four out of 48 sample pairs show "normal" (Zone III) Mo contents. The extremely low Mo content of Sierra Leone soils

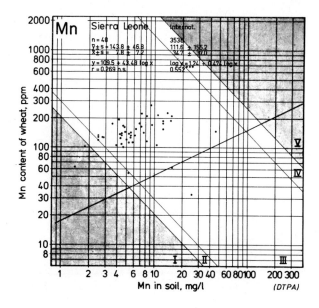

Fig. 534. Regression of Mn content of pot-grown wheat (y) on DTPA extractable soil Mn (x), Sierra Leone.

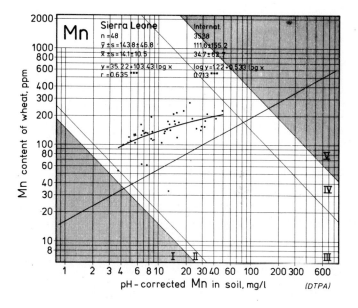

Fig. 535. Regression of Mn content of pot-grown wheat (y) on DTPA extractable soil Mn corrected for pH (x), Sierra Leone.

must be taken into account especially if crops with a high Mo requirement, e.g. nitrogen fixing legumes, are grown. For example, Haque and Bundu (1980) reported a good response of soybean to Mo in Sierra Leone.

Zinc. According to the analytical data on Sierra Leone soils and plants, the Zn level seems quite satisfactory irrespective of the method used for extracting Zn from the soils (Figs 538, 539, 41 and 43). With the exception of a few relatively high Zn values, the samples fall in the "normal" range (Zone III) and the analytical data obtained by plant and soil analyses are in agreement. At some sites a response to Zn could be expected.

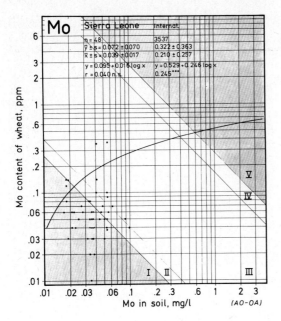

Fig. 536. Regression of Mo content of pot-grown wheat (y) on ammonium oxalate-oxalic acid extractable soil Mo (x), Sierra Leone.

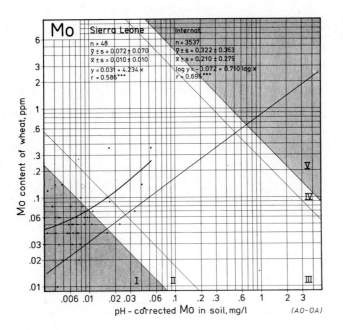

Fig. 537. Regression of Mo content of pot-grown wheat (y) on pH-corrected AO-OA extractable soil Mo (x), Sierra Leone.

9.5.4 Summary

The essential feature of the Sierra Leone soils is their low pH. The majority are coarse textured and have a relatively low CEC but are high in organic matter content. With the exception of nitrogen, the macronutrient contents of soils are very low.

Of the six micronutrients studied, the analytical data for Fe and Mn are usually "normal" but there are strong indications of Cu and Mo deficiency and some of B and Zn.

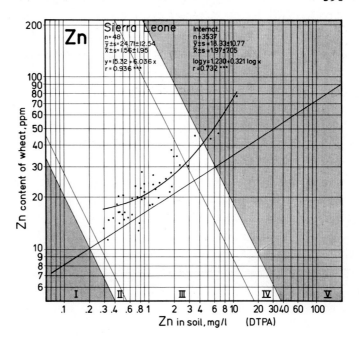

Fig. 538. Regression of Zn content of pot-grown wheat (y) on DTPA extractable soil Zn (x), Sierra Leone.

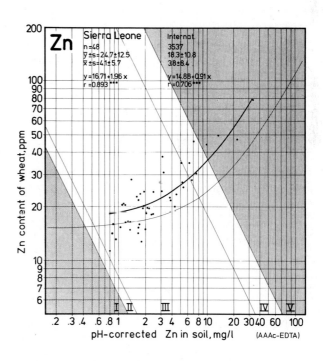

Fig. 539. Regression of Zn content of pot-grown wheat (y) on AAAc-EDTA extractable soil Zn corrected for pH (x), Sierra Leone.

9.6 Tanzania

9.6.1 General

The sample material of Tanzania consisting of 175 maize-soil and 5 wheat-soil sample pairs was collected from the following Regions: Arusha (25 sample pairs), Coast (7), Dodoma (8), Iringa (10), Kilimanjaro (27), Mara (16), Morogoro (13), Mtwara (13), Mwanza (10), Ruvuma (12), Shinyanga (3), Singida (13), Tabora (2), and Tanga (21). The approximate distribution of sampling sites is given in Fig. 540. With two exceptions the sampled crops were grown under rainfed conditions.

The soils vary widely in texture (from TI 13 to 88) but coarse and medium textures predominate (Fig. 541). Wide variations are characteristic also for pH, organic matter content and cation exchange capacity but soils with moderate acidity, medium organic matter content and low to medium CEC were most common. As will be seen later, the

Fig. 540. Sampling sites in Tanzania (dots = wheat fields, triangles = maize fields).

wide variation in the above soil properties are reflected as wide variations in the macro- and micronutrient contents of Tanzanian sample material. The electrical conductivity and $CaCO_3$ equivalent values and sodium contents are usually low (Appendixes 2—4).

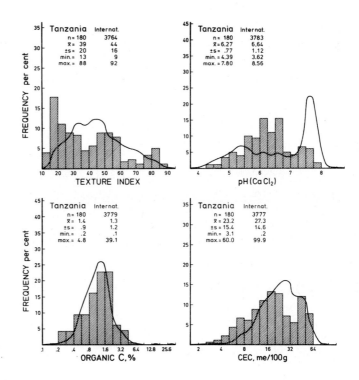

Fig. 541. Frequency distributions of texture, pH, organic carbon content, and cation exchange capacity in soils of Tanzania (columns). Curves show the international frequency of the same characters.

9.6.2 Macronutrients

Because only five wheat-soil sample pairs are included in the sample material, the following graphs show only the frequency distributions of macronutrient contents of maize and the respective soils. The averages for wheat and wheat soils are given in Appendix 2. It is noteworthy that only very low grain yields (200—1000 kg/ha) were expected from all original sampled wheats and from half of the sampled maize crops.

The total **nitrogen** contents of Tanzanian maize soils vary widely but are somewhat low when viewed in the international context of this study (Figs 542 and 6). The average N content of maize is lower than that of any other country. This must, at least in part, be due to the minimal amounts of nitrogen fertilizer applied (8 ± 17 kg N/ha) to the sampled crops.

Very wide variations are characteristic of the $NaHCO_3$ extractable **phosphorus** contents of Tanzanian soils as well as of the P contents of maize (Fig. 543). The average soil P content is somewhat higher than the respective international mean but the average P content of Tanzanian maize corresponds to the international mean. Only about 10 percent of the sampled maize crops had received phosphate fertilizer. The rates varied from 4 to 48 kg P/ha and for all sampled crops averaged 2 ± 6 kg P/ha which probably explains the relatively low P contents of maize. See also Fig. 7.

The **potassium** status of Tanzanian soils is generally high but varies largely from one soil to another (Figs 8 and 544). The K contents of Tanzanian maize are close to the

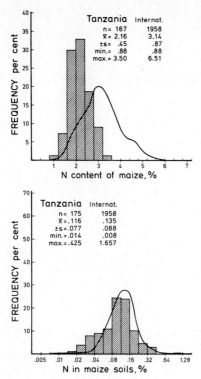
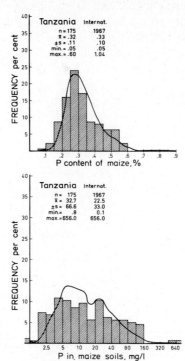
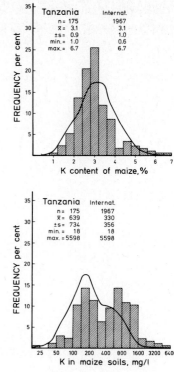

Fig. 542. Nitrogen, Tanzania.   *Fig. 543.* Phosphorus, Tanzania.   *Fig. 544.* Potassium, Tanzania.

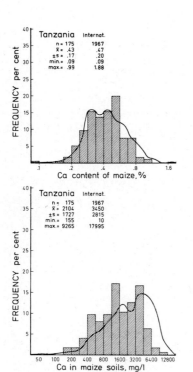
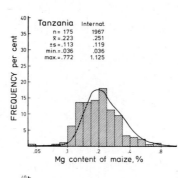
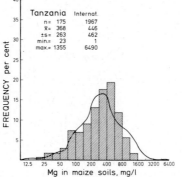

Figs 542–546. Frequency distributions of nitrogen, phosphorus, potassium, calcium and magnesium in original maize samples and respective soils (columns) of Tanzania. Curves show the international frequency of the same characters.

Fig. 545. Calcium, Tanzania.   *Fig. 546.* Magnesium, Tanzania.

international mean with correspondingly wide variations. No potassium fertilizer was applied to the sampled crops.

Wide variations are also typical for exchangeable **calcium** and **magnesium** contents of the Tanzanian soils and for Ca and Mg contents of maize (Figs 555 and 556). The average contents of these elements are somewhat lower than the respective international averages (Figs 9 and 10).

9.6.3 Micronutrients

Boron. With relatively few exceptions the B contents of soils and plants are within the "normal" B range (Zone III, Figs 547 and 548) and Tanzania locates near the centres of

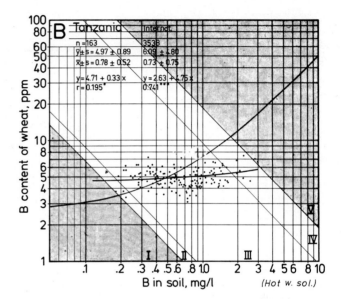

Fig. 547. Regression of B content of pot-grown wheat (y) on hot water soluble soil B (x), Tanzania. For details of summarized international background data, see Chapter 4.

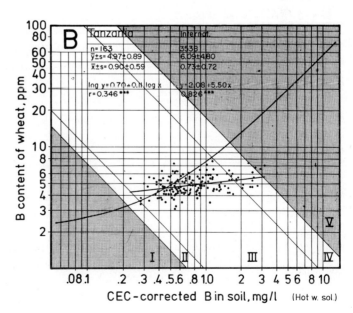

Fig. 548. Regression of B content of pot-grown wheat (y) on CEC-corrected (hot water soluble) soil B (x), Tanzania.

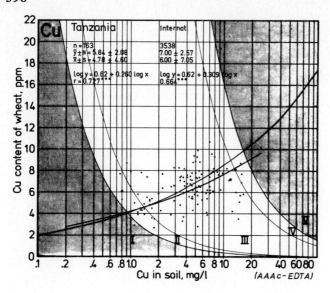

Fig. 549. Regression of Cu content of pot-grown wheat (y) on acid ammonium acetate-EDTA extractable soil Cu (x), Tanzania.

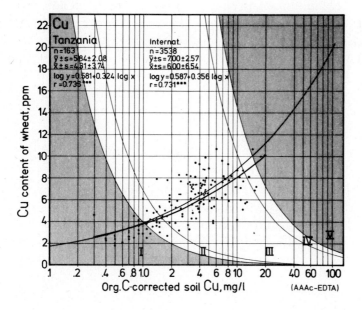

Fig. 550. Regression of Cu content of pot-grown wheat (y) on organic carbon-corrected (AAAc-EDTA extr.) soil Cu (x), Tanzania.

the "international B fields" (Figs 22 and 25). No extremely low or high values were recorded and, consequently, these analytical data do not indicate any widespread B problems in the country, although at some locations shortage of B may reduce the growth of crops sensitive to B deficiency.

Copper. Tanzania's national mean plant and soil Cu values are only slightly lower than the respective international averages (Figs 549 and 550) and its locations in the "international Cu fields" are almost central (Figs 27 and 29). However, variation among the Tanzanian samples is wide and several very low and a few relatively high Cu values occur. About 14 percent of the sample pairs fall in the lowest Cu Zone (I) and 23 percent

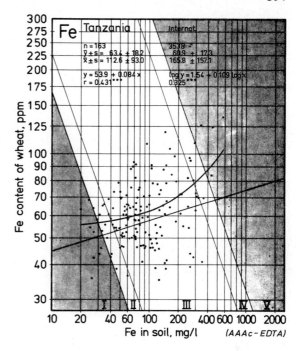

Fig. 551. Regression of Fe contents of pot-grown wheat (y) on acid ammonium acetate-EDTA extractable soil Fe (x), Tanzania.

in the two lowest Zones (I and II, Fig. 550). Low Cu values were relatively more frequent for samples which came from Mara than from other Regions. More than two-thirds of Mara samples show low (Zone I) Cu values indicating apparent shortage of Cu in this part of the country.

Iron. The average analytical data on Fe contents of Tanzanian sample material given in Fig. 31 and those of single sample pairs presented in Fig. 551 indicate that both the averages and the variations of Tanzanian plant and soil Fe values are of the same order of magnitude as those for the total international material in this study. Although several Fe points fall outside the "normal" range (Zone III) these are not in such extreme positions that severe disorders due to Fe should be expected.

Manganese. The non-significant plant Mn—soil Mn correlation (Fig. 552) becomes highly significant when he DTPA soil Mn values are corrected for pH (Fig. 553). No low (Zones I and II) Mn values were recorded from Tanzanian samples but some relatively high Mn values were measured from samples which came from sites with acid soils. These were somewhat more frequent for samples from Tanga and Singida than from other Regions.

Molybdenum. In the "international Mo fields" Tanzania stands clearly on the high side (Figs 15 and 20). However, the ranges of variation of Tanzanian plant and soil Mo values are exceptionally wide and both very low and very high Mo contents were recorded (Figs 554 and 555). Low (Zone I and II) Mo values occurred in samples from several Regions but were relatively more frequent in samples from sites with acid soils in Ruvuma, Mtwara, Morogoro and Dodoma. Almost all high (Zone IV and V) Mo values were obtained from Arusha and Kilimanjaro samples from sites where the soils are alkaline or only slightly acid.

Zinc. The Tanzanian national averages for plant and soil Zn correspond to the respective international mean values placing Tanzania at the centres of the "international

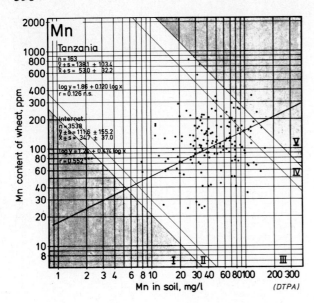

Fig. 552. Regression of Mn content of pot-grown wheat (y) on DTPA extractable soil Mn (x), Tanzania.

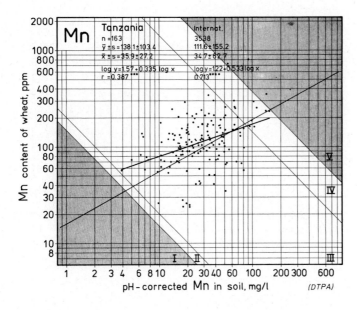

Fig. 553. Regression of Mn content of pot-grown wheat (y) on DTPA extractable soil Mn corrected for pH (x), Tanzania.

Zn fields" (Figs 41 and 43). As in the cases of most other nutrients, the variations in Zn contents are very wide (Figs 556 and 557). The results obtained by either of the extraction methods (DTPA and pH-corrected AAAc-EDTA) show a good correlation with the plant Zn analyses, but the former method gives relatively somewhat higher Zn values than the latter, especially at low Zn levels, i.e. more than twice as many plant Zn AAAc-EDTA Zn points fall in Zone I than in the case of DTPA extraction. Nevertheless, both methods give quite clear indications of a shortage of Zn at many locations in Tanzania. The lowest Zn values were measured from samples of Mwanza, Shinyanga, Singida, and Mara while most of the high Zn samples came from Kilimanjaro, Tanga and Iringa Regions.

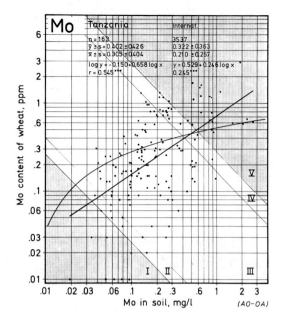

Fig. 554. Regression of Mo content of pot-grown wheat (y) on ammonium oxalate-oxalic acid extractable soil Mo (x), Tanzania.

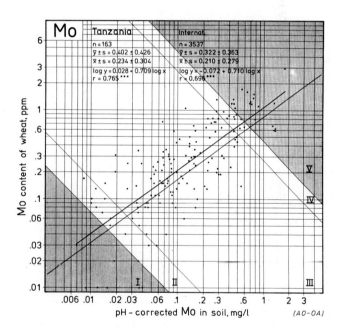

Fig. 555. Regression of Mo content of pot-grown wheat (y) on pH-corrected AO-OA extractable soil Mo (x), Tanzania.

9.6.4 Summary

None of the soils sampled from Tanzania for this study could be called a "typical Tanzanian" soil since soils varied greatly in all properties examined. However, soils with coarse to medium texture, moderate acidity, medium organic matter content, and low to medium CEC predominated.

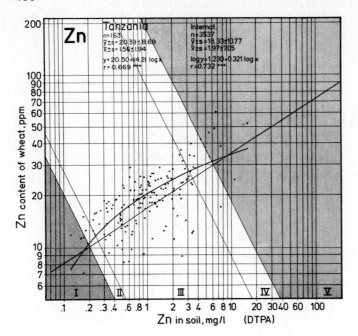

Fig. 556. Regression of Zn content of pot-grown wheat (y) on DTPA extractable soil Zn (x), Tanzania.

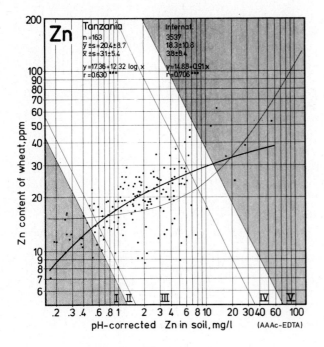

Fig. 557. Regression of Zn content of pot-grown wheat (y) on AAAc-EDTA extractable soil Zn corrected for pH (x), Tanzania.

Wide variations are characteristic also for the contents of all macronutrients, but on average the N, Mg, and Ca contents of soils are slightly on the low side while P and K contents are on the high side compared to other countries in this study.

The analytical data of the six micronutrients studied indicate that the most likely micronutrient problems in the country are those caused by shortage of Cu, Zn, B and Mo.

9.7 Zambia

9.7.1 General

The Zambian sample material consisting of 46 maize and respective soil samples was collected from the Central and Southern Provinces and effectively covers the maize growing areas of the country (Fig. 558). A great majority (41) of the sampled soils were classified as Nitosols. Two soils were classified as Arenosols and the remaining three as a Ferralsol, a Fluvisol and a Gleysol.

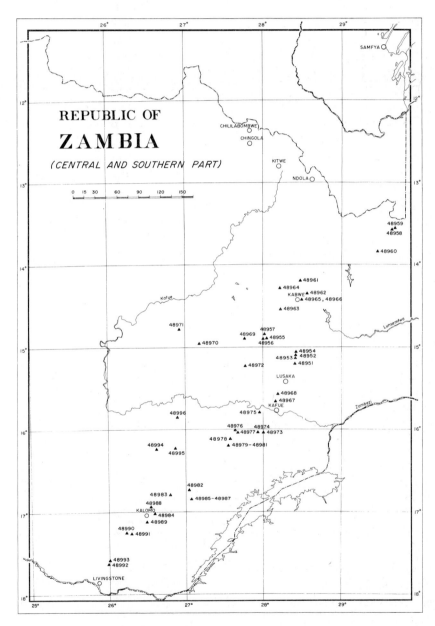

Fig. 558. Sampling sites in Zambia.

About two-thirds of the soils are coarse textured (TI < 30) and one-third have a medium texture (Fig. 559). The soil pH varies from 4.2 to 7.1 but acid soils predominate and the national mean pH is the second lowest (after Sierra Leone) among the 30 countries in this study. The national mean organic matter content is also among the lowest and that of CEC the absolute lowest recorded in this study. The values of electrical conductivity, $CaCO_3$ equivalent, and sodium content are also low (Appendixes 3 and 4).

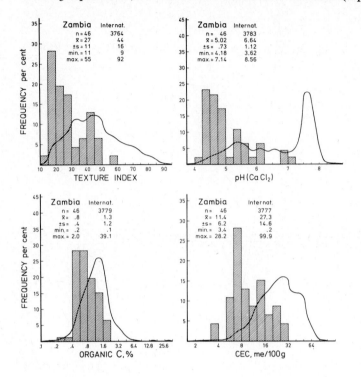

Fig. 559. Frequency distributions of texture, pH, organic carbon content, and cation exchange capacity in soils of Zambia (columns). Curves show the international frequency of the same characters.

9.7.2 Macronutrients

The average total **nitrogen** content of Zambian soils is the second lowest (after India) in the international scale (Fig. 6), but the N contents of maize are of the average international level (Fig. 560). This is because almost all sampled crops were heavily fertilized with nitrogen (Appendix 5).

The contents of $NaHCO_3$ extractable soil **phosphorus** are of the average international level but since the sampled crops were rather heavily fertilized with phosphates the P contents of maize are relatively high (Figs 7, 561 and Appendix 5).

The exchangeable **potassium** contents of Zambian soils vary considerably but are generally low and several very low values are included (Figs 8 and 562). Although moderate dressings of potassium fertilizers were applied to the sampled crops (Appendix 5) the K contents of maize remain at a very low level.

The Zambian national averages for exchangeable **calcium** and **magnesium** contents are among the lowest in this study (Figs 9, 10, 563 and 564). The Mg contents of maize are also relatively low but in relation to soil Ca the Ca contents of maize are high. As pointed out in Section 2.2.4 (Figs 12 and 14 and related text) plants seem to absorb these elements easily from coarse textured soils with low CEC and these properties are typical of Zambian soils.

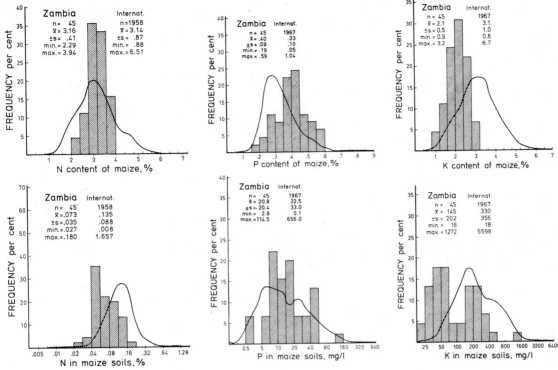

Fig. 560. Nitrogen, Zambia.

Fig. 561. Phosphorus, Zambia.

Fig. 562. Potassium, Zambia.

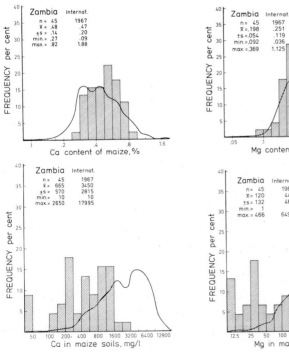

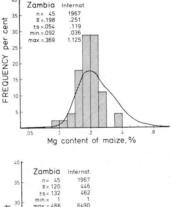

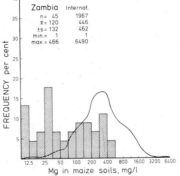

Figs 560–564. Frequency distributions of nitrogen, phosphorus, potassium, calcium and magnesium in original maize samples and respective soils (columns) of Zambia. Curves show the international frequency of the same characters.

Fig. 563. Calcium, Zambia.

Fig. 564. Magnesium, Zambia.

9.7.3 Micronutrients

Boron. The B contents of Zambian soils and pot-grown wheats vary within quite narrow limits (Figs 565 and 566). Almost all plant and soil B values are lower than the respective mean values for the total international material in this study, and the location of Zambia in the international scale (Figs 22 and 25) is low. Although no extremely low B contents were recorded in the Zambian samples, the analytical data indicate B shortage at several places in the country. B deficiency on cotton in Zambia was discovered in 1966 by Rothwell and co-workers (Anon. 1981). During recent years B fertilizers have commonly

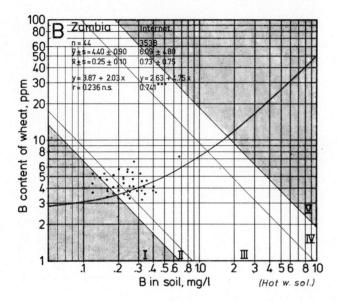

Fig. 565. Regression of B content of pot-grown wheat (y) on hot water soluble soil B (x), Zambia. For details of summarized international background data, see Chapter 4.

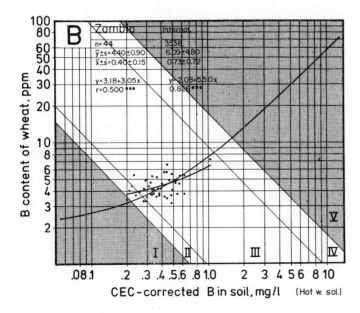

Fig. 566. Regression of B content of pot-grown wheat (y) on CEC-corrected (hot water soluble) soil B (x), Zambia.

been applied, often together with insecticides.

Copper. Zambia falls among the low-Cu countries in the "international Cu fields" (Figs 27 and 29). More than half of the Zambian sample pairs fall in the two lowest Cu Zones (I and II) and most of these are in Zone I (Figs 567 and 568) indicating apparent Cu shortages. Although no certain low-Cu areas could be geographically defined many of the samples lowest in Cu came from the Kalomo, Mkushi and Kabwe areas.

Iron. On average, the Fe contents of Zambian plants are of the average international level but the AAAc-EDTA extractable soil Fe values are low (Figs 569 and 31). Nevertheless, although several sample pairs fall in the low Fe Zones (I and II), indicating

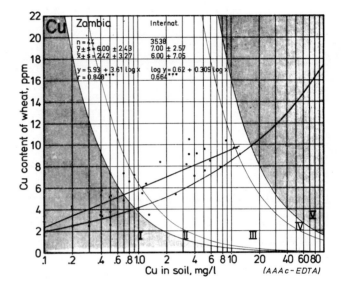

Fig. 567. Regression of Cu content of pot-grown wheat (y) on acid ammonium acetate-EDTA extractable soil Cu (x), Zambia.

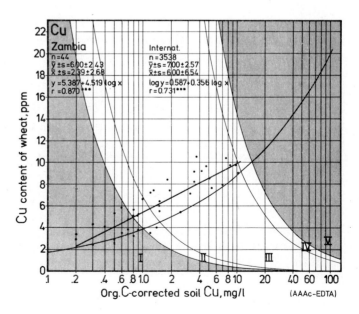

Fig. 568. Regression of Cu content of pot-grown wheat (y) on organic carbon-corrected (AAAc-EDTA extr.) soil Cu (x), Zambia.

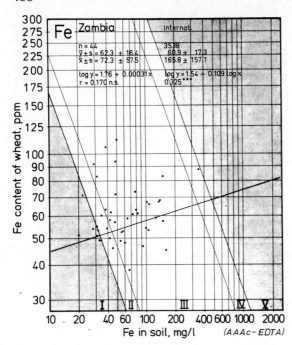

Fig. 569. Regression of Fe contents of pot-grown wheat (y) on acid ammonium acetate-EDTA extractable soil Fe (x), Zambia.

possible response to Fe of crops sensitive to Fe deficiency, a shortage of Fe is not likely to be among the primary micronutrient problems of the country.

Manganese. The Zambian mean for plant Mn content is higher than that of any other country in this study (Figs 33, 37, 570 and 571). It is five times that of the international mean and 25 times that of the lowest national mean (Malta). However, the Zambian national mean for DTPA extractable soil Mn is only the eighth highest, but moves up to second place when corrected for soil pH. Simultaneously, the plant Mn—soil Mn correlation is improved from non-significant (r = 0.266 n.s.) to a highly significant (r = 0.607***) level. More than half (55 %) of the Zambian plant-soil sample pairs fall in the highest Mn Zone (V) and only 30 percent of those show "normal" (Zone III) Mn contents (Fig. 571). According to these analytical data, excess of Mn is a factor which may cause disorder of plant growth in many parts of Zambia. The samples containing most Mn came from geographically scattered areas where very acid soils, $pH(CaCl_2) < 5$, existed.

Molybdenum. Unlike Mn, the Mo contents of Zambian plants are low. The national average is only one-third of the international mean and Zambia occupies one of the lowest positions in the "international Mo fields" (Figs 15 and 20). When pH correction of the AO-OA extractable soil Mo values takes place, the negative non-significant plant Mo—soil Mo correlation (r = —0.004 n.s.) becomes positive and highly significant (r = 0.656***, Figs 572 and 573). At the same time the Zambian national average for soil Mo decreases from 0.105 to 0.042 mg/l and Zambia's position in the "international Mo fields" moves closer to the international regression line. Half (50 %) of the plant-soil sample pairs are within the lowest Mo Zone (I), 9 percent in Zone II and 41 percent in the "normal" (Zone III) Mo range (Fig. 573). Problems due to the low Mo status of Zambian soils are likely to arise, especially if crops are grown which are sensitive to Mo deficiency, such as nitrogen fixing legumes.

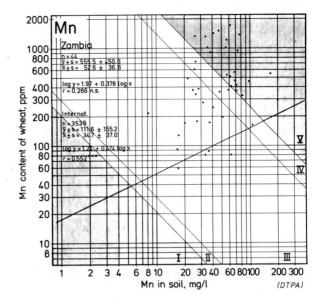

Fig. 570. Regression of Mn content of pot-grown wheat (y) on DTPA extractable soil Mn (x), Zambia.

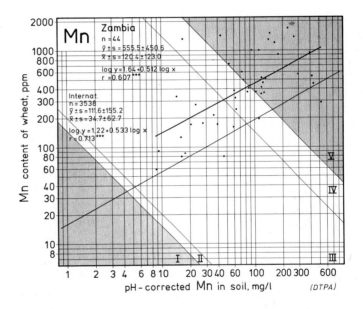

Fig. 571. Regression of Mn content of pot-grown wheat (y) on DTPA extractable soil Mn corrected for pH (x), Zambia.

Samples with low Mo contents occuring in geographically scattered sites with acid soils, $pH(CaCl_2) < 5$, are usually combined with high Mn contents. Obviously, in the long run, raising the soil pH by liming would be the best means of increasing the availability of Mo as well as reducing the risk of Mn toxicity.

Zinc. Irrespective of the method used for determining extractable soil Zn, the distributions of Zambian soil and plant Zn contents are much alike (Figs 574 and 575) and in both cases the plant Zn—soil Zn correlations are good. Almost all plant-soil Zn values are within the "normal" (Zone III) range, but many of them show such a low Zn content that disorders due to Zn shortage are likely (see also Section 2.7).

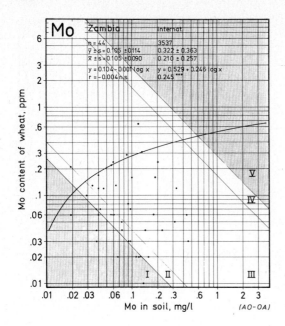

Fig. 572. Regression of Mo content of pot-grown wheat (y) on ammonium oxalate-oxalic acid extractable soil Mo (x), Zambia.

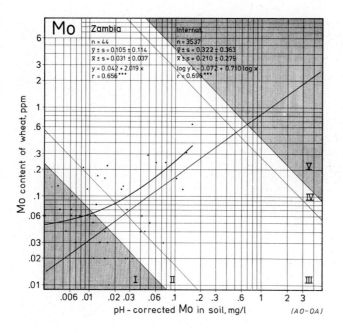

Fig. 573. Regression of Mo content of pot-grown wheat (y) on pH-corrected AO-OA extractable soil Mo (x), Zambia.

9.7.4 Summary

Most of the Zambian soils in this study are acid, coarse to medium textured, low in organic matter, and have a low cation exchange capacity.

The levels of macronutrients (N, K, Ca and Mg) are usually low but that of P corresponds to the average international level. The micronutrient content of Zambian soils

Fig. 574. Regression of Zn content of pot-grown wheat (y) on DTPA extractable soil Zn (x), Zambia.

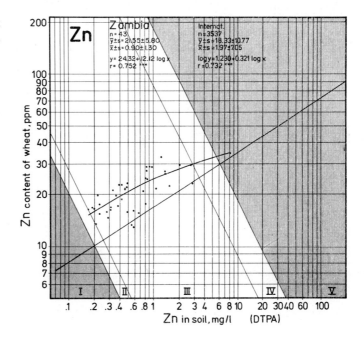

Fig. 575. Regression of Zn content of pot-grown wheat (y) on AAAc-EDTA extractable soil Zn corrected for pH (x), Zambia.

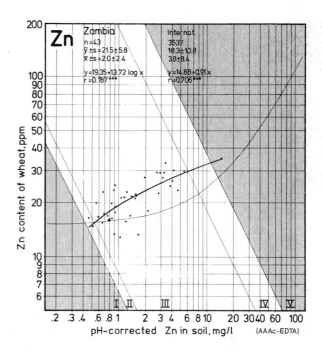

and plants varies considerably depending on the element. Only the contents of Zn are usually of the "normal" international level, but at several locations shortage of Zn is likely. Those of B and Fe are somewhat on the low side. Very low Mo and Cu and very high Mn values occur more frequently in Zambia than in most other countries.

10. Summary for Part II

The national data for each of the 30 participating countries on the most important soil characteristics, macronutrient contents of the original indicator plants (wheat and maize) and respective soils, as well as the micronutrient contents of pot-grown wheats and respective soils are presented in graphical forms in such a way that in the background of each graph the respective data for the whole international material are shown.

Analytical data for soil micronutrients are based on hot water extraction (B), acid ammonium acetate-EDTA extraction (Cu, Fe), DTPA extraction (Mn, Zn) and ammonium oxalate-oxalic acid extraction (Mo). Soil B values corrected for CEC, soil Cu values corrected for soil organic carbon content, Mo and Mn values corrected for pH, and AAAc-EDTA Zn values corrected for pH are also presented.

The nutrient analytical data for each country are briefly discussed and compared with the relevant international data as a whole and in the light of the soil characteristics typical of the country or area in question.

Data on four important soil characteristics are summarized by countries in Table 29, where the limits for the three categories for each characteristic are such that of the whole international soil sample material approximately 25 percent falls into the "low", 50 percent into the "medium" and 25 percent into the "high" class.

Coarse **texture** is typical of soils sampled from most African countries (except Ethiopia) and the majority of soils of Sri Lanka, India and Belgium fall into the same class. The highest percentage of fine textured soils was recorded from Brazil but soils of this type are also common in most Near East countries, Ethiopia and Thailand.

Low **soil pH** is most characteristic for Sierra Leone where 92 percent of the sampled soils have pH lower than 5.6. Relatively high percentages of acid soils were recorded from many other African countries as also in Finland, Brazil, Argentina, New Zealand and Korea. High soil alkalinity is most common in Near East countries, Pakistan and India.

The content of **organic matter** of soils varies considerably from one country to another. Soils of low organic matter content are especially common in India, Pakistan, Zambia and Nigeria while the soils sampled from New Zealand, Finland, Sierra Leone, Brazil, Argentina, Ethiopia and Thailand are generally rich in organic matter.

Low **cation exchange capacity** of soils is most typical for countries where the soils are coarse textured and/or low in organic matter such as Zambia, India, Nigeria, Sri Lanka, Ghana and Pakistan, while fine texture and/or high organic matter content generally contribute to high soil CEC, as in Egypt, New Zealand, Ethiopia, Thailand, Lebanon and Syria.

The macronutrient contents of soils of the 30 countries are summarized in Table 30 using the same low (25 %) -medium (50 %) -high (25 %) criteria as in Table 29.

The relative frequency of low total **nitrogen** content of soils is highest in countries with soils of low organic matter content (India, Pakistan, Zambia and Nigeria) and conversely, the high N soils are typically found in countries where soils have a high organic matter content (New Zealand, Finland, Argentina, Ethiopia, Hungary and Brazil).

Table 29. Relative frequencies (percentages) of low, medium and high values of soil texture (TI), pH, organic carbon content and cation exchange capacity in 30 countries.

Country	TI low <33	TI med. 33—55	TI high >55	pH(CaCl$_2$) low <5.6	pH(CaCl$_2$) med. 5.6—7.6	pH(CaCl$_2$) high >7.6	Org. C., % low <0.8	Org. C., % med. 0.8—1.5	Org. C., % high >1.5	CEC, me/100 g low <16	CEC, me/100 g med. 16—36	CEC, me/100 g high >36
Belgium	56	44	—	44	56	—	17	61	22	27	73	—
Finland	31	29	40	76	24	—	—	2	98	3	71	26
Hungary	10	68	22	12	73	15	6	46	48	5	71	24
Italy	19	50	31	5	83	12	6	78	16	14	69	16
Malta	4	84	12	—	88	12	—	80	20	—	100	—
New Zealand	6	61	32	63	37	—	—	—	100	—	18	82
Argentina	3	97	—	68	32	—	—	34	66	—	90	10
Brazil	4	20	76	75	25	—	4	24	72	7	76	17
Ecuador		no data		20	73	7	15	46	38	36	55	9
Mexico	18	46	36	7	48	45	41	56	3	13	47	39
Peru	46	43	11	10	83	7	13	51	36	6	83	11
India	58	34	8	1	39	60	82	18	—	76	15	10
Korea, Rep.	47	51	2	53	47	—	11	72	17	36	64	—
Nepal	28	68	4	32	68	—	32	56	12	24	74	2
Pakistan	13	85	2	—	13	87	72	28	0	51	48	0
Philippines	18	55	27	36	64	—	19	55	25	4	53	43
Sri Lanka	62	38	—	48	52	—	14	76	10	67	33	—
Thailand	6	38	56	12	84	4	1	45	54	2	41	57
Egypt	6	26	68	—	14	86	5	86	10	5	12	83
Iraq	3	63	35	—	27	73	47	49	4	5	85	11
Lebanon	—	38	62	—	62	38	25	75	—	—	44	56
Syria	13	34	53	—	63	37	32	47	21	5	42	53
Turkey	10	57	33	3	52	46	31	64	5	7	59	34
Ethiopia	4	33	63	51	48	1	—	34	66	—	20	80
Ghana	67	33	—	35	65	—	10	67	24	66	33	1
Malawi	47	50	3	69	31	—	12	58	30	49	48	3
Nigeria	73	20	7	46	52	2	54	37	10	74	22	4
Sierra Leone	62	36	2	92	8	—	—	6	94	24	74	2
Tanzania	46	33	21	18	78	4	27	32	41	41	36	23
Zambia	70	30	—	78	22	—	59	35	7	80	20	—

High NaHCO$_3$ extractable **phosphorus** contents of soils occur most commonly in countries with traditionally high use of phosphates where the soil P reserves have been built up during past decades. These include most European countries, New Zealand, Korea and Lebanon. Soils of low P content are most typical for Pakistan, Ghana, Iraq and India.

High exchangeable (CH$_3$COONH$_4$) **potassium** contents were recorded most frequently from soils of Argentina, Egypt, Mexico and Ethiopia. Low K values again were relatively most common in Nepal, Ghana, Sierra Leone, Zambia, Nigeria, Sri Lanka and India. Unlike in the case of phosphorus, in countries with the traditionally highest K fertilizer consumption (New Zealand, Belgium, Hungary, Finland, Korea, Italy, Sri Lanka, Lebanon and Brazil) the soils are not generally higher in exchangeable K than soils in other countries. This is apparently due to the high uptake of K by plants preventing the build-up of soil K reserves.

Exchangeable (CH$_3$COONH$_4$) **calcium** and **magnesium** contents of soils are generally

Table 30. Relative frequencies (percentages) of low, medium and high values of nitrogen, phosphorus, potassium, calcium and magnesium in soils of 30 countries.

Country	N, %			P, mg/l			K, mg/l			Ca, mg/l			Mg, mg/l		
	low <.083	med. .083—.16	high. >.16	low <6	med. 6—25	high >25	low <140	med. 140—450	high >450	low <1500	med. 1500—6000	high >6000	low <190	med. 190—550	high >550
Belgium	2	93	5	—	—	100	14	86	—	59	41	—	100	—	—
Finland	—	10	90	—	7	93	22	76	1	55	44	1	68	22	10
Hungary	4	37	59	0	38	62	17	77	6	6	56	38	18	60	23
Italy	3	65	33	10	51	39	31	50	19	8	50	42	38	46	17
Malta	—	80	20	—	4	96	—	64	36	—	28	72	—	100	—
New Zealand	—	—	100	5	37	58	21	42	37	26	66	8	53	11	37
Argentina	—	29	71	1	64	35	—	2	98	7	93	—	0	99	1
Brazil	3	39	58	27	70	3	46	51	3	79	21	—	49	49	1
Ecuador	13	38	50	13	67	20	7	87	7	40	60	—	27	40	33
Mexico	34	62	4	19	64	17	4	30	66	9	53	38	5	41	54
Peru	9	43	49	16	54	29	30	63	7	7	66	27	19	69	13
India	73	27	—	46	49	5	51	47	2	25	67	8	24	60	16
Korea, Rep.	11	62	27	1	32	67	45	55	—	76	24	—	44	56	—
Nepal	22	62	16	26	48	26	92	8	—	52	48	—	48	52	—
Pakistan	57	43	0	61	36	2	25	69	5	0	98	2	11	81	8
Philippines	14	57	29	18	60	22	45	45	10	21	62	17	15	32	53
Sri Lanka	10	86	5	38	48	14	52	48	—	76	24	—	62	33	5
Thailand	1	60	39	38	55	7	33	61	6	11	43	45	18	56	26
Egypt	6	83	11	10	72	18	1	22	78	—	22	78	—	5	95
Iraq	35	60	5	50	48	2	1	83	16	—	25	75	—	42	58
Lebanon	31	69	—	6	31	63	13	63	25	—	12	88	6	63	31
Syria	26	55	18	29	50	21	3	55	42	—	21	79	3	24	74
Turkey	39	55	6	27	65	8	7	67	27	1	44	56	11	55	34
Ethiopia	13	26	61	45	44	11	3	31	65	25	54	21	7	57	36
Ghana	19	54	27	53	45	2	74	26	—	81	19	—	56	40	
Malawi	20	69	11	9	43	48	34	58	8	56	44	—	54	42	4
Nigeria	57	37	7	33	50	18	55	38	7	84	14	2	70	24	6
Sierra Leone	—	50	50	26	68	6	70	30	—	94	6	—	94	6	—
Tanzania	35	47	18	41	28	30	3	52	44	57	39	4	32	48	19
Zambia	62	36	2	9	65	26	63	33	4	91	9	—	76	24	—

low in African countries (except Ethiopia), Belgium, Finland, Brazil, Korea, Nepal and Sri Lanka and high in Near East countries and Mexico. Several very low Mg values indicating apparent deficiency were recorded particularly from Belgium, Zambia, Sierra Leone, Sri Lanka, New Zealand and Finland.

To facilitate the comparison of the relative abundance of the six micronutrients in different countries, the frequency distributions of analytical micronutrient data in five zones are summarized in Table 31. The zone limits are based on both the plant and the soil micronutrient contents (plant micronutrient content x soil micronutrient content) and adjusted so that each of the two lowest Zones (I and II) contain 5 percent, the "normal" Zone (III) 80 percent and the two highest Zones (IV and V) 5 percent each of the whole international material (see the uppermost line in the Table). The zone limits are statistically defined and therefore do not necessarily refer to critical deficiency of toxicity limits. Especially for B and Zn, a considerable number of plant-soil analytical data falling in the lower part of Zone III would still be suspect of deficiency.

Table 31. Relative distributions (percentages) of the six micronutrients in five plant x soil content zones in various countries.[1]

Country	Zone n	Boron					Copper					Iron				
		I	II	III	IV	V	I	II	III	IV	V	I	II	III	IV	V
Whole material	3538	5	5	80	5	5	5	5	80	5	5	5	5	80	5	5
Belgium	36	—	—	92	8	—	—	—	94	3	3	—	—	72	17	11
Finland	90	3	1	93	1	1	6	13	80	1	—	—	—	30	26	44
Hungary	201	0	1	87	9	2	—	—	99	1	—	—	—	97	2	1
Italy	170	—	—	92	5	3	—	—	68	10	22	1	1	86	7	5
Malta	25	—	—	72	20	8	—	—	92	8	—	60	8	32	—	—
New Zealand	35	—	—	97	3	—	17	11	49	—	23	—	—	43	26	31
Argentina	208	—	—	99	0	0	4	19	77	—	—	—	—	95	3	3
Brazil	58	—	—	100	—	—	—	—	48	12	40	—	2	91	5	2
Mexico	242	—	2	74	12	13	2	3	93	1	0	27	15	57	0	—
Peru	68	1	3	87	7	1	1	1	93	3	1	—	3	90	6	1
India	258	12	8	68	6	6	—	—	97	2	1	3	9	74	4	11
Korea, Rep.	90	3	13	79	2	2	1	7	89	2	1	2	—	92	2	3
Nepal	35	46	23	31	—	—	—	3	89	—	9	—	—	71	9	20
Pakistan	237	2	1	75	11	11	—	—	97	3	0	1	8	88	1	1
Philippines	194	30	25	45	—	—	—	1	35	27	38	—	—	64	16	19
Sri Lanka	18	—	11	89	—	—	—	—	94	6	—	—	11	83	6	—
Thailand	150	10	13	77	—	—	3	1	91	2	2	4	3	86	3	4
Egypt	198	—	—	94	4	2	—	—	67	26	7	—	1	86	11	3
Iraq	150	3	3	45	13	36	—	—	99	—	1	2	7	91	1	—
Lebanon	16	—	—	100	—	—	—	—	100	—	—	13	—	88	—	—
Syria	38	3	—	76	16	5	—	3	95	3	—	8	13	79	—	—
Turkey	298	1	1	84	6	8	—	—	94	5	1	14	9	74	1	1
Ethiopia	125	4	6	90	—	—	21	6	74	—	—	—	1	90	6	4
Ghana	93	—	—	98	2	—	23	24	54	—	—	1	—	99	—	—
Malawi	97	9	14	76	—	—	7	6	86	—	1	—	4	93	2	1
Nigeria	153	11	14	75	—	—	12	20	67	1	—	5	12	81	2	1
Sierra Leone	48	—	2	98	—	—	69	21	10	—	—	—	4	92	4	—
Tanzania	163	—	1	88	9	2	14	9	73	2	2	6	7	81	3	2
Zambia	44	2	11	86	—	—	36	16	48	—	—	16	11	70	2	—

[1] Extraction methods for soil micronutrients:

    B: hot water extraction + CEC correction
    Cu: AAAc-EDTA + org. C correction
    Fe: AAAc-EDTA
    Mn: DTPA + pH correction
    Mo: AO-OA + pH correction
    Zn: DTPA

Although **boron** deficiency can be suspected at some locations in almost every country it seems to be relatively most common in Far East countries, Nepal, Philippines, India and Thailand as well as in Nigeria and Malawi. Problems due to excess B are most likely to arise in Iraq, Mexico, Pakistan and Turkey, often at irrigated sites.

The analytical data on **copper** indicate that deficiency of this micronutrient is relatively common in all African countries studied but also many soils of New Zealand and some of Finland are low in available Cu. Elsewhere, shortage of Cu seems to be rare.

Table 31 (cont.)

Country	Zone n	Manganese					Molybdenum					Zinc				
		I	II	III	IV	V	I	II	III	IV	V	I	II	III	IV	V
Whole material	3538	5	5	80	5	5	5	5	80	5	5	5	5	80	5	5
Belgium	36	3	—	83	11	3	—	—	100	—	—	—	—	17	31	53
Finland	90	1	1	94	2	1	—	1	94	3	1	—	—	71	18	11
Hungary	201	4	4	85	3	3	1	3	93	1	1	—	—	96	2	1
Italy	170	12	9	79	—	—	—	2	85	7	6	3	4	82	5	6
Malta	25	80	16	4	—	—	—	—	80	16	4	—	—	20	36	44
New Zealand	35	—	—	91	3	6	9	29	63	—	—	—	—	60	34	6
Argentina	208	—	—	92	5	3	—	—	98	0	1	—	—	90	5	4
Brazil	58	2	—	47	19	33	45	26	29	—	—	—	—	84	7	9
Mexico	242	6	4	87	1	2	1	1	84	6	8	2	5	86	2	4
Peru	68	1	4	89	1	4	1	3	84	3	9	—	—	87	4	9
India	258	14	11	75	—	—	—	2	88	6	5	11	7	81	0	1
Korea, Rep.	90	8	6	66	7	14	4	10	81	2	2	—	—	78	9	13
Nepal	35	—	9	91	—	—	20	14	66	—	—	3	3	91	—	3
Pakistan	237	11	15	74	—	—	—	—	62	17	22	8	12	79	—	1
Philippines	194	1	1	80	12	7	3	9	85	1	3	1	—	80	11	8
Sri Lanka	18	—	—	72	11	17	—	6	94	—	—	—	—	83	11	6
Thailand	150	—	—	91	7	3	1	2	88	6	3	1	2	87	5	4
Egypt	198	8	13	79	—	—	—	—	86	13	1	—	2	95	1	2
Iraq	150	2	5	93	—	—	—	1	70	14	15	34	23	43	1	—
Lebanon	16	13	—	88	—	—	—	—	94	6	—	6	6	88	—	—
Syria	38	13	11	76	—	—	—	—	92	3	5	5	11	76	3	5
Turkey	298	2	5	92	0	—	—	2	92	2	4	17	18	64	1	1
Ethiopia	125	—	—	62	19	18	—	4	89	2	6	—	1	65	15	19
Ghana	93	—	—	80	12	9	16	22	62	—	—	—	2	95	2	1
Malawi	97	—	—	45	26	29	7	24	69	—	—	1	1	89	9	—
Nigeria	153	1	3	71	14	12	20	12	65	3	1	—	1	87	5	8
Sierra Leone	48	—	2	98	—	—	81	10	8	—	—	—	—	88	4	8
Tanzania	163	—	—	91	7	2	6	7	69	7	12	3	2	85	5	5
Zambia	44	—	—	30	16	55	50	9	41	—	—	—	2	95	—	2

Exceptionally high Cu values were frequently found in samples from the Philippines, Brazil, Italy and Tonga[1] but only occasionally elsewhere.

Although the highest percentage of low (Zone I) **iron** values were recorded from Malta, most of the lowest single plant x soil Fe values originated in Mexico and Turkey. Indications of low Fe availability were also obtained occasionally from several other countries.

Low availability of **manganese** is usually associated with alkaline soils and excess of Mn with acid soils. The most probable cases of Mn deficiency are therefore to be found in countries with soils of generally high pH. The highest percentage of low (Zone I) Mn values were recorded from Malta where all sampled soils had a pH of 7.48 or higher. Among other countries with a relatively high frequency of alkaline soils and high probability of Mn deficiency are India, Pakistan, Syria, Italy, Egypt and Lebanon. Indications of Mn deficiency were very rarely obtained from African countries, where

[1] Tonga samples are included in New Zealand material.

acid soils predominate, and indeed in many of them an excess of Mn is a more likely problem as it could be in other countries with acidic soils such as Brazil, Sri Lanka and Korea.

Unlike Mn, a deficiency of **molybdenum** seems to be most widespread in countries with acid soils such as most of the African countries, especially Sierra Leone, Zambia, Nigeria and Ghana. Low Mo values were most frequently recorded also from samples from Brazil, Nepal and New Zealand and high Mo values from samples obtained at irrigated sites in Pakistan and Iraq.

According to the present analytical data deficiency of **zinc** can be suspected somewhere in almost every country, Belgium and Malta being the most likely exceptions. It seems to be most widespread in Iraq, Turkey, India and Pakistan but in several other countries such as Syria, Lebanon, Mexico, Italy, Nepal, Tanzania and Thailand the data indicate some shortages of Zn. High plant Zn x soil Zn values were recorded chiefly in samples from Belgium, Malta, Korea and Ethiopia.

When interpreting the nutrient data by countries, attention has been drawn to certain soil characteristics and other factors (such as texture, pH, organic matter content, CEC, fertilization, varietal differences in original indicator plants and irrigation practices) affecting the nutrient contents of soils and/or plants.

It should be stressed that the limited number of soil and plant samples collected from the countries participating in this investigation impose a restriction on the application of the findings. These findings cannot be regarded as being applicable to the whole agricultural areas of a country, but rather should be considered as general guidelines to draw attention to the nature of problems likely to arise and to the future work necessary to resolve them.

References

ALMEIDA, A.M.R. & S'FREDO, G.J. Leaf curl and dwarfing of soybean associated with manganese toxicity. Fitopatol., Brasil. 4,2: 333—335.
1979

ALTEN, F., WANDROWSKY, B. & KNIPPENBERG, E. Beitrag zur Humusbestimmung. Erg. Agric. Chem. 4: 61—69.
1935

ANDERSON, O.E. & BOSWELL, F.C. Boron and manganese effects on cotton yield, lint quality, and earliness of harvest. Agron. J. 60: 488—493.
1968

ANON. Small packs of boron for Zambian cotton. Micronutrient News 1, 2: 2.
1981

BASCOMB, C.L. Rapid method for the determination of cation exchange capacity of calcareous and non-calcareous soils. J. Sci.Fd.Agric. 15: 821—823.
1964

BASSON, W.D., BÖHMER, R.G. & STANTON, D.A. An automated procedure for the determination of boron in plant tissue. Analyst 94: 1135—1141.
1969

BERGER, K.C. Micronutrient shortages, Micronutrient deficiencies in the United States. Agric. Food. Chem. 10: 178—181.
1962

BERGER, K.C. & TRUOG, E. Boron tests and determination for soils and plants. Soil Sci. 57: 25—36.
1944

BERGMANN, W. & NEUBERT, P. Pflanzendiagnose und Pflanzenanalyse. 711 p., Jena.
1976

BOULD, C. & HEWITT, E.J. Mineral nutrition of plants in soils and in culture media. Plant Physiology (ed. F.S. Steward) Vol. III: 15—133.
1963

BRANDENBURG, E. & KORONOWSKI, P. Die nichtparasitären Krankheiten, VIII. Bor.Handbuch der Pflanzenkrankheiten. Verlag Paul Parey, Berlin. 2.Teil: 132—170.
1969

BROWN, A.L., QUICK, J. & EDDINGS, J.L. A comparison of analytical methods for soil zinc. Soil.Sci.Soc. Amer.Proc. 35: 105—107.
1971

CAHOON, G.A. Handbook of reference methods for soil testing. The council of soil testing and plant analysis. 101 p. Athens, Georgia.
1974

CAWSE, P.A. Deposition of trace elements from the atmosphere in the U K. Inorganic pollution and agriculture. Ministry Agric. Fish.Food (U K), Reference Book 326: 22—46.
1980

COTTENIE, A. Soil and plant testing as a basis of fertilizer recommendation. FAO Soils Bull. 38/2: 100 p, FAO, Rome.
1980

DE KOCK, P.C. Iron nutrition of plants at high pH. Soil Sci. 79: 167—175.
1955

DELVER, P. Saline soils in the Lower Mesopotamian Plain. Rep.Iraq, Ministry Agric. Tech.Bull. 7: 54 p.
1960

DUDAL, R. Definitions of soil units for the soil map of the world. World Soil Resources Rep. 33: 72 p.
1968

EL-DUJAILI, A. & ISMAIL, H.N. Reclamation, improvement and management of samt-affected and waterlogged soils in Iraq. Salinity seminar, Baghdad Dec. 5—14, 1970: 157—193. FAO, Rome.
1971

ELONEN, P. Particle size analysis of soil. Acta Agr. Fenn. 122: 1—122.
1971

ELSEEWI, A.A. & EMALKY, A.E. Boron distribution in soils and waters of Egypt. Soil Sci.Soc. Amer. J. 43: 297—300.
1979

FAO. Fertilizer demonstration and distribution programme, Nigeria. Project findings and recommendations, Vol. 2, Niger State: 104 p. FAO, Rome.
1979

FAO. 1979 FAO fertilizer yearbook 29: 143 p. FAO, Rome.
1980a

FAO. Fertilizer demonstration and pilot scheme distribution 1978—79. The Philippines. AG: DP/PHI/ 77/009, Terminal Report: 129 p. FAO, Rome.
1980b

FAO/UNESCO. Soil map of the world. Southeast Asia. Sheet IX. Edition I/1976.
1976

FIREMAN, M. & KRAUS, Y. Salinity control in irrigated agriculture, Tahal, Israel. 46 p.
1965

FRANCK, E.v. & FINCK. A. Evaluation of critical zinc contents of oats and wheat. Z.Pfl.ernähr. Bodenk. 143(1): 38—46.
1980

GERICKE, S. & KURMIES, B. Die kolorimetrische Phosphorsäurebestimmung mit Ammonium Vanadat-Molybdat und ihre Anwendung in der Pflanzenanalyse. Z. Pfl.ernähr.Düng. Bodenk. 59: 235—247.
1952

GERTSCH, M.E. Fertilizer demand in Ethiopia. FAO Joint. Proj. ETH-054-A(SIS) Subrep. Ethiopian Gov.Tech. Agency, Addis Ababa. 111 p.
1972

GRANICK, S. Iron metabolism in animals and plants. Trace Elements (ed. C.A. Lamb et al). Academic Press, New York.
1958

GRAVEN, E.H., ATTOE, O.J. & SMITH, D. Effect of liming and flooding on manganese toxicity in alfalfa. Soil Sci. Soc.Amer.Proc. 29: 702—706.
1965

GUPTA, U.C. Boron nutrition of crops. Adv.Agron. 31: 273—307.
1979

GUPTA, U.C. & MUNRO, D.C. The boron content of tissues and roots of rutabagas and of soil associated with brown heart condition. Soil Sci. Soc.Amer.Proc. 33: 424—426.
1969

HAQUE, I. & BUNDU, H.S. Effects of inoculation, N, Mo and mulch on soyabean in Sierra Leone. Commun. in Soil Sci. & Plant Anal. 11 (5): 477—483.
1980

HAQUE, I. & GODFREY-SAM-AGGREY, W. Nutritional survey of coffee and cacoa growes in Sierra Leone. Commun. in Soil Sci. & Plant Anal. 11(5): 485—505.
1980

HONG, C.W. The fertility status of Korean soils. ASPAC Food & Fert.Technol.Center, Techn. Bull. 10: 56—87.
1972

JACKSON, M.L. Chemical composition of soils. Chemistry of the Soil (Ed. F.E. Bear). Reinhold, New York. 71:141.
1964

JOHN, M.K., CHUAH, H.H. & NEUFELD, J.H. Application of improved azomethine-H method to determination of boron in soils and plants. Anal. Lett. 8: 559—568.
1975

KANWAR, J.S. & RANDHAWA, N.S. Micronutrient Research in Soils and Plants in India. A review. Indian Council of Agric. Res. Tech. Bull. (Agric.) No. 50.
1974

KORONOWSKI, P. Die Nichtparasitären Krankheiten,
1969 Handbuch der Pflanzenkrankheiten, 2. Teil; X. Mangan: 211—246.
KURKI, M. Suomen peltojen viljavuuden kehityksestä.
1979 41 p. Viljavuuspalvelu Oy, Helsinki.
LAKANEN, E. & ERVIÖ, R. A comparison of eight
1971 extractants for the determination of plant available micronutrients in soils. Acta Agr.Fenn. 123: 223—232.
LINDSAY, W.L. & NORVELL, W.A. Development of a
1969 DTPA micronutrient soil test. Agron.Abstr. p. 84.
LINDSAY, W.L. & NORVELL, W.A. Development of a
1978 DTPA soil test for zinc, iron, manganese, and copper. Soil Sci Soc.Amer.J. 42: 421—428.
LOIZIDES, P. Report of UNSF FAO Damascus Agri-
1967 cultural Research Station Project, Iron deficiency and its control in the Syrian Arab Republic, 7. p. FAO, Rome.
MEHLICH, A. Uniformity of expressing soil test results.
1972 A case for calculating results on a volume basis. Comm.Soil Sci.Plant Analysis. 3: 417—424.
MESTANZA, S.S. & LAINEZ, C. The correction of boron
1970 deficiency in cacao in Ecuador. Trop. Aric. 47: 57—61.
MITCHELL, R.L. Trace element problems on Scottish
1974 soils. Neth.J.agric.Sci. 22: 295—304.
MOKADY, R. The effect of partial pressure of oxygen in
1961 field soils on lime induced chlorosis. Plant and Soil 15: 377—386.
MORTVEDT, J.J. & OSBORN, G. Boron concentration
1965 adjacent to fertilizer granules in soil and its effect on root growth. Soil Sci.Soc.Amer.Proc. 29: 187—191.
OLSEN, S.R., COLE, C.V., WATANABE, F.S. & DEAN,
1954 L.A. Estimation of available phosphorus in soils by extraction with sodium bicarbonate. U.S.Dept. Agric., Circ. 939.
OSINAME, O.A. Research aspects in soil fertility and
1976 fertilizer use in Western State. Rep. FAO/ NORAD and Fed. Dept. Agric. Seminar on fertilizer development in Nigeria: 24—29. FAO, Rome.
PANIN, M.S. & PREDTEČENSKIJ, S.A. Zur Frage der
1971 Cu-Dynamik im System Boden-Sommerweizen. Chimija v sel'skom Choz. 9,8: 20—23.
PARK, C.S. & PARK, N.J. Studies on the available boron
1966 content in Korean upland soils. The Research Rept. Office of Rural Development. 9: 163—174.
PEECH, M. Hydrogen-ion activity in Methods of Soil
1965 Analysis Part 2; C.A. Black, ed. pp. 914—926.
RANDHAWA, N.S. & TAKKAR, P.N. Micronutrient re-
1975 search in India. The present status and future projections. Repr.Fert. News May 1975: 9 p.
RATHORE, G.S., GUPTA, G.P., KHAMPARIA, R.S. &
1978 SINHA, S.B. Response of wheat to zinc application in alluvial soils of Morena District, Madhya Pradesh. J.Ind.Soc.Soil Sci. 26: 58—62.
REISENAUER, H.M., WALSH, L.M. & HOEFT, R.G.
1973 Testing soils for sulphur, boron, molybdenum and chlorine, p. 173—200. L.M. Walsh and J.D. Beaton: Soil Testing and Plant Analysis. Soil Sci.Soc.Amer. Inc., Madison.

RINNE, S.-L., SILLANPÄÄ, M., HUOKUNA E. & HII-
1974a VOLA, S.-L. Effects of heavy nitrogen fertilization on potassium, calcium, magnesium and phosphorus contents in ley grasses. Ann. Agric. Fenn. 13: 96—108.
RINNE, S.-L., SILLANPÄÄ, M., HUOKUNA E. & HII-
1974b VOLA, S.-L. Effects of heavy nitrogen fertilization on iron, manganese, sodium, zinc, copper, strontium, molybdenum and cogalt contents in ley grasses. Ann. Agric. Fenn. 13: 109—118.
SAKAL, R., SINHA, H., SINGH, A.P. & THAKUR, K.N.
1979 A critical limit for the response of rice and wheat to applied zinc in Tarai soils. J.Agric.Sci. 93: 419—422.
SEDBERRY, J.E. Jr., PETERSON, F.J., WILSON, F.E.,
1980 MENGEL, D.B., SCHILLING, P.E. & BRUP- BACHER, R.H. Influence of soil reaction and applications of zinc on yields and zinc contents of rice plants. Commun. Soil. Sci. Pl. Anal. 11 (3): 283—295.
SILLANPÄÄ, M. Trace elements in Finnish soils as
1962a related to soil texture and organic matter content. J.Sci.Agric. Soc. Finland. 34: 34—40.
SILLANPÄÄ, M. On the effect of some soil factors on the
1962b solubility of trace elements. Agrogeol. publ. 81: 1—24.
SILLANPÄÄ, M. Trace elements in soils and agriculture.
1972a Soils Bull. 17: 67 p., FAO, Rome.
SILLANPÄÄ, M. Distribution of trace elements in peat
1972b profiles. Proc. 4th Intern. Peat Congr. Otaniemi. Vol 5, p. 185—191.
SILLANPÄÄ, M. Unbalanced use of fertilizers as a
1974 possible cause of soil degradation. Rep. Expert Consultation on Soil Degradation. FAO, Rome 10—14 June 1974: 12—13.
SILLANPÄÄ, M. Preliminary results of the world wide
1981 microelement study. Mineral elements' 80. A nordic symposium on soil-plant-animal-man inter- relationship. Proc.Part. I: 63—73. Kemira, Hel- sinki.
SINGH, C.P., PRASAD, R.N., SINHA, H. & PRASAD, B.
1977 Evaluation of the critical limit and extractants for the determination of available zinc in cal- careous soils. Beiträge zur Tropischen Landwirt- schaft u. Veterinärmedizin. 15: 131—136.
SIPPOLA, J. & ERVIÖ, R. Determination of boron in
1977 soils and plants by the azomethine-H method. Finn.Chem.Lett. 1977: 138—140.
SIPPOLA, J. & TARES, T. The soluble content of mineral
1978 elements in cultivated Finnish soils. Acta Agric. Scand., Suppl. 20: 11—25.
SIPPOLA, J., YLÄRANTA, T. & JANSSON, H. Macro-
1978 nutrient contents of wheat during the growing season. Ann.Agric.Fenn. 17: 158—162.
SMILDE, K.W. Minor elements in the nutrition of cereals.
1976 Semaine d'étude céréaliculture, Gembloux: 303— 312.
SPRAQUE, H.B. Why do plants starve? Hunger signs in
1964 crops. (ed. H.B. Sprague) p. 1—24.
STANTON, R.E. & HARDWICK, A.J. The colorimetric
1967 determination of molybdenum in soils and sedi- ments by zinc dithiol. Analyst 92: 387—390.

TÄHTINEN, H. Copper content of the soil and the
1971 effect of copper fertilization. Acta Agr.Fenn.
123: 136—142.
TAINIO, A. Maan kuparipitoisuus ja kuparilannoituksen
1960 vaikutus. Koetoim. ja käyt. 17: 3.
TAKKAR, P.N., MANN, M.S. & RANDHAWA, N.S.
1974 A note on relationship of zinc contents in wheat
plants exhibiting different degree of zinc deficiency
symptoms with wheat yield. Indian J.Agric.Sci.
(ref. Randhawa, N.S., Takkar, P.N. 1975 Fert.
News, May 1975).
TAKKAR, P.N. & MANN, M.S. Toxic levels of soil and
1978 plant zinc for maize and wheat. Plant and Soil
49: 667—669.
TAMM, O. Eine Methode zur Bestimmung der an-
1922 organischen Komponenten des Gelkomplexes im
Boden. Medd.Stat. Skogsförsöksanst. 19: 387—
404.
TARES, T. & SIPPOLA, J. Changes in pH, in electrical
1978 conductivity and in the extractable amounts of
mineral elements in soil, and the utilization and
losses of the elements in some field experiments.
Acta Agric.Scand., Suppl. 20: 90—113.
TÖLGYESI, G. & MIKÓ, Z. T. (Correlations between
1977 yield and mineral uptake in maize). Eng. summary.
Növénytermelés 26, 2/3: 169—175.
TOLLENAR, D. Boron deficiency in cacao, bananas, and
1966 other crops on volcanic soils of Ecuador. Neth.
J.Agric.Sci. 14: 138—151.
TRIERWEILER, J.F. & LINDSAY, W.L. EDTA-
1969 ammonium carbonate soil test for zinc. Soil Sci.
Soc.Amer.Proc. 33: 49—54.

VIETS, F.G.Jr. & LINDSAY, W.L. Testing soils for zinc,
1973 copper, manganese, and iron. p. 153—172. L.M.
Walsh and J.D. Beaton: Soil testing and plant
analysis. Soil Sci.Soc.Amer.Inc., Madison.
WADSWORTH, G.A. & WEBBER, J. Deposition of
1980 minerals and trace elements in rainfall. Inorganic
pollution and agriculture. Ministry Agric.Fish.
Food (U K), Reference Book 326: 47—55.
WATANABE, F.S. & OLSEN, S.R. Test of an ascorbic
1965 acid method for determining phosphorus in
water and $NaHCO_3$ extracts from soil. Soil Sci.
Soc.Amer.Proc. 29: 677—678.
WEIR, R.G. & MILHAM, P.J. Use of plant analysis to
1978 asses zinc status of maize seedling. N.Z. Dept.Sci.
Ind.Res. 134: 547—552.
WHITNEY, D.A. Micronutrient soil tests — zinc, iron,
1980 manganese, and copper. Recommended chemical
soil test procedures for the North Central Region.
Agricultural Exp.Sta. & U.S.Dept.Agric.Bull. 499:
18—21.
WHITNEY, D.A., ELLIS, R. Jr., MURPHY, L.S. &
1973 HERRON, G. Identifying and correcting zinc and
iron deficiency in field crops. Kansas Coop. Ext.
Ser. Leaflet L-360, Manhattan, KS.
YLÄRANTA, T., JANSSON, H. & SIPPOLA, J. Seasonal
1979 variation in micronutrient contents of wheat. Ann.
Agric.Fenn. 18: 218—224.
YLÄRANTA, T. & SILLANPÄÄ, M. (Unpublished data
on micronutrient contents of various plant species).
Institute of Soil Science, Jokioinen, Finland.

APPENDIX 1. **Instructions (condensed) for taking plant and soil samples.**

Selection of sampling sites:
— The main wheat and maize growing areas of the country should be included in such a way that samples from the major soil types will be represented.
— Areas where deficiencies of one or more trace elements are suspected or known to exist should be included in proportion to the total area to be sampled.
— To minimize the risk of contamination, sampling sites close to roads or other dusty places must not be chosen.
— The locations of sampling sites must be indicated on a suitable large scale working map in such a way that the exact sites can later be relocated easily. From these large scale working maps the sampling sites and numbers should be transferred to small scale maps covering the whole country.

Plant sampling:
— To avoid contamination by soil, plant samples must always be taken with clean hands before taking soil samples.
— Size of plant sample: not less than 350 g fresh material collected from various parts of the sampling plot of about 10 × 10 m.
— Timing: Wheat samples should be taken at *mid-tillering stage* of the plant growth (i.e. for spring wheat about 35—40 days after planting depending on weather conditions, for winter wheat the corresponding stage of growth). Maize samples at *5-6 leaf stage* (i.e. approximately 35—40 days after planting).
— Plant part to be sampled: Wheat — the upper half of the plant (use scissors). Maize — two uppermost fully expanded leaf blades of each plant.
— Equipment: Clean hands, clean scissors, clean knife (rusted equipment should not be used).
— Bags: Paper bags, which will be made available, should be used when collecting the plant samples in the field and during transportation from field to a laboratory (or to any clean dust-free room) for drying. Every effort should be made to avoid contamination.
— Drying: All plant samples must be air-dried in a dust-free room, or even oven-dried at 60—65 °C.
— Milling: No milling should be done.

Soil sampling:
— Equipment: Stainless steel soil bores (which will be made available) only should be used to avoid trace element contamination due to equipment.
— Sampling depth: Full plough layer (appr. 0—20 cm).
— Size of the soil sample: 0.6 dm^3, i.e. full pre-numbered carton box composed of a minimum of 6—10 subsamples collected from various parts of the plant-soil sampling plot of about 10 × 10 m.
— Drying: Soil samples must be air-dried in a dust-free area.
— Contamination: Every effort should be made to avoid contamination at all stages.

APPENDIX 2. National mean values of basic analytical data on **original wheat** plants and respective soils. For analytical methods, see Section 1.3. Only basic (uncorrected) results for soil micronutrients are given: hot w.sol. B, AAAc-EDTA extr. Cu and Fe, DTPA extr. Mn and Zn, AO-OA extr. Mo.

Country		GENERAL SOIL PROPERTIES										SOIL MACRONUTRIENTS				
		Particle size distr.			Tex. ind.	CEC	pH		El. cond.	$CaCO_3$ equiv.	Org. C	Volume weight	N (total)	P	K	Ca	M
		<.002	.002—.06	.06—2	>2 mm			H_2O	$CaCl_2$							(extractable)	
		%	%	%	%		me/100 g			10^{-4}S/cm	%	%	g/cm³	%	mg/l	mg/l	mg/l	mg/l	mg/l
EUROPE AND OCEANIA																	
BELGIUM	mean	12	72	15	0	36	16.4	6.55	6.08	1.2	.4	.9	1.23	.103	1272	83.7	213	1868	92
(n = 21)	± s	4	11	13	0	4	2.3	.60	.66	.3	1.5	.2	.03	.013	150	23.3	58	983	29
	min.	8	44	3	0	29	12.8	5.70	5.00	.7	.0	.6	1.19	.081	983	45.6	122	805	46
	max.	19	84	45	0	43	21.1	7.90	7.40	1.9	7.1	1.2	1.28	.129	1552	121.6	369	5045	146
FINLAND	mean	28	48	25	0	45	31.8	5.73	5.21	1.6	.0	3.9	.99	.288	2711	61.0	205	1521	207
(n = 94)	± s	21	19	21	0	18	13.9	.52	.52	.7	.0	2.8	.13	.160	1058	32.5	100	969	213
	min.	1	11	2	0	16	12.5	4.60	4.10	.6	.0	1.4	.66	.106	1181	14.2	40	190	10
	max.	81	89	74	0	86	89.7	7.30	6.90	3.9	.4	17.5	1.22	1.023	7169	208.0	632	8800	1075
HUNGARY	mean	29	52	19	0	47	29.8	7.25	6.83	2.2	4.1	1.6	1.20	.183	2176	32.7	217	4910	437
(n = 144)	± s	12	15	20	0	11	9.9	.79	.83	1.1	6.6	.5	.07	.054	592	20.3	122	1893	310
	min.	4	1	1	0	15	6.9	4.80	4.25	.5	.0	.5	1.06	.052	753	5.6	46	520	30
	max.	69	76	94	0	78	57.1	8.35	7.90	8.3	45.0	3.6	1.45	.331	3511	130.0	843	8850	1826
ITALY	mean	36	37	25	3	49	27.6	7.62	7.22	6.0	11.4	1.3	1.19	.142	1684	19.6	334	6598	341
(n = 118)	± s	19	13	18	11	17	10.9	.58	.57	9.0	11.5	.7	.10	.057	643	17.8	279	4229	306
	min.	9	6	1	0	19	9.4	5.61	4.96	.6	.0	.5	.96	.065	798	3.2	40	820	45
	max.	85	79	93	58	89	74.9	8.28	7.82	57.0	57.4	5.7	1.75	.422	4696	116.4	1720	21930	1889
MALTA	mean	28	39	33	0	43	22.2	7.91	7.54	5.9	58.6	1.3	1.17	.145	1711	80.4	365	6518	271
(n = 25)	± s	10	7	10	0	9	4.4	.10	.05	2.4	5.7	.3	.02	.034	414	51.3	226	744	47
	min.	18	25	20	0	32	17.9	7.78	7.48	3.2	52.0	.8	1.13	.115	1364	20.7	186	5640	207
	max.	51	53	51	0	61	29.8	8.08	7.64	9.1	68.4	2.1	1.20	.212	2536	171.8	789	8090	335
NEW ZEALAND	mean	30	58	13	0	48	35.1	5.87	5.33	1.3	.0	3.2	1.04	.340	3435	24.7	210	1916	169
(n = 14)	± s	15	14	11	0	12	14.9	.25	.26	.5	.0	1.3	.08	.139	1072	7.8	165	753	176
	min.	16	30	1	0	33	20.5	5.49	5.02	.6	.0	2.0	.87	.221	2443	15.2	68	1170	48
	max.	69	78	43	0	78	66.9	6.41	5.90	2.2	.0	6.2	1.14	.701	6096	41.6	699	3610	638
LATIN AMERICA																			
ARGENTINA	mean	22	59	19	0	42	27.2	6.11	5.49	1.4	.0	1.9	1.04	.199	2067	24.4	784	1993	296
(n = 119)	± s	4	12	13	0	5	5.3	.31	.34	2.2	.1	.7	.07	.066	713	22.8	198	632	72
	min.	11	24	2	0	25	17.0	5.42	4.85	.5	.0	.8	.89	.111	1175	5.4	420	1020	175
	max.	31	73	64	0	50	45.3	7.41	6.96	24.4	.7	4.1	1.20	.425	4309	200.0	1649	5480	584
BRAZIL	mean	59	21	20	0	67	27.1	5.66	5.10	.9	.0	2.0	1.12	.184	2028	10.9	170	1004	227
(n = 71)	± s	19	10	20	0	17	8.5	.50	.57	.3	.0	.8	.10	.057	578	9.6	111	782	137
	min.	7	3	2	0	17	6.4	4.55	3.97	.4	.0	.6	.92	.051	742	2.4	37	180	40
	max.	83	50	90	0	88	46.8	6.77	6.30	2.0	.0	4.7	1.46	.325	3778	74.2	632	5730	1031
ECUADOR	mean						21.5	6.76	6.22	10.7	.1	1.4	1.20	.209	2349	17.7	280	1884	421
(n = 16)	± s						9.6	.73	.82	6.6	.2	.6	.11	.162	1905	16.3	93	1082	287
	min.						9.7	5.33	4.66	3.1	.0	.6	.98	.049	664	2.4	134	590	78
	max.						42.1	8.00	7.65	27.2	.6	2.3	1.45	.747	8844	67.6	472	4110	1038

	SOIL MICRONUTRIENTS						NUTRIENT CONTENTS OF ORIGINAL WHEAT PLANTS									
	B	Cu (extractable)	Fe	Mn	Mo	Zn	N	P	K	Ca	Mg	B	Cu	Fe	Mn	Mo	Zn
	mg/l	mg/l	mg/l	mg/l	mg/l	mg/l	%	%	%	%	%	ppm	ppm	ppm	ppm	ppm	ppm
	.54	5.1	262	48.7	.253	5.17	3.95	.501	3.95	.358	0.088	4.18	7.3		40.7	.30	29.7
	.18	2.7	76	22.2	.087	2.10	.61	.102	.55	.076	.018	1.62	1.1		25.3	.20	10.5
	.26	2.4	130	18.4	.089	2.79	2.68	.330	2.94	.239	.057	2.43	4.7		14.6	.09	22.4
	1.09	13.1	425	94.4	.489	10.53	5.04	.668	4.76	.493	.132	9.43	9.0		124.2	.83	72.8
	.56	4.3	563	21.4	.540	2.77	5.03	.429	4.33	.399	.142	8.98	7.1		69.9	.37	27.0
	.29	2.6	361	14.3	.357	2.64	.76	.093	.59	.094	.038	5.86	1.7		49.5	.33	5.6
	.14	.9	130	1.0	.086	.34	2.64	.230	3.26	.224	.076	3.48	4.3		19.5	.01	17.4
	1.88	16.5	1995	63.2	2.071	13.13	7.17	.752	6.70	.652	.285	37.30	13.1		423.1	1.88	43.3
	.98	5.5	148	34.1	.133	1.18	5.46	.490	4.15	.529	.177	4.28	8.5		75.8	.29	25.9
	.46	2.5	83	27.9	.100	1.09	.78	.127	.86	.107	.048	1.46	1.8		31.3	.25	7.0
	.11	1.6	61	8.9	.030	.34	2.72	.256	1.50	.304	.105	2.27	5.1		26.7	.03	14.2
	2.68	14.6	605	150.0	.634	8.75	7.45	.829	6.55	.769	.399	12.87	13.6		175.9	1.87	48.4
	.87	11.3	197	20.8	.213	2.31	4.55	.424	3.55	.505	.140	4.47	8.6		54.3	.74	30.6
	.60	13.8	119	24.3	.215	6.18	1.03	.114	.75	.214	.081	1.87	2.7		40.4	.69	15.6
	.24	1.4	30	4.0	.007	.17	1.99	.175	2.05	.182	.065	2.41	4.0		13.0	.04	10.7
	4.17	87.5	690	222.3	1.311	47.57	7.22	.781	5.30	1.302	.876	15.36	20.1		268.0	5.12	128.3
	1.14	7.2	35	7.5	.112	10.29	3.07	.533	4.21	.491	.102	3.40	7.4		37.0	.89	57.9
	.53	2.5	11	2.0	.057	12.51	.75	.148	.84	.132	.013	.70	1.5		6.7	.36	11.9
	.63	5.0	22	4.9	.021	2.02	1.63	.422	2.87	.308	.075	2.45	4.5		26.0	.40	38.8
	2.20	12.9	52	11.3	.217	39.40	4.70	.800	5.93	.836	.126	4.75	9.5		51.2	1.70	81.1
	.67	2.3	290	16.6	.148	1.45	4.68	.392	3.48	.384	.153	4.56	8.4		54.7	.15	26.5
	.23	1.6	233	12.0	.104	1.66	.59	.049	.68	.128	.038	1.08	1.8		30.9	.09	8.1
	.38	.6	112	6.3	.035	.28	3.78	.321	2.47	.262	.109	2.29	4.2		20.9	.03	16.8
	1.20	6.7	960	51.2	.419	5.18	5.51	.484	5.12	.672	.249	6.04	11.4		141.4	.35	42.7
	.72	3.2	207	39.7	.387	2.03	4.71	.419	4.74	.385	.158	3.98	7.6		64.3	.71	27.1
	.22	.7	73	11.5	.222	1.66	.88	.134	.69	.084	.032	1.58	1.9		27.4	.67	6.5
	.42	1.7	105	13.1	.132	.41	2.78	.168	3.21	.208	.096	1.83	2.7		18.2	.69	15.5
	1.82	5.4	460	75.2	1.949	12.04	6.42	.800	6.42	.591	.244	10.96	12.5		137.5	5.37	49.4
	.64	17.5	166	108.1	.199	3.81	3.96	.418	4.42	.338	.223	6.15	14.5		107.3	.24	34.9
	.19	16.0	111	85.3	.072	14.88	.91	.117	1.20	.111	.056	1.96	11.0		43.5	.19	12.0
	.26	1.4	42	2.4	.049	.20	2.10	.227	1.49	.158	.140	2.88	6.9		37.4	.02	17.6
	1.21	81.1	700	286.0	.458	123.90	6.04	.752	6.83	.661	.449	12.66	99.9		231.3	.95	77.7
	.42			24.1	.249	1.41	3.47	.432	4.31	.247	.145	2.33	7.7		34.3	.40	23.8
	.20			20.8	.232	.82	1.07	.058	.84	.053	.050	.62	1.7		18.8	.40	6.1
	.16			1.2	.048	.24	2.21	.326	3.35	.170	.096	1.45	5.1		15.0	.13	12.5
	.95			76.7	.870	2.90	6.43	.547	5.95	.321	.267	3.90	10.8		91.8	1.50	34.9

APPENDIX 2 (cont.)

Country		Particle size distr.				Tex. ind.	CEC	pH		El. cond.	CaCO₃ equiv.	Org. C	Volume weight	N (total)	P	K	Ca (extractable)	
		<.002 %	.002–.06 %	.06–2 %	>2 mm %	%	me/100 g	H₂O	CaCl₂	10^{-4}S/cm	%	%	g/cm³	%	mg/l	mg/l	mg/l	mg/l	mg
MEXICO (n = 100)	mean	36	40	25	0	50	30.7	8.11	7.64	5.1	10.1	.8	1.19	.100	1183	14.7	693	7023	7
	± s	17	15	19	0	14	12.2	.32	.28	6.1	15.3	.3	.06	.029	325	28.5	318	3020	3
	min.	9	7	2	0	21	11.0	6.77	6.48	.7	.1	.3	1.02	.044	563	2.4	168	1970	1
	max.	77	78	82	3	83	80.3	8.85	8.09	42.8	59.8	1.8	1.44	.209	2358	271.4	2019	19980	15
PERU (n = 13)	mean	25	32	29	14	38	25.3	6.99	6.58	1.8	4.6	1.2	1.23	.144	1786	10.6	178	4067	3
	± s	10	14	11	18	12	6.9	.94	.92	.7	6.0	.5	.15	.056	809	8.2	65	2016	2
	min.	14	13	6	0	22	15.3	5.33	4.98	.8	.0	.5	1.01	.084	893	2.8	78	1220	1
	max.	46	57	52	44	61	35.0	8.00	7.57	3.0	16.7	2.2	1.48	.285	3959	27.4	273	6490	10
FAR EAST																			
INDIA (n = 188)	mean	18	47	35	0	36	18.4	8.06	7.56	2.9	2.8	.5	1.26	.071	881	9.3	186	3530	4
	± s	15	17	23	1	14	16.2	.52	.50	2.7	7.3	.2	.08	.021	232	8.1	119	2795	3
	min.	2	9	1	0	16	5.1	6.02	5.14	.3	.0	.2	.95	.009	95	1.2	42	380	
	max.	62	91	86	8	73	76.5	9.10	8.32	23.1	40.4	1.3	1.50	.142	1722	54.0	856	12040	329
KOREA (n = 50)	mean	20	47	33	0	37	16.9	6.10	5.61	1.7	.2	1.2	1.15	.118	1327	45.4	131	1068	1
	± s	10	11	19	0	11	4.2	.70	.73	.8	.2	.4	.07	.041	398	28.2	48	312	
	min.	5	18	5	0	18	8.2	5.00	4.20	.5	.0	.8	.98	.035	451	6.7	45	400	
	max.	41	68	77	0	56	34.3	7.30	6.85	4.1	.9	2.6	1.32	.270	2641	116.3	256	2220	35
NEPAL (n = 50)	mean	18	54	27	0	37	19.7	6.75	6.16	1.1	1.0	1.0	1.14	.119	1332	21.6	80	1766	20
	± s	8	14	17	1	9	5.7	.84	.91	.5	3.8	.4	.11	.042	390	26.6	48	1173	1
	min.	5	22	3	0	23	9.5	5.19	4.50	.3	.0	.4	.84	.061	770	1.2	23	220	3
	max.	40	76	67	10	57	36.5	8.20	7.59	2.5	26.3	2.0	1.36	.222	2191	121.4	323	4660	45
PAKISTAN (n = 156)	mean	20	60	20	0	40	17.3	8.38	7.83	5.0	6.9	.7	1.23	.082	.993	7.1	245	4516	36
	± s	9	13	16	0	8	4.6	.27	.21	7.7	4.0	.2	.08	.026	275	6.9	127	770	15
	min.	4	10	1	0	19	7.6	7.89	7.40	1.0	.7	.2	1.04	.033	465	1.4	55	1200	7
	max.	50	79	82	0	65	32.4	9.25	8.56	57.5	18.1	1.4	1.43	.169	1785	51.6	945	7240	92
NEAR EAST																			
EGYPT (n = 100)	mean	47	37	15	0	60	44.5	8.01	7.68	7.8	4.4	1.2	1.19	.132	1559	18.7	722	7463	158
	± s	15	12	21	0	14	10.3	.19	.13	5.5	1.8	.3	.06	.030	316	13.8	308	1319	52
	min.	5	1	2	0	15	10.2	7.57	7.38	1.6	1.3	.4	1.08	.061	825	4.2	228	1640	39
	max.	78	60	94	0	84	58.5	8.43	8.05	26.5	16.3	1.8	1.51	.221	2537	85.8	2097	9920	294
IRAQ (n = 119)	mean	35	51	13	0	52	26.9	8.04	7.69	8.3	25.2	.8	1.17	.097	1130	8.3	341	7315	67
	± s	12	11	13	3	10	7.4	.15	.15	10.1	6.4	.3	.07	.035	400	10.2	141	1608	33
	min.	7	9	1	0	18	11.0	7.68	7.22	1.2	2.1	.3	1.04	.031	354	1.2	130	4400	19
	max.	63	79	84	28	74	43.8	8.44	8.00	53.5	43.5	1.8	1.38	.239	2993	69.7	1028	15431	155
SYRIA (n = 20)	mean	45	35	18	3	57	37.8	7.94	7.53	3.8	27.5	1.1	1.23	.121	1466	19.5	465	9159	91
	± s	19	15	20	6	17	13.6	.21	.13	4.2	15.7	.8	.07	.078	896	24.6	226	3391	58
	min.	7	9	2	0	18	9.3	7.61	7.36	1.6	6.8	.3	1.14	.040	558	2.0	166	4900	10
	max.	74	58	84	20	80	62.9	8.51	7.84	18.2	59.2	3.9	1.39	.379	4442	111.9	965	17700	262

	SOIL MICRONUTRIENTS						NUTRIENT CONTENTS OF ORIGINAL WHEAT PLANTS										
Na	B	Cu (extractable)	Fe	Mn	Mo	Zn	N	P	K	Ca	Mg	B	Cu	Fe	Mn	Mo	Zn
mg/l	mg/l	mg/l	mg/l	mg/l	mg/l	mg/l	%	%	%	%	%	ppm	ppm	ppm	ppm	ppm	ppm
326	1.16	3.9	49	12.9	.200	.98	4.88	.378	4.77	.402	.185	10.81	11.1		100.8	1.85	37.8
352	.72	1.5	31	6.4	.168	1.75	.87	.120	.67	.131	.037	10.47	2.1		38.3	1.19	12.4
4	.19	.9	10	4.8	.028	.22	2.28	.182	2.91	.231	.118	2.74	5.4		48.7	.11	14.0
2358	3.45	8.7	170	44.5	1.080	14.83	6.46	1.020	6.30	.912	.307	83.53	17.4		223.5	6.18	76.9
23	.54	4.2	134	27.5	.224	3.66	2.94	.302	2.89	.332	.131	4.37	9.2		48.8	.88	37.3
17	.29	1.7	59	13.9	.136	8.07	.62	.095	.77	.124	.045	2.14	3.6		17.0	1.16	19.6
7	.22	1.9	65	9.1	.053	.42	1.84	.167	1.38	.157	.085	1.59	3.9		26.7	.11	15.4
70	1.20	7.9	239	52.8	.481	30.21	4.24	.486	4.34	.490	.194	8.42	18.4		93.0	4.22	76.1
175	.50	4.3	145	13.4	.122	1.16	3.31	.323	3.83	.356	.187	4.83	8.8		53.9	1.11	23.3
197	.40	2.4	170	6.1	.107	3.04	1.51	.131	1.30	.159	.072	4.50	3.7		29.1	.98	9.2
1	.07	.6	28	3.8	.014	.10	.60	.048	.94	.113	.052	1.68	2.5		8.6	.04	3.1
1295	2.10	18.9	885	48.7	.936	38.83	6.14	.608	6.22	.926	.407	42.78	18.0		160.1	6.31	55.7
32	.39	4.5	217	41.6	.295	10.97	5.27	.430	3.98	.386	.184	4.15	6.9		85.7	.41	33.0
39	.31	4.1	192	36.3	.518	34.33	1.07	.129	.72	.092	.048	3.16	2.3		45.2	.29	11.0
5	.08	.6	35	3.5	.066	.47	2.99	.203	2.20	.244	.108	1.40	3.4		25.5	.07	18.1
198	1.92	24.6	905	149.5	3.241	186.45	7.09	.777	5.77	.591	.314	18.55	16.4		244.4	1.36	74.1
27	.19	5.7	334	21.5	.084	.75	4.19	.400	3.70	.477	.196	2.99	10.7		69.1	.43	33.8
24	.07	7.2	237	15.8	.039	1.08	1.07	.111	.89	.134	.059	.97	2.8		29.3	.44	9.6
0	.10	1.0	48	3.7	.023	.11	2.24	.217	2.23	.230	.109	1.75	5.7		21.9	.03	18.6
100	.43	40.4	1125	74.1	.204	6.72	6.34	.756	6.48	.756	.358	8.28	17.8		140.5	2.08	60.9
349	.69	4.8	127	8.5	.198	.82	4.56	.372	4.81	.393	.189	7.60	12.6		78.0	2.24	27.3
554	.68	2.3	64	2.3	.164	1.47	.82	.100	.90	.134	.044	6.18	10.2		33.4	1.17	8.2
5	.08	1.3	28	3.6	.022	.10	2.19	.120	2.17	.134	.106	1.74	5.9		25.7	.35	11.8
2765	4.42	22.7	371	14.3	1.049	12.10	6.25	.719	6.84	.780	.358	33.78	100.6		228.2	7.99	61.0
746	.87	12.1	195	13.4	.144	1.31	4.31	.356	4.04	.355	.184	5.93	11.1		56.8	.94	31.1
686	.37	3.9	89	4.5	.045	.92	.58	.071	.64	.108	.031	3.39	2.0		20.8	.33	7.9
98	.30	3.3	49	7.3	.042	.46	2.81	.191	2.98	.184	.130	2.99	7.0		27.4	.41	17.0
3398	2.74	26.4	458	39.9	.304	7.50	5.76	.584	5.80	.839	.280	32.40	17.5		164.1	2.16	61.5
329	1.36	4.8	110	9.1	.123	.28	3.39	.300	3.48	.375	.162	12.89	9.6		84.1	1.19	21.2
439	1.37	1.2	45	2.9	.114	.32	.79	.082	.61	.094	.045	14.63	12.2		40.9	.98	5.9
4	.16	2.1	32	3.6	.019	.07	1.90	.172	1.83	.192	.075	2.36	4.6		26.4	.05	9.9
2224	8.90	9.8	215	19.1	.656	3.56	5.16	.524	4.84	.776	.272	95.44	99.9		261.4	4.45	46.0
116	1.04	5.1	83	14.9	.114	1.21	4.14	.375	3.98	.413	.162	5.44	8.5		74.5	.74	26.5
139	.67	2.2	46	13.1	.073	1.73	.76	.087	.70	.100	.051	4.47	1.5		37.2	.54	7.0
20	.31	1.9	29	5.8	.024	.10	2.82	.239	3.04	.220	.109	2.65	5.4		28.8	.03	12.0
486	3.18	12.8	185	67.0	.290	7.45	5.55	.560	5.57	.624	.315	22.27	11.4		148.7	2.11	42.7

APPENDIX 2 (cont.)

Country		Particle size distr.				Tex. ind.	CEC	pH		El. cond.	CaCO$_3$ equiv.	Org. C	Volume weight	N (total)	P	K	Ca (extractable)	
		<.002	.002—.06	.06—2	>2 mm			H$_2$O	CaCl$_2$									
							me/100g			10^{-4}S/cm	%	%	g/cm^3	%	mg/l	mg/l	mg/l	mg/l	mg
TURKEY (n = 249)	mean	36	35	29	0	50	32.2	7.90	7.52	1.6	11.8	.9	1.21	.094	1124	10.6	411	6408	49
	±s	15	10	16	0	13	10.6	.46	.45	1.1	11.5	.3	.07	.034	377	7.9	210	2140	30
	min.	3	12	1	0	17	9.7	5.70	5.30	.5	.0	.2	1.00	.026	357	1.0	58	660	
	max.	77	67	75	0	84	64.2	8.90	8.55	12.0	54.2	2.8	1.47	.288	3192	69.6	1338	13160	175
AFRICA																			
ETHIOPIA (n = 54)	mean	58	32	10	0	68	56.7	6.63	5.97	.8	.2	2.0	1.12	.176	1905	10.0	740	5662	71
	±s	14	9	9	1	12	12.5	.75	.90	.5	.4	.9	.08	.093	905	9.6	260	3261	39
	min.	15	20	1	0	28	34.8	5.50	4.60	.3	.0	.8	.92	.056	678	3.0	127	1250	18
	max.	76	63	47	9	82	79.0	8.15	7.60	2.2	1.8	3.6	1.25	.360	3826	57.8	1336	11550	143
NIGERIA (n = 42)	mean	23	36	41	0	38	16.4	6.54	6.12	2.5	1.3	.7	1.29	.083	1058	16.7	207	2009	34
	±s	21	15	23	0	18	10.9	1.06	1.06	2.4	5.3	.4	.11	.041	499	13.1	134	1952	36
	min.	2	1	6	0	12	5.1	4.57	4.13	.5	.0	.1	1.06	.030	486	1.2	50	320	4
	max.	69	67	97	0	77	44.9	9.17	7.96	10.5	26.8	2.3	1.62	.227	3132	56.4	633	7000	179
TANZANIA (n = 5)	mean	38	13	49	0	47	19.0	5.66	4.99	.6	.0	1.9	1.07	.153	1635	3.2	65	420	8
	±s	2	2	2	0	2	3.7	.19	.26	.3	.0	.4	.03	.042	433	1.2	29	407	4
	min.	36	10	45	0	45	14.8	5.48	4.79	.4	.0	1.5	1.04	.104	1100	1.9	20	110	4
	max.	41	15	52	0	50	23.2	5.96	5.41	1.1	.0	2.5	1.10	.202	2141	4.9	97	1110	15
WHOLE INTERNAT. MATERIAL (n = 1768)	mean	31	45	24	0	47	27.8	7.38	6.91	3.5	7.4	1.3	1.18	.133	1506	20.2	365	4671	48
	±s	18	17	19	4	15	13.6	1.02	1.07	5.4	12.0	1.1	.11	.084	779	24.7	283	3076	43
	min.	1	1	1	0	12	5.1	4.55	3.97	.3	.0	.1	.66	.009	95	1.0	20	110	
	max.	85	91	97	58	89	89.7	9.25	8.56	57.5	68.4	17.5	1.75	1.023	8844	271.4	2097	21930	329

427

	SOIL MICRONUTRIENTS						NUTRIENT CONTENTS OF ORIGINAL WHEAT PLANTS									
	B	Cu (extractable)	Fe	Mn	Mo	Zn	N	P	K	Ca	Mg	B	Cu	Fe	Mn	Mo	Zn
	mg/l	mg/l	mg/l	mg/l	mg/l	mg/l	%	%	%	%	%	ppm	ppm	ppm	ppm	ppm	ppm
	1.16	5.3	89	15.2	.111	.39	4.12	.295	3.33	.568	.174	9.30	8.6		85.3	.84	18.1
	1.13	2.4	66	12.9	.126	.34	.78	.087	.61	.218	.069	12.67	2.1		37.0	.97	6.2
	.06	.9	13	3.4	.020	.12	1.02	.141	1.80	.214	.087	1.61	4.9		28.0	.03	7.6
	9.99	15.5	558	84.0	.956	3.65	6.73	.572	5.36	1.677	.948	99.19	16.6		225.8	8.57	41.7
	.55	6.0	237	85.8	.537	2.07	3.05	.314	3.77	.307	.108	2.51	6.7		65.1	1.34	25.4
	.19	2.2	133	51.0	.398	1.99	.77	.077	.81	.085	.017	.49	1.6		29.5	1.87	5.5
	.28	1.1	74	11.5	.106	.49	1.94	.226	2.59	.158	.075	1.39	1.5		24.6	.02	12.9
	1.09	14.3	700	193.0	2.600	9.90	5.43	.557	6.43	.553	.162	3.94	10.0		166.0	10.04	43.9
	.25	2.1	162	27.3	.175	.78	4.28	.343	4.54	.461	.231	3.31	9.7		114.4	.71	33.3
	.24	1.7	161	17.2	.244	.51	.72	.112	.71	.167	.072	1.28	2.0		65.3	.93	12.7
	.03	.2	22	4.5	.016	.28	2.44	.206	2.98	.260	.126	1.47	5.7		47.8	.07	16.6
	1.31	7.0	870	81.0	1.248	3.12	5.95	.732	6.01	1.102	.483	7.00	14.0		364.7	4.13	84.8
	.36	4.0	58	25.0	.094	.48	4.54	.296	4.54	.386	.219	2.24	10.8		60.6	.25	42.9
	.05	1.2	19	10.7	.034	.14	.39	.090	.52	.091	.057	.32	1.1		19.7	.29	17.6
	.28	2.7	38	17.4	.063	.26	3.98	.199	3.99	.251	.141	1.92	9.8		48.3	.04	28.5
	.40	5.2	88	43.2	.150	.65	5.04	.415	5.38	.470	.274	2.74	12.6		95.6	.76	69.2
	.81	6.1	172	25.2	.204	1.77	4.28	.375	4.03	.428	.172	6.56	9.4		73.6	.94	27.4
	.75	6.4	175	33.1	.229	7.30	1.15	.125	.97	.166	.060	7.80	5.7		39.8	1.03	11.3
	.03	.2	10	1.0	.007	.07	.60	.048	.94	.113	.044	1.39	1.5		8.6	.01	3.1
	9.99	87.5	1995	286.0	3.241	186.45	7.45	1.020	6.84	1.677	.948	99.19	100.6		423.1	11.04	128.3

APPENDIX 3. National mean values of basic analytical data on **original maize** plants and respective soils. For analytical methods, see Appendix 2 and Section 1.3.

Country			GENERAL SOIL PROPERTIES											SOIL MACRONUTRIENTS			
		Particle size distr.				Tex. ind.	CEC	pH H₂O CaCl₂		El. cond.	CaCO₃ equiv.	Org. C	Volume weight	N (total)	P	K	Ca (extractable)
		<.002 %	.002—.06 %	.06—2 %	>2 mm %		me 100 g			10⁻⁴S cm	%	%	g/cm³	%	mg/l	mg/l	mg/l mg/l

EUROPE AND OCEANIA

Country		<.002	.002—.06	.06—2	>2mm	Tex	CEC	pH H₂O	pH CaCl₂	El.cond	CaCO₃	Org.C	Vol.wt	N	P	K	Ca
BELGIUM (n = 20)	mean	7	35	59	0	23	19.8	5.91	5.52	2.5	.2	1.5	1.26	.134	1674	119.8	232	1240
	± s	5	24	29	0	9	3.0	.71	.77	.6	.5	.4	.07	.023	271	39.3	94	866
	min.	2	12	6	0	15	13.4	4.85	4.45	1.2	.0	.6	1.15	.088	1033	43.2	108	490
	max.	15	79	85	0	39	24.5	7.30	7.05	3.9	2.4	2.1	1.37	.171	2223	187.0	406	4050
HUNGARY (n = 106)	mean	30	52	18	0	47	29.5	7.32	6.96	2.7	3.6	1.6	1.22	.169	2026	38.9	241	4858
	± s	13	15	21	0	13	11.6	.89	.95	1.1	5.9	.6	.06	.061	655	34.0	136	2094
	min.	3	12	1	0	16	4.6	5.00	4.20	.5	.0	.4	.99	.036	508	7.8	50	165
	max.	77	73	83	0	84	64.2	8.40	8.00	8.7	42.2	3.9	1.42	.404	4252	322.0	956	8270
ITALY (n = 70)	mean	28	43	27	2	44	25.5	7.37	7.00	3.6	8.4	1.3	1.16	.178	2051	32.5	210	4990
	± s	16	12	17	7	15	10.8	.76	.76	3.3	10.2	.6	.08	.141	1739	23.0	189	2566
	min.	9	3	1	0	23	9.9	5.18	4.65	1.0	.0	.7	.97	.096	1110	2.8	24	730
	max.	71	76	67	48	80	58.9	8.10	7.77	22.0	37.8	4.3	1.31	1.169	14729	105.0	1121	11130
NEW ZEALAND (n = 24)	mean	45	39	16	0	58	60.0	5.83	5.42	2.6	.1	6.2	.92	.481	3944	40.0	500	3215
	± s	20	12	12	0	16	17.2	.73	.71	1.1	.5	9.0	.16	.340	1418	33.1	317	2025
	min.	11	21	2	0	32	29.7	4.00	3.62	.8	.0	2.1	.47	.211	2287	4.8	76	1070
	max.	75	63	35	0	82	99.7	7.07	6.77	5.2	2.6	39.1	1.09	1.657	8378	157.4	1063	7120

LATIN AMERICA

Country		<.002	.002—.06	.06—2	>2mm	Tex	CEC	pH H₂O	pH CaCl₂	El.cond	CaCO₃	Org.C	Vol.wt	N	P	K	Ca
ARGENTINA (n = 90)	mean	23	67	10	0	44	28.4	6.08	5.45	1.6	.0	1.9	1.06	.207	2189	23.7	790	2090
	± s	3	6	7	0	4	5.6	.29	.27	.7	.1	.8	.03	.074	758	16.1	196	309
	min.	15	46	2	0	22	20.4	5.24	4.79	.7	.0	1.1	.96	.124	1289	.7	479	1100
	max.	36	81	29	0	54	44.9	6.72	6.08	4.3	.4	4.3	1.16	.433	4434	105.4	1649	2990
MEXICO (n = 147)	mean	35	36	29	0	49	32.8	7.37	6.89	2.2	1.3	1.0	1.18	.100	1154	19.4	548	4363
	± s	21	15	22	2	18	16.0	.93	.93	1.7	2.4	.4	.11	.045	435	17.5	338	2519
	min.	1	6	1	0	10	.2	4.60	4.10	.3	.0	.1	.88	.011	176	2.5	51	290
	max.	76	75	82	19	82	79.1	8.80	8.25	9.2	24.8	3.3	1.60	.342	3068	90.4	1918	14880
PERU (n = 57)	mean	24	29	38	10	36	26.1	7.30	6.92	3.0	3.8	1.5	1.24	.175	2117	20.9	230	4556
	± s	15	10	16	14	14	11.3	.82	.82	2.2	7.3	.7	.13	.076	794	16.8	144	2491
	min.	2	12	1	0	9	13.5	4.54	4.00	.4	.0	.3	.98	.053	664	2.4	56	280
	max.	61	54	73	55	69	66.1	8.14	7.70	13.7	47.0	3.4	1.58	.420	4402	79.8	777	10840

FAR EAST

Country		<.002	.002—.06	.06—2	>2mm	Tex	CEC	pH H₂O	pH CaCl₂	El.cond	CaCO₃	Org.C	Vol.wt	N	P	K	Ca
INDIA (n = 107)	mean	11	55	33	0	31	13.6	8.07	7.55	1.8	5.2	.6	1.26	.073	910	8.6	126	2660
	± s	5	18	19	0	6	3.5	.52	.49	1.8	11.0	.2	.08	.017	171	7.1	78	1271
	min.	3	11	5	0	15	5.1	6.04	5.54	.5	.0	.2	1.01	.042	571	1.9	41	470
	max.	28	87	86	0	48	26.0	8.90	8.55	15.4	39.6	1.1	1.43	.128	1443	46.2	517	5330
KOREA, REP. OF (n = 50)	mean	18	44	38	0	35	17.8	6.02	5.58	2.6	1.3	1.3	1.10	.144	1553	44.0	190	1390
	± s	10	15	22	0	11	5.2	1.06	1.09	1.7	6.1	.5	.11	.043	382	32.2	104	831
	min.	3	18	4	0	19	6.7	4.40	4.10	.7	.0	.4	.88	.041	529	5.0	52	300
	max.	42	81	76	0	56	36.0	8.00	7.50	10.0	42.9	2.3	1.31	.231	2318	140.0	492	4000

	SOIL MICRONUTRIENTS						NUTRIENT CONTENTS OF ORIGINAL MAIZE PLANTS									
	B	Cu (extractable)	Fe	Mn	Mo	Zn	N	P	K	Ca	Mg	B	Cu	Fe	Mn	Mo	Zn
mg/l	mg/l	mg/l	mg/l	mg/l	mg/l	mg/l	%	%	%	%	%	ppm	ppm	ppm	ppm	ppm	ppm
9	.48	4.5	401	31.0	.170	19.23	5.01	.535	4.35	.618	.184	4.61	10.5		137.6	.34	168.1
9	.25	2.2	180	31.0	.049	11.72	.28	.104	.51	.111	.040	.85	2.1		107.7	.22	106.8
0	.23	2.2	115	4.0	.101	2.75	4.43	.369	3.23	.415	.083	3.20	7.6		43.7	.10	22.8
1	1.09	9.3	750	104.6	.250	44.06	5.54	.778	5.02	.864	.250	6.66	15.3		517.1	1.09	403.7
9	1.09	5.2	145	44.2	.138	1.03	4.72	.502	4.13	.768	.385	6.14	15.0		115.8	.37	30.8
1	.48	2.4	85	33.2	.130	.91	.50	.105	.89	.262	.138	1.74	4.1		56.2	.41	11.9
2	.12	.6	52	7.5	.040	.34	2.93	.262	1.48	.326	.144	3.05	7.7		43.1	.01	16.5
6	3.09	13.1	567	154.0	1.161	5.96	5.87	1.038	6.02	1.880	1.073	10.69	25.6		429.1	2.86	77.9
1	.68	19.2	236	21.8	.184	2.63	4.08	.363	3.09	.685	.292	9.53	14.5		76.5	.45	32.9
7	.36	20.2	120	21.5	.162	3.16	.69	.079	.96	.172	.103	3.88	4.0		31.2	.49	11.4
2	.15	2.5	88	2.7	.047	.39	2.11	.150	.76	.330	.145	4.02	2.5		19.4	.03	15.6
4	1.85	96.0	820	127.0	1.224	18.97	5.73	.591	5.22	1.204	.711	24.06	27.4		144.8	2.72	64.1
6	.77	16.3	589	67.0	.195	3.70	3.68	.355	3.52	.340	.179	10.55	9.5		64.7	.08	40.3
7	.43	17.6	452	50.6	.192	3.52	.89	.088	.80	.064	.053	4.07	2.4		34.5	.10	23.0
4	.31	.3	75	2.7	.036	.38	1.71	.229	2.13	.237	.106	3.71	5.4		26.1	.01	15.8
7	1.78	42.9	2275	237.0	.940	18.29	5.40	.507	4.85	.471	.291	19.14	13.6		144.9	.50	102.8
8	.78	3.2	201	71.8	.357	2.97	3.81	.326	4.10	.596	.242	7.45	12.0		85.6	.46	25.4
7	.24	.7	74	18.6	.218	4.35	.46	.073	.51	.291	.094	3.37	9.8		27.3	.48	4.2
4	.41	1.9	110	31.8	.182	.61	2.52	.189	2.85	.184	.117	3.22	4.7		34.2	.05	14.2
0	1.91	7.0	506	123.9	1.633	28.69	4.95	.537	5.61	1.218	.474	23.07	99.6		153.7	3.26	39.1
7	1.30	4.0	80	35.4	.183	1.73	2.86	.308	2.71	.383	.186	10.90	9.6		65.1	.73	28.5
3	1.68	4.8	66	29.9	.160	3.62	.70	.080	.70	.140	.051	7.30	2.5		26.9	.65	20.7
2	.07	.3	14	.9	.020	.16	1.40	.131	.98	.163	.087	2.91	3.2		16.6	.01	6.2
8	9.99	54.4	535	163.2	1.151	26.95	4.67	.629	4.32	1.024	.336	43.13	15.5		180.6	4.76	224.5
9	.67	5.8	167	25.8	.237	2.26	3.24	.332	2.85	.511	.241	8.05	11.6		54.7	1.02	30.7
9	.46	3.9	92	25.3	.292	4.61	.56	.089	1.32	.205	.126	3.34	2.9		15.6	.96	13.2
2	.21	.4	30	6.3	.046	.29	1.99	.107	.85	.213	.093	3.90	6.6		25.9	.07	16.0
7	2.60	23.0	394	138.0	1.853	34.35	5.33	.514	4.68	1.024	.723	21.54	19.2		107.6	5.24	104.2
0	.37	4.4	183	16.0	.105	.90	2.62	.310	2.35	.556	.313	7.99	11.8		63.6	2.70	25.2
3	.26	7.8	227	10.3	.067	4.02	.62	.067	.99	.259	.133	4.04	4.4		27.7	2.75	18.9
3	.05	.7	31	4.5	.010	.09	1.57	.156	.96	.194	.057	1.88	5.7		24.9	.03	8.6
6	1.61	82.0	930	55.9	.482	41.89	4.34	.480	4.34	1.159	.823	25.90	33.3		180.7	16.76	195.2
0	.34	3.0	134	39.1	.196	1.67	3.95	.369	3.39	.442	.270	6.48	10.4		103.5	.37	34.2
6	.20	1.5	59	37.5	.109	1.02	.41	.076	.78	.115	.085	6.26	2.4		93.5	.37	14.3
2	.14	.5	25	3.3	.018	.32	2.90	.213	1.87	.211	.147	3.24	6.0		21.8	.01	18.5
4	1.20	6.2	310	148.0	.624	5.68	4.69	.507	5.30	.777	.471	31.03	15.8		453.8	1.94	109.0

429

APPENDIX 3 (cont.)

Country			Particle size distr.			Tex. ind.	CEC	pH H$_2$O	pH CaCl$_2$	El. cond.	CaCO$_3$ equiv.	Org. C	Volume weight	N (total)	P	K (extractable)	Ca (extractable)
		<.002 %	.002—.06 %	.06—2 %	>2 mm %	%	me 100 g			10^{-4}S cm	%	%	g/cm^3	%	mg/l	mg/l	mg/l	mg/l
PAKISTAN (n = 86)	mean	18	64	19	0	39	14.9	8.22	7.67	2.3	6.0	.6	1.22	.077	937	6.3	188	4102
	± s	6	10	11	0	6	5.1	.31	.28	1.5	4.3	.3	.06	.022	264	6.0	98	839
	min.	7	36	1	0	28	7.5	6.85	6.15	.7	.3	.2	1.08	.008	97	.8	58	2230
	max.	37	79	48	0	55	41.2	8.90	8.10	12.5	20.0	3.2	1.37	.139	1771	41.0	510	7160
PHILIPPINES (n = 197)	mean	33	41	26	0	48	35.6	6.50	5.99	1.5	.7	1.2	1.09	.139	1471	18.5	212	3839
	± s	21	16	19	2	18	12.8	.76	.81	1.2	1.9	.7	.10	.063	562	18.9	212	2752
	min.	3	9	0	0	17	6.7	4.56	4.23	.4	.0	.3	.80	.039	487	1.5	18	100
	max.	87	77	81	22	91	67.2	8.10	7.56	11.6	16.0	4.8	1.42	.527	4313	138.6	1997	12800
SRI LANKA (n = 21)	mean	23	11	64	2	31	14.2	6.21	5.62	1.0	.1	1.2	1.28	.122	1553	14.0	144	1159
	± s	11	5	14	8	10	4.3	.75	.85	.4	.2	.7	.08	.061	727	12.3	59	886
	min.	3	3	40	0	13	6.8	5.15	4.38	.5	.0	.5	1.07	.075	1028	2.2	21	300
	max.	47	20	94	33	55	24.6	8.02	7.60	1.8	.8	3.9	1.45	.363	4400	48.4	258	3700
THAILAND (n = 150)	mean	44	34	21	1	56	41.8	7.31	6.80	1.3	4.8	1.6	1.21	.157	1884	11.7	216	6444
	± s	16	12	15	4	14	17.0	.76	.80	.9	10.1	.5	.07	.047	513	18.8	131	4244
	min.	6	9	1	0	12	11.8	5.30	4.50	.3	.0	.7	1.00	.076	1045	.1	40	600
	max.	81	63	84	25	85	82.8	8.20	7.70	10.5	50.0	3.7	1.43	.380	4134	165.2	849	15390

NEAR EAST

Country			Particle size distr.			Tex. ind.	CEC	pH H$_2$O	pH CaCl$_2$	El. cond.	CaCO$_3$ equiv.	Org. C	Volume weight	N (total)	P	K (extractable)	Ca (extractable)
EGYPT (n = 100)	mean	45	35	20	0	57	41.8	8.29	7.81	4.8	4.6	1.2	1.17	.117	1358	15.8	635	6348
	± s	14	13	21	0	14	10.6	.19	.12	3.6	2.9	.3	.06	.028	291	13.4	306	986
	min.	5	3	1	0	15	5.9	7.75	7.60	1.2	1.0	.4	1.07	.044	642	3.4	82	3450
	max.	82	54	92	0	87	57.2	8.95	8.20	32.0	24.1	1.8	1.46	.208	2314	101.1	2325	8098
IRAQ (n = 31)	mean	33	56	11	0	51	23.6	8.16	7.83	13.5	25.8	.7	1.16	.087	1011	7.5	304	6456
	± s	8	9	9	0	7	5.4	.16	.17	16.7	4.7	.2	.08	.032	405	4.6	126	2851
	min.	14	34	1	0	36	13.6	7.87	7.65	1.4	15.6	.3	1.00	.041	464	.7	120	4200
	max.	49	72	39	0	63	40.8	8.65	8.55	73.0	42.5	1.1	1.34	.197	2634	20.9	645	17995
LEBANON (n = 16)	mean	47	35	17	0	60	38.3	7.80	7.44	3.9	17.1	.9	1.20	.097	1158	42.5	354	7789
	± s	17	16	11	0	13	9.4	.37	.29	3.6	20.7	.3	.04	.023	287	31.0	234	1919
	min.	15	19	1	0	38	23.8	7.10	6.95	1.5	.4	.4	1.10	.061	736	5.3	101	5000
	max.	78	70	33	0	84	61.9	8.10	7.75	13.5	67.1	1.3	1.26	.137	1722	104.6	1070	11950
SYRIA (n = 18)	mean	36	40	18	6	50	39.0	8.05	7.66	3.4	23.8	1.3	1.18	.134	1494	16.6	418	7618
	± s	18	18	15	13	15	12.7	.25	.24	2.6	15.3	1.2	.12	.101	877	19.1	249	2763
	min.	9	9	4	0	23	22.2	7.28	6.90	1.3	1.4	1.5	.86	.023	304	4.4	136	3310
	max.	75	72	66	43	82	74.3	8.38	7.92	12.6	61.4	5.6	1.34	.480	4128	82.0	887	15910
TURKEY (n = 50)	mean	31	36	33	0	45	28.3	7.36	6.79	1.5	3.5	1.2	1.20	.119	1407	17.5	213	5411
	± s	15	12	19	0	14	12.2	.76	.74	.9	5.5	.5	.06	.047	507	14.4	115	2282
	min.	10	12	1	0	25	7.7	5.60	5.20	.3	.0	.4	1.08	.047	560	3.0	46	1290
	max.	79	59	72	0	85	58.2	8.30	7.60	4.7	24.8	2.7	1.40	.270	2905	73.0	556	10375

	SOIL MICRONUTRIENTS						NUTRIENT CONTENTS OF ORIGINAL MAIZE PLANTS										
	B	Cu	Fe (extractable)	Mn	Mo	Zn	N	P	K	Ca	Mg	B	Cu	Fe	Mn	Mo	Zn
mg/l	mg/l	mg/l	mg/l	mg/l	mg/l	mg/l	%	%	%	%	%	ppm	ppm	ppm	ppm	ppm	ppm
2	.65	4.4	98	12.2	.146	.60	3.13	.314	3.58	.498	.280	13.85	13.7		86.0	1.91	31.0
6	.26	1.5	40	7.3	.079	.57	.76	.071	.95	.234	.127	7.97	4.6		36.9	1.19	9.2
8	.26	1.7	32	5.6	.037	.17	1.37	.152	1.50	.216	.124	4.37	5.9		24.7	.36	15.9
9	1.45	9.5	200	57.3	.505	3.04	4.67	.522	5.52	1.620	.652	49.42	26.3		218.4	9.01	59.2
6	.28	13.4	273	45.0	.228	3.26	2.99	.287	2.46	.524	.340	6.93	12.3		54.9	.56	32.6
7	.15	10.6	131	35.8	.189	15.89	.51	.071	.81	.162	.174	3.35	2.6		26.7	.76	17.0
1	.07	1.4	63	2.9	.036	.22	1.60	.156	.58	.223	.102	3.08	5.8		19.5	.02	12.9
1	.86	93.0	840	226.0	1.585	185.20	4.50	.600	4.39	1.114	1.125	26.57	20.8		221.1	6.36	183.0
0	.34	3.9	156	64.6	.196	3.78	2.82	.322	3.51	.412	.231	6.22	9.8		94.2	.17	27.8
3	.19	2.7	104	52.5	.132	11.27	.76	.097	.80	.126	.098	1.91	2.4		39.1	.27	11.8
0	.16	.9	50	3.8	.065	.55	1.56	.167	2.02	.216	.106	4.00	4.9		26.8	.02	15.6
6	.85	10.7	470	203.0	.510	52.81	4.17	.537	4.80	.704	.499	10.34	13.2		167.9	1.27	57.9
8	.42	6.3	140	52.5	.305	1.60	3.27	.294	3.01	.321	.182	6.21	10.5		49.6	.46	27.9
0	.15	8.8	104	34.2	.454	2.66	.52	.073	.60	.096	.059	1.24	1.9		12.2	.44	6.4
0	.16	.2	22	10.0	.026	.20	1.52	.143	1.30	.157	.081	3.78	4.0		26.4	.03	12.8
1	.83	99.7	775	162.0	3.326	20.51	4.63	.474	4.75	.637	.424	9.70	15.0		88.6	2.66	54.3
5	1.00	11.0	191	10.6	.140	1.04	3.02	.302	3.31	.330	.263	13.63	10.9		59.6	.74	31.6
5	.39	3.3	97	4.1	.042	1.12	.45	.054	.49	.075	.071	5.21	2.1		12.8	.21	6.6
8	.32	2.5	44	5.6	.023	.29	1.72	.166	2.37	.131	.156	7.69	5.6		32.0	.30	15.1
3	2.86	21.3	675	36.9	.224	9.18	3.83	.527	4.64	.593	.691	40.18	15.6		105.7	1.40	49.8
9	2.08	4.7	121	9.6	.134	.27	3.14	.276	3.52	.487	.291	38.23	14.0		102.5	1.27	22.8
2	2.24	1.2	38	3.7	.096	.10	.40	.060	.45	.084	.089	26.53	4.7		33.6	.71	8.9
0	.37	2.5	38	3.6	.034	.13	2.50	.165	2.34	.337	.160	7.16	8.7		44.5	.23	9.7
8	10.02	7.7	220	21.2	.439	.60	4.12	.405	4.45	.654	.483	100.04	24.5		193.4	2.90	44.6
4	.54	6.0	98	17.9	.077	.85	3.43	.359	3.32	.497	.219	10.43	13.0		85.8	.30	26.2
1	.16	2.0	37	11.5	.068	.42	.73	.061	.54	.153	.075	3.94	3.4		24.1	.32	5.1
5	.34	4.1	24	6.8	.014	.21	1.83	.246	2.21	.220	.143	6.38	7.9		57.3	.05	14.7
6	.97	11.5	156	37.9	.250	1.67	4.69	.425	4.15	.854	.406	22.12	22.4		139.2	1.33	34.1
6	1.12	4.5	91	15.1	.125	1.21	2.98	.284	3.10	.499	.263	19.62	9.8		86.3	.76	32.1
4	.80	1.5	86	20.9	.121	2.60	.83	.070	1.10	.151	.099	15.05	2.8		34.7	.75	16.1
9	.22	2.0	25	2.8	.027	.24	1.41	.198	2.18	.282	.159	4.79	4.7		8.3	.12	14.5
8	3.45	6.4	380	97.5	.529	11.36	4.22	.422	5.30	.854	.476	56.72	14.1		135.7	3.17	80.6
0	.78	7.7	210	46.1	.168	1.72	3.28	.303	2.74	.569	.290	7.82	12.2		64.9	.34	26.6
9	.45	4.5	103	38.5	.120	2.73	.63	.060	.79	.140	.131	2.85	2.9		19.6	.35	8.2
6	.08	1.8	65	9.6	.035	.27	1.59	.173	1.00	.326	.134	3.62	6.6		33.8	.03	15.6
5	2.20	20.2	540	167.0	.504	16.16	4.56	.460	4.18	.895	.614	18.25	19.0		111.6	1.64	47.9

APPENDIX 3 (cont.)

Country			Particle size distr.				Tex. ind.	CEC	pH		El. cond.	CaCO₃ equiv.	Org. C	Volume weight	N (total)	P	K	Ca (extractable)	
			<.002 %	.002–.06 %	.06–2 %	>2 mm %	%	me/100 g	H_2O	$CaCl_2$	10^{-4}S cm	%	%	g/cm³	%	mg/l	mg/l	mg/l	mg/l
AFRICA																				
ETHIOPIA	mean		45	32	23	0	56	41.5	6.23	5.56	1.0	.1	2.2	1.01	.200	1987	18.7	549	2292
(n = 71)	± s		20	12	15	1	17	10.9	.75	.81	.6	.4	.8	.11	.068	611	43.0	390	1608
	min.		5	12	2	0	20	16.2	4.60	4.10	.3	.0	.9	.76	.064	775	1.2	87	510
	max.		79	62	66	11	84	64.5	8.30	7.65	3.2	1.8	4.0	1.25	.394	3767	311.0	2077	7680
GHANA	mean		18	19	58	5	30	14.7	6.36	5.85	.9	.1	1.3	1.30	.132	1692	8.2	129	1145
(n = 93)	± s		8	12	18	10	9	6.0	.55	.66	.4	.1	.5	.10	.056	663	10.7	70	520
	min.		4	7	17	0	13	4.9	4.80	4.35	.3	.0	.6	1.08	.053	727	1.8	34	71
	max.		37	50	88	56	55	41.2	7.37	7.05	2.3	.5	2.7	1.52	.333	3770	95.8	321	3150
MALAWI	mean		24	17	59	0	35	17.7	5.91	5.38	2.6	.1	1.3	1.29	.115	1461	39.4	227	1534
(n = 100)	± s		11	8	14	1	10	8.3	.56	.51	3.0	.2	.6	.12	.041	457	26.3	144	889
	min.		4	8	8	0	15	5.9	4.89	4.58	.3	.0	.3	.76	.039	298	2.6	38	200
	max.		49	47	87	8	61	41.1	7.30	6.80	16.7	.8	3.3	1.51	.257	2864	113.4	693	3838
NIGERIA	mean		14	20	64	1	26	14.0	6.23	5.62	1.1	.1	1.0	1.36	.092	1211	21.7	191	1099
(n = 103)	± s		12	16	24	7	12	6.7	.94	1.09	.8	.6	.6	.15	.049	564	41.4	175	1107
	min.		3	2	5	0	13	2.4	4.70	3.95	.2	.0	.2	1.03	.041	607	1.2	35	65
	max.		78	63	93	45	83	47.1	8.20	7.60	4.8	5.4	4.9	1.77	.392	4500	303.2	1042	4880
SIERRA LEONE	mean		21	18	61	0	33	20.9	5.72	4.90	.8	.0	2.5	1.09	.198	2086	12.3	108	585
(n = 49)	± s		9	10	15	0	8	7.6	.50	.58	.6	.2	1.0	.09	.091	768	21.1	55	547
	min.		11	6	25	0	22	10.7	4.85	4.10	.3	.0	.8	.88	.103	1231	3.0	47	100
	max.		48	46	82	0	58	44.7	7.50	7.10	3.0	.9	5.7	1.22	.492	4512	152.0	280	2840
TANZANIA	mean		29	16	55	0	39	23.4	6.83	6.30	1.0	.4	1.3	1.26	.116	1356	32.7	639	2104
(n = 175)	± s		22	9	26	4	20	15.6	.66	.75	.6	.8	.9	.19	.077	768	66.6	737	1727
	min.		2	2	3	0	13	3.1	5.20	4.39	.1	.0	.2	.90	.014	237	.8	18	155
	max.		85	46	93	54	88	60.0	8.30	7.80	4.7	7.6	4.8	1.72	.425	4747	656.0	5598	9265
ZAMBIA	mean		15	18	66	0	27	11.4	5.72	5.02	1.1	.0	.8	1.36	.073	957	20.8	145	665
(n = 45)	± s		11	9	17	2	11	6.2	.63	.73	1.7	.1	.4	.12	.035	385	20.4	202	570
	min.		1	1	28	0	11	3.4	4.85	4.18	.2	.0	.2	1.10	.027	448	2.8	18	10
	max.		44	41	97	11	55	28.2	7.64	7.14	11.5	.5	2.0	1.66	.180	2136	114.5	1272	2650
WHOLE	mean		28	36	35	1	42	27.0	6.92	6.40	2.1	2.8	1.3	1.20	.135	1547	22.5	330	3450
INTERNAT.	± s		18	20	26	5	17	15.4	1.05	1.11	3.2	7.1	1.3	.14	.088	793	33.0	356	2815
MATERIAL	min.		1	1	0	0	9	.2	4.00	3.62	.1	.0	.1	.47	.008	97	.1	18	10
(n = 1976)	max.		87	87	97	56	91	99.7	8.95	8.55	73.0	67.1	39.1	1.77	1.657	14729	656.0	5598	17995

	SOIL MICRONUTRIENTS							NUTRIENT CONTENTS OF ORIGINAL MAIZE PLANTS									
Na	B	Cu	Fe	Mn	Mo	Zn	N	P	K	Ca	Mg	B	Cu	Fe	Mn	Mo	Zn
			(extractable)														
g/l	mg/l	mg/l	mg/l	mg/l	mg/l	mg/l	%	%	%	%	%	ppm	ppm	ppm	ppm	ppm	ppm
31	.54	3.3	199	89.7	.771	4.88	3.90	.304	3.65	.396	.210	5.67	10.6		78.8	.71	45.7
59	.26	2.9	123	50.4	.580	6.52	.87	.120	.77	.112	.069	1.80	3.5		32.4	1.76	30.5
0	.16	.1	49	6.5	.093	.23	1.59	.126	2.14	.191	.074	3.10	2.0		32.3	.03	15.5
93	1.24	11.5	825	213.5	3.560	43.63	6.51	.814	5.16	.707	.449	10.21	17.0		238.5	14.33	156.2
6	.57	1.9	94	62.3	.081	1.06	2.80	.318	3.10	.329	.203	9.15	9.4		68.0	.38	23.9
9	.22	1.2	37	27.4	.028	.95	.41	.071	.46	.057	.043	2.05	1.9		17.6	.43	6.3
0	.24	.4	32	13.6	.033	.16	1.58	.174	1.87	.159	.119	4.36	3.9		35.8	.01	12.8
71	1.37	5.7	225	138.1	.160	4.36	3.84	.513	4.43	.498	.316	15.84	13.6		119.9	2.68	43.1
4	.29	3.3	131	86.4	.184	1.59	2.72	.393	2.92	.477	.215	4.04	12.7		94.2	.40	17.0
4	.11	2.7	67	53.7	.101	.94	.80	.153	.98	.142	.078	.84	4.5		61.0	.49	4.5
0	.12	.3	39	18.7	.049	.17	1.18	.206	1.00	.180	.096	2.61	5.4		18.2	.03	9.2
19	.69	22.3	316	378.4	.748	4.10	4.32	.926	5.09	.939	.595	6.30	27.0		322.8	2.98	32.9
11	.30	1.9	89	37.3	.112	3.33	2.48	.315	3.42	.437	.232	5.38	10.8		76.1	1.53	29.6
8	.17	1.7	65	32.4	.064	8.04	.52	.096	.80	.145	.075	1.58	2.8		35.0	2.79	12.9
1	.05	.3	15	2.9	.011	.17	1.31	.138	1.78	.174	.080	2.17	5.9		28.9	.03	13.3
56	.86	13.4	376	203.0	.384	57.19	3.80	.593	6.02	.789	.530	10.48	22.1		232.8	21.03	81.3
6	.38	1.1	123	7.8	.039	1.52	2.83	.266	3.65	.289	.168	6.59	10.2		62.1	.28	49.7
6	.14	.6	71	7.1	.017	1.91	.53	.064	.72	.073	.033	.91	3.1		14.4	.17	41.9
2	.17	.3	33	1.4	.017	.29	1.72	.166	1.61	.169	.111	4.42	3.8		39.5	.09	15.0
25	.81	3.1	400	45.5	.099	11.15	4.14	.472	5.20	.454	.267	8.60	25.0		115.7	.93	256.3
29	.80	4.6	114	52.1	.307	1.64	2.16	.317	3.05	.434	.223	16.49	10.8		103.0	1.63	80.7
74	.53	4.5	96	31.8	.396	1.99	.45	.109	.93	.168	.113	10.86	4.4		60.4	1.49	128.9
0	.15	.2	20	7.9	.019	.12	.88	.050	1.03	.091	.036	4.06	2.8		22.1	.03	13.0
570	2.97	24.8	570	161.2	2.724	15.95	3.50	.596	6.71	.990	.772	99.99	25.0		316.3	7.60	915.6
4	.25	2.4	72	53.4	.110	.88	3.16	.400	2.09	.485	.200	4.02	13.5		128.3	.29	24.3
5	.10	3.3	57	36.7	.094	1.26	.41	.095	.49	.139	.052	1.00	3.6		116.8	.29	11.0
0	.10	.1	20	8.1	.019	.17	2.29	.189	.94	.274	.111	2.78	7.3		27.5	.02	12.0
34	.66	12.5	350	200.0	.460	8.19	3.94	.587	3.15	.815	.369	7.62	21.2		488.5	1.34	78.7
86	.65	6.0	160	43.2	.212	2.14	3.14	.330	3.13	.470	.251	9.24	11.6		77.6	.86	35.7
260	.71	7.9	139	38.4	.273	6.49	.87	.104	.96	.205	.119	8.00	4.2		47.8	1.35	47.2
0	.05	.1	14	.9	.010	.09	.88	.050	.58	.091	.036	1.88	2.0		8.3	.01	6.2
058	10.02	99.7	2275	378.4	3.560	185.20	6.51	1.038	6.71	1.880	1.125	100.04	99.6		517.1	21.03	915.6

APPENDIX 4. National mean values of basic analytical data on **pot-grown wheat** plants and respective soils. For analytical methods, see Appendix 2 and Section 1.3.

Country			GENERAL SOIL PROPERTIES											SOIL MACRONUTRIENTS				
		Particle size distr.				Tex. ind.	CEC	pH		El. cond.	$CaCO_3$ equiv.	Org. C	Volume weight		N (total)	P	K	Ca (extractable)
		<.002 %	.002–.06 %	.06–2 %	>2 mm %		me 100 g	H_2O	$CaCl_2$	$10^{-4}S$ cm	%	%	g/cm³	%	mg/l	mg/l	mg/l	mg/l

EUROPE AND OCEANIA

BELGIUM (n = 36)	mean	9	53	38	0	29	18.2	6.20	5.76	1.8	.3	1.2	1.25	.119	1483	103.4	225	1541
	± s	5	27	31	0	10	3.3	.72	.76	.8	1.2	.4	.05	.024	291	37.5	78	1000
	min.	2	12	3	0	15	12.8	4.85	4.45	.7	.0	.6	1.15	.081	983	45.6	108	490
	max.	19	84	85	0	43	24.5	7.90	7.40	3.9	7.1	2.1	1.37	.171	2223	187.0	406	5045
FINLAND (n = 90)	mean	28	48	25	0	45	32.0	5.72	5.20	1.6	.0	3.9	.99	.291	2725	60.6	206	1522
	± s	21	18	21	0	18	14.1	.52	.52	.7	.0	2.9	.13	.164	1079	32.7	101	988
	min.	1	11	2	0	16	12.5	4.60	4.10	.6	.0	1.4	.66	.106	1181	14.2	40	190
	max.	81	89	74	0	86	89.7	7.30	6.90	3.9	.4	17.5	1.22	1.023	7169	208.0	632	8800
HUNGARY (n = 201)	mean	29	52	19	0	46	29.6	7.23	6.83	2.4	3.6	1.6	1.21	.177	2116	34.1	226	4775
	± s	12	15	21	0	12	10.5	.83	.87	1.2	6.0	.6	.07	.055	604	21.2	136	1944
	min.	3	1	1	0	15	4.6	4.80	4.20	.5	.0	.4	1.05	.036	508	5.6	46	165
	max.	77	76	94	0	84	59.2	8.40	8.00	8.7	45.0	3.6	1.45	.331	3511	130.0	956	8850
ITALY (n = 170)	mean	32	39	27	3	46	26.1	7.50	7.10	5.2	9.7	1.3	1.18	.151	1783	24.1	293	5857
	± s	18	14	17	10	16	10.3	.69	.67	7.9	11.0	.6	.09	.096	1179	20.3	258	3905
	min.	9	3	1	0	19	9.4	5.18	4.65	.6	.0	.5	.96	.065	798	2.8	24	730
	max.	85	79	93	58	89	74.9	8.28	7.82	57.0	57.4	5.7	1.75	1.169	14729	116.4	1720	21930
MALTA (n = 25)	mean	28	39	33	0	43	22.2	7.91	7.54	5.9	58.6	1.3	1.17	.145	1711	80.4	365	6518
	± s	10	7	10	0	9	4.4	.10	.05	2.4	5.7	.3	.02	.034	414	51.3	226	744
	min.	18	25	20	0	32	17.9	7.78	7.48	3.2	52.0	.8	1.13	.115	1364	20.7	186	5640
	max.	51	53	51	0	61	29.8	8.08	7.64	9.1	68.4	2.1	1.20	.212	2536	171.8	789	8090
NEW ZEALAND (n = 35)	mean	40	46	15	0	55	49.0	5.93	5.46	2.0	.1	4.1	.99	.398	3660	30.8	403	2849
	± s	19	16	11	0	15	19.1	.51	.50	1.1	.4	4.8	.12	.254	1265	18.4	309	1803
	min.	11	21	1	0	32	20.5	4.94	4.50	.6	.0	2.0	.51	.211	2287	4.8	68	1070
	max.	75	78	43	0	82	99.9	7.07	6.77	4.5	2.6	30.8	1.14	1.657	8378	72.2	1063	7120

LATIN AMERICA

ARGENTINA (n = 208)	mean	23	63	15	0	43	27.6	6.10	5.48	1.5	.0	1.9	1.05	.200	2101	24.2	785	2030
	± s	3	10	12	0	5	5.4	.30	.31	1.7	.1	.7	.06	.068	716	20.2	197	518
	min.	11	24	2	0	22	17.0	5.24	4.79	.5	.0	.8	.89	.111	1175	.7	420	1020
	max.	36	81	64	0	54	45.3	7.41	6.96	24.4	.7	4.3	1.20	.433	4434	200.0	1649	5480
BRAZIL (n = 158)	mean	58	21	21	0	66	26.5	5.62	5.06	.9	.0	2.0	1.11	.180	1969	11.3	162	892
	± s	19	10	20	0	17	8.3	.50	.57	.3	.0	.9	.10	.057	561	10.3	99	486
	min.	7	3	2	0	17	6.4	4.55	3.97	.4	.0	.6	.92	.051	742	3.2	37	180
	max.	83	50	90	0	88	45.6	6.77	6.27	2.0	.0	4.7	1.46	.325	3778	74.2	588	2000
MEXICO (n = 142)	mean	36	38	26	0	50	32.4	7.69	7.22	3.5	5.0	.9	1.18	.101	1174	17.6	615	5516
	± s	19	15	21	1	16	14.4	.81	.80	4.3	10.9	.4	.09	.039	390	22.9	334	3001
	min.	2	7	1	0	12	2.3	4.60	4.10	.3	.0	.3	.88	.024	351	2.4	69	290
	max.	77	78	82	9	83	80.3	8.85	8.25	42.8	59.8	3.3	1.46	.342	3068	271.4	2019	19980

	SOIL MICRONUTRIENTS						MICRONUTRIENTS IN POT-GROWN WHEAT PLANTS							
a	B	Cu (extractable)	Fe	Mn	Mo	Zn	DM yield per pot	B	Cu	Fe	Mn	Mo	Zn	Country
/l	mg/l	mg/l	mg/l	mg/l	mg/l	mg/l	mg	ppm	ppm	ppm	ppm	ppm	ppm	
18	.52	4.8	327	39.8	.217	12.59	1189	7.0	7.3	54.0	82	.39	36.5	BELGIUM
15	.22	2.5	148	28.3	.085	11.23	180	1.1	2.8	11.2	75	.24	21.4	
0	.23	2.3	115	4.0	.089	2.84	826	5.3	3.6	33.6	21	.03	14.8	
71	1.09	13.1	750	104.6	.489	44.06	1542	9.5	20.6	77.4	326	.96	102.3	
13	.55	4.3	569	21.2	.540	2.73	1171	5.6	6.0	64.1	107	.27	23.9	FINLAND
8	.28	2.7	367	14.1	.357	2.64	184	1.3	1.7	15.7	76	.23	8.4	
2	.14	.9	130	1.0	.086	.34	648	3.4	3.1	37.8	26	.03	12.0	
47	1.88	16.5	1995	63.2	2.071	13.13	1506	10.9	11.5	132.0	597	1.19	67.4	
48	1.02	5.4	151	38.5	.141	1.17	1194	6.3	6.1	56.6	66	.24	15.1	HUNGARY
30	.48	2.5	85	31.4	.122	1.05	207	1.5	1.6	11.0	105	.16	5.9	
0	.11	.6	60	7.5	.030	.34	616	3.9	2.8	33.7	12	.01	5.1	
46	3.09	14.6	605	154.0	1.161	8.75	1795	13.0	14.6	104.3	1295	.82	56.5	
36	.82	13.7	212	21.6	.213	2.44	1266	5.9	9.0	59.8	55	.41	17.9	ITALY
48	.54	16.1	126	23.8	.201	5.36	222	2.0	2.9	12.0	29	.38	7.9	
2	.15	1.4	30	2.7	.007	.17	618	3.3	4.4	38.5	7	.03	7.1	
395	4.17	96.0	820	222.3	1.311	47.57	1837	21.4	18.0	114.5	187	3.88	59.8	
202	1.14	7.2	35	7.5	.112	10.29	1425	6.4	9.7	48.7	22	.64	38.5	MALTA
143	.53	2.5	11	2.0	.057	12.51	144	1.5	1.6	8.2	9	.28	13.9	
52	.63	5.0	22	4.9	.021	2.02	1094	4.5	7.5	35.9	9	.27	20.6	
441	2.20	12.9	52	11.3	.217	39.40	1650	11.0	12.8	63.8	45	1.42	65.1	
40	.76	12.0	472	51.3	.179	2.99	1304	5.3	6.4	62.8	121	.07	22.6	NEW ZEALAND
26	.37	15.9	405	48.3	.169	3.24	293	1.3	2.6	18.6	54	.08	9.8	
11	.31	.6	88	6.3	.035	.28	817	3.3	2.1	31.3	34	.02	9.6	
117	1.78	42.9	2275	237.0	.940	18.29	2211	8.9	10.7	117.1	284	.47	46.4	
23	.74	3.2	203	53.6	.375	2.42	1372	5.2	4.0	59.1	112	.18	16.7	ARGENTINA
40	.22	.7	72	21.9	.220	3.15	196	1.0	1.1	13.6	51	.26	5.6	
4	.41	1.7	105	13.1	.132	.41	799	3.3	1.9	32.5	23	.02	7.8	
292	1.91	7.0	506	123.9	1.949	28.69	1906	9.6	8.0	103.6	301	3.15	47.7	
4	.63	16.0	166	107.4	.198	3.96	1168	4.5	9.2	60.7	288	.03	24.0	BRAZIL
3	.17	15.2	116	87.4	.072	16.39	182	.9	3.2	11.2	319	.04	13.1	
0	.26	1.4	42	2.4	.049	.20	599	3.3	3.5	42.5	23	.01	10.1	
18	1.10	81.1	700	286.0	.458	123.90	1526	7.3	16.8	92.4	1709	.23	81.4	
277	1.26	4.0	67	25.9	.191	1.45	1413	7.3	6.4	50.6	71	.42	16.1	MEXICO
413	1.39	3.8	56	24.9	.164	3.05	297	4.6	2.2	13.1	67	.37	8.2	
2	.12	.3	10	.9	.020	.11	619	2.8	1.7	29.3	13	.01	6.2	
748	9.99	54.4	535	163.2	1.151	26.95	2116	44.4	14.0	115.4	647	2.85	64.3	

APPENDIX 4 (cont.)

Country		Particle size distr. <.002 %	.002—.06 %	.06—2 %	>2 mm %	Tex. ind.	CEC me/100 g	pH H₂O	pH CaCl₂	El. cond. 10⁻⁴S cm	CaCO₃ equiv. %	Org. C %	Volume weight g/cm³	N (total) %	N (total) mg/l	P mg/l	K (extractable) mg/l	Ca (extractable) mg/l	m
PERU (n = 68)	mean	23	30	36	11	35	25.0	7.24	6.85	2.8	4.0	1.4	1.24	.168	2041	19.2	214	4310	3
	±s	13	11	15	15	13	9.0	.85	.85	2.1	7.1	.6	.14	.074	808	16.2	130	2258	
	min.	2	12	1	0	9	13.5	4.54	4.00	.4	.0	.3	.98	.053	664	2.4	56	280	
	max.	55	57	73	55	65	56.4	8.14	7.65	13.7	47.0	3.4	1.58	.420	4402	79.8	777	10840	10
FAR EAST																			
INDIA (n = 258)	mean	16	51	32	0	35	17.4	8.04	7.53	2.3	4.0	.6	1.26	.071	886	8.8	163	3291	3
	±s	14	18	21	0	12	13.9	.53	.51	2.2	9.4	.2	.08	.020	211	7.8	109	2505	3
	min.	2	11	1	0	16	5.1	6.02	5.14	.3	.0	.2	.95	.009	95	1.4	41	380	
	max.	62	91	85	8	73	76.5	9.10	8.32	16.7	40.4	1.3	1.50	.142	1722	54.0	856	12040	32
KOREA, REP. (n = 90)	mean	18	45	37	0	36	17.1	6.05	5.59	2.2	.8	1.2	1.13	.131	1442	46.0	161	1242	2
	±s	10	13	21	0	11	4.5	.90	.93	1.4	4.6	.4	.10	.044	408	31.0	89	662	2
	min.	3	18	4	0	18	6.7	4.40	4.10	.5	.0	.3	.88	.035	451	5.0	45	300	
	max.	42	81	77	0	56	36.0	8.00	7.50	10.0	42.9	2.6	1.32	.270	2641	140.0	492	4000	18
NEPAL (n = 35)	mean	19	57	22	0	39	20.9	6.69	6.06	1.1	.4	1.1	1.11	.124	1342	22.0	86	1693	2
	±s	9	12	13	0	8	5.9	.83	.94	.5	.9	.5	.11	.042	356	29.7	52	1083	1
	min.	9	23	3	0	25	11.9	5.19	4.50	.3	.0	.5	.84	.071	862	1.2	31	220	4
	max.	40	76	56	0	57	36.5	8.14	7.54	2.5	4.3	2.0	1.27	.222	2032	121.4	323	4660	45
PAKISTAN (n = 237)	mean	19	61	19	0	40	16.6	8.33	7.77	4.1	6.5	.6	1.23	.080	975	6.9	226	4379	33
	±s	8	12	15	0	8	4.9	.30	.25	6.4	4.2	.3	.08	.025	274	6.6	121	824	14
	min.	4	10	1	0	19	7.5	6.85	6.15	.7	.3	.2	1.04	.008	97	1.4	55	1200	5
	max.	50	79	82	0	65	41.2	9.25	8.56	57.5	20.0	3.2	1.43	.169	1785	51.6	945	7240	92
PHILIPPINES (n = 194)	mean	32	41	27	0	47	35.5	6.49	5.99	1.5	.7	1.2	1.09	.139	1477	18.6	212	3842	80
	±s	21	16	19	0	18	12.8	.76	.82	1.2	1.9	.7	.10	.063	564	19.0	213	2769	86
	min.	3	9	0	0	17	6.7	4.56	4.23	.4	.0	.3	.80	.039	487	1.5	18	100	3
	max.	87	77	81	0	91	67.2	8.10	7.56	11.6	16.0	4.8	1.42	.527	4313	138.6	1997	12800	649
SRI LANKA (n = 18)	mean	21	11	66	2	31	13.6	6.34	5.77	1.0	.1	1.0	1.29	.108	1383	12.9	145	1286	19
	±s	11	5	14	8	10	3.8	.73	.82	.4	.2	.4	.08	.026	324	12.7	62	896	14
	min.	3	3	40	0	13	6.8	5.32	4.56	.5	.0	.5	1.07	.075	1028	2.2	21	310	3
	max.	47	20	94	33	55	22.1	8.02	7.60	1.8	.8	2.1	1.45	.155	2107	48.4	258	3700	57
THAILAND (n = 150)	mean	44	34	21	1	56	41.8	7.31	6.80	1.3	4.8	1.6	1.21	.157	1884	11.6	216	6444	44
	±s	16	12	15	4	14	17.0	.76	.80	.9	10.1	.5	.07	.047	513	18.8	131	4244	30
	min.	6	9	1	0	12	11.8	5.30	4.50	.3	.0	.7	1.00	.076	1045	.1	40	600	3
	max.	81	63	84	25	85	82.8	8.20	7.70	10.5	50.0	3.7	1.43	.380	4134	165.2	849	15390	1762
NEAR EAST																			
EGYPT (n = 198)	mean	46	36	18	0	58	43.2	8.15	7.74	6.3	4.4	1.2	1.18	.125	1459	17.3	679	6911	1395
	±s	15	12	21	0	14	10.5	.23	.14	4.9	2.4	.3	.06	.030	320	13.7	310	1294	493
	min.	5	1	1	0	15	5.9	7.57	7.38	1.2	1.0	.4	1.07	.044	642	3.4	82	1640	363
	max.	82	60	94	0	87	58.5	8.95	8.20	32.0	24.1	1.8	1.51	.221	2537	101.1	2325	9920	2946
IRAQ (n = 150)	mean	34	52	13	0	51	26.2	8.06	7.72	9.4	25.3	.8	1.17	.095	1106	8.1	333	7138	680
	±s	11	11	12	2	10	7.1	.16	.16	11.9	6.1	.3	.07	.035	402	9.3	138	1951	317
	min.	7	9	1	0	18	11.0	7.68	7.22	1.2	2.1	.3	1.00	.031	354	.7	120	4200	190
	max.	63	79	84	28	74	43.8	8.65	8.55	73.0	43.5	1.8	1.38	.239	2993	69.7	1028	17995	1555

	SOIL MICRONUTRIENTS						MICRONUTRIENTS IN POT-GROWN WHEAT PLANTS							
a	B	Cu (extractable)	Fe	Mn	Mo	Zn	DM yield per pot	B	Cu	Fe	Mn	Mo	Zn	Country
/l	mg/l	mg/l	mg/l	mg/l	mg/l	mg/l	mg	ppm	ppm	ppm	ppm	ppm	ppm	
6	.65	5.6	161	25.9	.238	2.56	1339	5.7	7.2	62.1	78	.44	21.5	PERU
4	.44	3.6	88	23.9	.272	5.45	212	1.1	2.0	14.8	92	.39	13.0	
2	.21	.4	30	6.3	.046	.29	759	3.8	3.3	39.6	24	.02	7.7	
7	2.60	23.0	394	138.0	1.853	34.35	1911	10.4	11.8	108.1	593	1.74	81.6	
8	.42	4.3	165	14.5	.112	.82	1215	6.1	7.5	63.4	48	.36	13.7	INDIA
9	.35	5.3	200	8.3	.095	2.65	215	2.8	2.3	25.8	30	.26	5.6	
1	.05	.6	29	3.8	.010	.09	602	2.9	2.3	33.6	11	.02	5.9	
5	2.10	82.0	930	55.9	.936	41.89	1774	22.7	16.5	298.0	232	1.41	58.3	
21	.37	3.8	177	39.4	.247	6.78	1180	4.8	6.5	54.0	199	.22	25.5	KOREA, REP.
1	27	3.3	155	36.5	.395	25.91	229	1.5	1.9	12.2	295	.20	21.9	
2	.08	.5	25	3.3	.018	.32	666	3.0	3.1	36.4	14	.02	10.2	
8	1.92	24.6	905	149.5	3.241	186.45	1763	10.9	11.2	95.1	1625	1.03	204.4	
29	.19	6.9	362	22.7	.088	.88	1220	4.1	8.6	60.4	80	.13	16.9	NEPAL
23	.07	8.2	254	16.8	.039	1.24	200	1.0	2.2	14.0	37	.13	4.9	
0	.10	1.7	105	5.4	.025	.15	637	2.5	3.3	39.0	28	.01	9.1	
96	.43	40.4	1125	74.1	.204	6.72	1657	6.5	12.3	111.0	175	.60	29.0	
65	.68	4.7	116	9.9	.181	.75	1255	8.8	8.3	57.6	47	.68	14.2	PAKISTAN
75	.57	2.1	58	5.1	.143	1.24	235	7.9	2.0	15.6	21	.40	5.7	
5	.08	1.3	28	3.6	.022	.10	534	3.1	4.2	31.9	13	.08	5.5	
65	4.42	22.7	371	57.3	1.049	12.10	1787	57.5	13.6	111.3	179	2.40	52.3	
56	.28	13.4	273	44.8	.228	3.29	1204	4.3	10.5	79.1	185	.22	28.5	PHILIPPINES
87	.15	10.7	132	35.8	.190	16.01	242	.8	1.9	19.5	137	.24	14.3	
1	.07	1.4	63	2.9	.036	.22	617	2.5	3.7	45.5	27	.01	9.5	
81	.86	93.0	840	226.0	1.585	185.20	1780	8.1	15.4	156.9	1248	1.57	166.2	
12	.32	4.1	154	67.1	.208	4.26	1189	3.9	7.5	61.2	332	.15	25.6	SRI LANKA
14	.16	2.8	104	50.3	.139	12.16	255	.8	1.6	12.3	351	.16	15.5	
4	.16	1.3	50	8.5	.065	.55	664	2.3	5.1	41.6	44	.03	12.6	
66	.81	10.7	470	203.0	.510	52.81	1659	5.4	10.1	95.0	1380	.60	83.3	
18	.42	6.3	140	52.5	.305	1.60	1577	4.1	6.6	67.0	114	.26	18.1	THAILAND
40	.15	8.8	104	34.2	.454	2.66	299	.8	1.7	19.2	76	.20	9.2	
0	.16	.2	22	10.0	.026	.20	647	2.8	1.7	39.8	39	.01	7.2	
41	.83	99.9	775	162.0	3.326	20.51	2339	6.7	14.6	169.9	579	1.07	60.4	
72	.93	11.5	193	12.0	.142	1.18	1460	5.5	8.3	61.5	41	.51	14.2	EGYPT
80	.38	3.7	93	4.5	.044	1.04	217	1.6	1.5	13.2	14	.18	5.2	
48	.30	2.5	44	5.6	.023	.29	623	2.8	3.6	30.5	10	.14	5.7	
98	2.86	26.4	675	39.9	.304	9.18	1995	18.8	12.5	115.8	96	1.14	43.5	
414	1.51	4.7	112	9.2	.125	.28	1357	13.5	8.7	65.3	68	.56	11.1	IRAQ
519	1.61	1.2	44	3.0	.110	.29	187	14.9	2.2	17.1	23	.40	3.0	
4	.16	2.1	32	3.6	.019	.07	491	3.0	5.0	38.9	27	.02	6.3	
058	9.99	9.8	220	21.2	.656	3.56	1695	99.9	20.7	160.0	170	1.63	32.0	

APPENDIX 4 (cont.)

Country		Particle size distr.				Tex. ind.	CEC	pH H₂O	pH CaCl₂	El. cond.	CaCO₃ equiv.	Org. C	Volume weight	N (total)	P	K	Ca (extractable)	Mg
		<.002 %	.002–.06 %	.06–2 %	>2 mm %	%	me/100 g			10⁻⁴S cm	%	%	g/cm³	%	mg/l	mg/l	mg/l	mg/l
LEBANON (n=16)	mean	47	35	17	0	60	38.3	7.80	7.44	3.9	17.1	.9	1.20	.097	1158	42.5	354	7789	5
	±s	17	16	11	0	13	9.4	.37	.29	3.6	20.7	.3	.04	.024	287	31.0	234	1919	4
	min.	15	19	1	0	38	23.8	7.10	6.95	1.5	.4	.4	1.10	.061	736	5.3	101	5000	1
	max.	78	70	33	0	84	61.9	8.10	7.75	13.5	67.1	1.3	1.26	.137	1722	104.6	1070	11950	15
SYRIA (n=38)	mean	40	38	18	4	54	38.4	7.99	7.59	3.6	25.8	1.2	1.21	.127	1480	18.1	443	8429	9
	±s	19	16	18	10	16	13.0	.23	.20	3.5	15.5	1.0	.10	.089	875	21.9	235	3166	5
	min.	7	9	2	0	18	9.3	7.28	6.90	1.3	1.4	.3	.86	.023	305	2.0	136	3310	1
	max.	75	72	84	43	82	74.3	8.51	7.92	18.2	61.4	5.6	1.39	.480	4442	111.9	965	17700	26
TURKEY (n=298)	mean	35	35	29	0	49	31.6	7.81	7.40	1.6	10.5	.9	1.21	.098	1171	11.6	378	6240	4
	±s	15	10	17	0	13	11.0	.56	.58	1.1	11.1	.4	.07	.038	415	9.6	211	2196	3
	min.	3	12	1	0	17	7.7	5.60	5.20	.3	.0	.2	1.00	.026	357	1.0	46	660	
	max.	79	67	75	0	85	64.2	8.90	8.55	12.0	54.2	2.8	1.47	.288	3192	73.0	1338	13160	17

AFRICA

Country		Particle size distr.				Tex. ind.	CEC	pH H₂O	pH CaCl₂	El. cond.	CaCO₃ equiv.	Org. C	Volume weight	N (total)	P	K	Ca (extractable)	Mg
ETHIOPIA (n=125)	mean	50	32	17	0	62	48.2	6.42	5.76	.9	.2	2.1	1.06	.189	1954	15.3	644	3831	5
	±s	19	11	14	1	16	13.9	.77	.87	.6	.4	.9	.11	.081	758	33.3	356	3022	
	min.	5	12	1	0	20	16.2	4.60	4.10	.3	.0	.8	.76	.056	678	1.2	87	510	
	max.	79	63	66	11	92	79.0	8.30	7.65	3.2	1.8	4.0	1.25	.394	3826	311.0	2077	11550	17
GHANA (n=93)	mean	18	19	58	5	30	14.7	6.35	5.84	.9	.1	1.3	1.30	.131	1676	8.1	128	1148	2
	±s	8	12	18	10	9	6.0	.55	.65	.4	.1	.5	.10	.056	667	10.7	68	523	1
	min.	4	7	17	0	13	4.9	4.80	4.35	.3	.0	.6	1.08	.053	727	1.8	34	71	
	max.	37	50	88	56	55	41.2	7.37	7.05	2.3	.5	2.7	1.52	.333	3770	95.8	302	3150	14
MALAWI (n=97)	mean	24	17	59	0	35	17.9	5.91	5.38	2.6	.1	1.4	1.29	.116	1468	39.8	229	1539	2
	±s	11	8	14	1	10	8.3	.56	.51	3.0	.2	.6	.12	.042	462	26.4	145	895	1
	min.	4	8	8	0	15	5.9	4.89	4.58	.3	.0	.3	.76	.039	298	2.6	38	200	
	max.	49	47	87	6	61	41.1	7.30	6.80	16.7	.8	3.3	1.51	.257	2864	113.4	693	3838	7
NIGERIA (n=153)	mean	16	25	58	1	29	14.5	6.32	5.75	1.4	.4	.9	1.34	.088	1141	18.6	173	1225	2
	±s	15	17	25	6	15	8.3	.93	1.05	1.5	2.9	.6	.14	.046	523	33.2	146	1384	29
	min.	2	1	5	0	12	2.4	4.57	3.95	.2	.0	.1	1.03	.030	486	1.2	35	50	1
	max.	75	67	97	45	82	47.1	9.17	7.96	10.5	26.8	4.9	1.77	.392	4500	303.2	1042	7000	246
SIERRA LEONE (n=48)	mean	21	18	60	0	33	21.1	5.73	4.92	.8	.0	2.5	1.09	.201	2107	12.5	110	592	
	±s	9	11	15	0	9	7.6	.51	.59	.6	.2	1.0	.09	.092	767	21.3	54	550	6
	min.	11	6	25	0	22	10.7	4.85	4.10	.3	.0	.8	.88	.103	1231	4.4	47	100	1
	max.	48	46	82	0	58	44.7	7.50	7.10	3.0	.9	5.7	1.22	.492	4512	152.0	280	2840	41
TANZANIA (n=163)	mean	30	16	55	0	40	23.0	6.78	6.25	.9	.4	1.3	1.26	.112	1314	32.0	556	2036	36
	±s	22	9	26	1	21	15.2	.66	.75	.5	.8	.8	.18	.067	675	68.0	550	1746	26
	min.	2	2	3	0	13	3.1	5.20	4.39	.1	.0	.1	.95	.014	237	.8	18	110	2
	max.	85	46	92	15	88	60.0	8.30	7.80	3.2	7.6	4.0	1.72	.409	4747	656.0	3867	9625	135
ZAMBIA (n=44)	mean	15	18	67	0	27	11.4	5.74	5.03	.8	.0	.8	1.36	.073	961	20.4	145	675	12
	±s	11	9	17	2	11	6.3	.63	.75	.7	.1	.4	.12	.035	386	20.2	204	573	
	min.	1	1	28	0	11	3.4	4.85	4.18	.2	.0	.2	1.10	.027	448	2.8	18	10	
	max.	44	41	97	11	55	28.2	7.64	7.14	3.4	.5	2.0	1.66	.180	2136	114.5	1272	2650	46
WHOLE INTERNAT. MATERIAL (n=3538)	mean	30	40	30	1	45	27.6	7.14	6.64	2.7	5.1	1.3	1.19	.133	1513	21.1	349	4059	46
	±s	18	19	24	4	16	14.6	1.06	1.12	4.5	10.1	1.1	.13	.083	767	29.0	307	3030	45
	min.	1	1	0	0	9	2.3	4.40	3.95	.1	.0	.1	.51	.008	95	.1	18	10	
	max.	87	91	97	58	92	99.9	9.25	8.56	73.0	68.4	30.8	1.77	1.657	14729	656.0	3867	21930	649

439

	SOIL MICRONUTRIENTS						MICRONUTRIENTS IN POT-GROWN WHEAT PLANTS							
a	B	Cu	Fe (extractable)	Mn	Mo	Zn	DM yield per pot	B	Cu	Fe	Mn	Mo	Zn	Country
/l	mg/l	mg/l	mg/l	mg/l	mg/l	mg/l	mg	ppm	ppm	ppm	ppm	ppm	ppm	
44	.54	6.0	98	17.9	.077	.85	1426	5.1	7.7	65.6	62	.23	16.7	LEBANON
31	.16	2.0	37	11.5	.068	.42	267	.7	1.4	13.2	21	.18	6.6	
15	.34	4.1	24	6.8	.014	.21	928	4.1	6.1	47.1	26	.01	8.4	
26	.97	11.5	156	37.9	.250	1.67	1912	6.8	11.1	95.0	93	.72	33.4	
21	1.08	4.8	87	15.0	.119	1.21	1288	6.6	7.2	62.5	48	.36	14.8	SYRIA
30	.73	1.9	67	17.0	.097	2.15	193	2.5	2.0	11.1	17	.39	7.6	
20	.22	1.9	25	2.8	.024	.10	663	3.0	3.7	45.8	8	.03	7.5	
38	3.45	12.8	380	97.5	.529	11.36	1601	15.5	13.3	94.8	91	1.85	40.0	
58	1.10	5.7	110	20.4	.120	.62	1253	6.6	7.5	54.2	59	.25	12.0	TURKEY
33	1.06	3.0	86	22.7	.126	1.25	200	5.5	1.7	12.6	23	.26	4.4	
3	.06	.9	13	3.4	.020	.12	590	3.1	3.8	32.0	16	.01	5.5	
25	9.99	20.2	558	167.0	.956	16.16	1871	79.9	13.5	113.9	157	2.00	40.7	
31	.54	4.6	214	86.5	.657	3.60	1320	4.8	4.8	59.2	123	.27	23.1	ETHIOPIA
52	.23	2.9	129	50.9	.519	5.21	267	.9	1.9	18.4	84	.89	13.1	
0	.16	.1	49	6.5	.093	.23	682	2.8	.9	37.2	28	.01	6.6	
493	1.24	14.3	825	213.5	3.560	43.63	1998	7.8	10.0	149.8	433	9.79	73.0	
6	.57	1.9	94	62.5	.082	1.06	1244	6.0	4.8	65.7	163	.10	19.3	GHANA
9	.22	1.2	37	27.3	.029	.96	248	.9	1.5	13.4	104	.10	6.0	
0	.24	.4	32	13.6	.026	.16	686	3.9	2.0	43.4	45	.01	9.7	
71	1.37	5.7	225	138.1	.160	4.36	1756	8.4	8.5	116.0	601	.75	46.0	
4	.29	3.3	130	85.4	.184	1.61	1245	4.3	5.9	65.1	357	.09	22.6	MALAWI
4	.11	2.7	66	54.2	.102	.94	240	1.0	1.9	19.0	368	.09	7.1	
0	.12	.3	39	18.7	.049	.17	719	2.4	1.9	36.4	47	.01	9.0	
19	.69	22.3	316	378.4	.748	4.10	1931	7.0	14.1	169.4	2009	.53	42.0	
55	.27	1.9	110	35.3	.126	2.70	1246	4.5	5.3	60.2	205	.19	23.6	NIGERIA
241	.17	1.7	105	29.8	.141	7.39	254	1.1	1.8	15.5	213	.29	15.1	
0	.03	.2	15	2.8	.011	.18	640	2.4	1.5	34.5	20	.01	9.1	
880	1.31	13.4	870	203.0	1.248	57.19	1819	7.6	12.1	135.8	1660	2.14	120.4	
6	.39	1.1	124	7.8	.039	1.56	1241	5.1	3.7	73.3	144	.07	24.7	SIERRA LEONE
6	.14	.6	71	7.2	.017	1.95	200	.8	1.2	17.2	47	.07	12.5	
2	.17	.3	33	1.4	.017	.29	856	3.1	1.8	45.2	33	.01	11.4	
25	.81	3.1	400	45.5	.099	11.15	1644	7.4	7.5	115.1	272	.38	79.9	
27	.78	4.8	113	53.0	.305	1.56	1353	5.0	5.8	63.4	138	.40	20.4	TANZANIA
70	.52	4.6	93	32.2	.404	1.94	249	.9	2.1	18.2	103	.43	8.7	
0	.15	.2	20	7.9	.019	.12	710	3.1	1.8	35.2	24	.01	7.1	
570	2.97	24.8	570	161.2	2.724	15.95	1928	7.7	10.7	135.8	813	2.93	63.4	
4	.25	2.4	72	52.6	.105	24.32	982	4.4	6.0	62.3	555	.11	21.5	ZAMBIA
5	.10	3.3	57	36.6	.090	12.12	242	.9	2.4	16.4	451	.11	5.8	
0	.10	.1	20	8.1	.019	.17	608	3.2	2.4	38.5	59	.01	13.0	
34	.66	12.5	350	200.0	.460	8.19	1583	7.3	10.5	111.0	1703	.65	35.7	
126	.73	6.0	166	34.7	.210	1.97	1293	6.1	7.0	60.9	112	.32	18.3	WHOLE
320	.75	7.0	157	36.9	.257	7.05	254	4.8	2.6	17.3	155	.36	10.8	INTERNAT.
0	.03	.1	10	.9	.007	.07	491	2.3	.9	29.3	7	.01	5.1	MATERIAL
058	9.99	99.9	2275	378.4	3.560	186.45	2339	99.9	20.7	298.0	2009	9.79	204.4	

APPENDIX 5. Indicative data on fertilizer application to the original indicator crops in different countries[1].

Country	Number of sites		Nitrogen kg N/ha		Phosphorus kg P/ha		Potassium kg K/ha	
	Wheat fields	Maize fields	Wheat fields	Maize fields	Wheat fields	Maize fields	Wheat fields	Maize fields
Belgium	21	20	98 ± 21	118 ± 66	42 ± 9	45 ± 26	130 ± 29	122 ± 90
Finland	94		76 ± 24		43 ± 26		63 ± 19	
Hungary	144	106	140 ± 74	136 ± 58	52 ± 55	52 ± 33	98 ± 96	104 ± 56
Italy	118	70	105 ± 66	186 ± 127	42 ± 29	51 ± 44	29 ± 31	80 ± 98
Malta	25		60 ± 0		26 ± 0		83 ± 0	
New Zealand	14	24	9 ± 16	55 ± 67	16 ± 7	30 ± 29	7 ± 12	44 ± 25
Argentina	119	90	3 ± 14	6 ± 16	0 ± 0	6 ± 18	0 ± 0	0 ± 0
Brazil	71		19 ± 11		49 ± 30		33 ± 18	
Ecuador	16		44 ± 13		23 ± 8		13 ± 6	
Mexico	100	147	122 ± 54	40 ± 49	9 ± 13	4 ± 7	1 ± 5	0 ± 1
Peru	13	57	38 ± 35	46 ± 58	10 ± 13	8 ± 13	12 ± 19	9 ± 17
India	188	107	75 ± 51	106 ± 57	15 ± 15	18 ± 12	11 ± 20	9 ± 19
Korea, Rep.	50	50	100 ± 25	107 ± 31	31 ± 7	35 ± 14	58 ± 10	77 ± 80
Nepal	50		68 ± 70		15 ± 8		16 ± 20	
Pakistan	156	86	58 ± 57	83 ± 90	9 ± 15	12 ± 29	0 ± 3	2 ± 8
Philippines		197		8 ± 19		1 ± 4		1 ± 4
Sri Lanka		21		24 ± 26		10 ± 9		14 ± 12
Thailand		150		0 ± 0		0 ± 0		0 ± 0
Egypt	100	100	129 ± 31	161 ± 39	9 ± 7	2 ± 6	0 ± 0	0 ± 0
Iraq	119	31	5 ± 9	27 ± 24	2 ± 4	10 ± 10	1 ± 3	0 ± 1
Lebanon		16		53 ± 99		33 ± 54		0 ± 0
Syria	20	18	52 ± 59	117 ± 87	7 ± 11	7 ± 12	0 ± 0	0 ± 0
Turkey	249	50	34 ± 30	61 ± 50	18 ± 13	6 ± 11	0 ± 1	0 ± 0
Ethiopia	54	71	11 ± 10	7 ± 11	10 ± 10	5 ± 7	0 ± 0	0 ± 0
Ghana		93		48 ± 10		11 ± 11		14 ± 12
Malawi		100		81 ± 63		17 ± 14		11 ± 26
Nigeria	42	103	85 ± 27	13 ± 23	17 ± 8	4 ± 6	4 ± 11	6 ± 10
Sierra Leone		49		5 ± 22		2 ± 7		3 ± 15
Tanzania	5	175	16 ± 36	8 ± 17	2 ± 4	2 ± 6	0 ± 0	0 ± 0
Zambia		45		190 ± 84		35 ± 35		25 ± 29
Internat. average	1768	1976	66 ± 61	27 ± 37	21 ± 27	6 ± 10	21 ± 44	7 ± 18

[1] The data are based on information given by co-operators and sample collectors in "Field Information Forms". Because of several missing figures the data are not complete and, therefore, must be considered only as indicative. It should be pointed out that in many cases the samples come from fertilizer trial fields and therefore, the indicated fertilization level may be considerable higher than elsewhere in the country.

APPENDIX 6. Average general soil properties and macronutrient contents of soils classified by FAO/Unesco soil units. For analytical methods, see Section 1.3.

		Texture index	pH(CaCl$_2$)	El. cond. 10^{-4} S cm	CaCO$_3$ equiv. %	CEC me 100 g	Org. C %	N total %	P mg/l	K mg/l	Ca mg/l	Mg mg/l	Na mg/l
											(extractable)		
Acrisols (n = 73)	x̄	42	5.19	1.2	0.0	19.4	1.1	0.122	15.0	156	1208	238	14
	s	16	0.49	0.9	0.0	11.1	0.7	0.059	14.7	97	653	174	15
Andosols (n = 8)	x̄	33	5.26	1.9	0.4	35.5	2.6	0.297	16.7	388	1581	111	14
	s	14	0.63	0.9	0.7	13.0	1.9	0.157	6.8	226	803	94	13
Arenosols (n = 22)	x̄	19	5.13	1.6	0.2	13.7	1.1	0.107	82.2	163	793	83	13
	s	8	0.83	1.0	0.2	8.1	0.6	0.053	58.8	89	682	86	11
Cambisols (n = 246)	x̄	40	6.41	1.8	2.1	22.2	1.8	0.159	29.5	179	2549	271	46
	s	13	1.27	1.0	5.7	12.3	2.2	0.136	29.4	104	1688	186	77
Chernozems (n = 48)	x̄	47	7.20	2.3	5.6	28.1	1.6	0.183	30.4	174	5350	289	16
	s	7	0.70	0.7	5.1	4.3	0.4	0.044	15.4	49	1609	171	18
Ferralsols (n = 127)	x̄	32	5.26	0.9	0.0	14.2	1.3	0.116	12.6	112	813	136	9
	s	13	0.74	0.5	0.1	5.3	0.5	0.049	18.2	72	632	89	7
Fluvisols (n = 470)	x̄	51	7.44	5.2	7.6	33.4	1.1	0.123	18.1	475	5859	904	397
	s	16	0.68	6.8	9.6	13.9	0.5	0.073	21.2	327	2539	613	543
Gleysols (n = 31)	x̄	36	5.62	2.1	0.8	21.3	1.7	0.179	47.2	206	1600	269	17
	s	14	1.00	1.1	2.3	12.7	1.2	0.116	40.8	131	1504	276	17
Halosols (n = 60)	x̄	50	7.78	12.3	21.0	23.7	0.8	0.097	10.7	265	6017	780	713
	s	9	0.29	11.8	11.2	6.0	0.3	0.036	13.1	142	1964	336	767
Histosols (n = 7)	x̄	—	4.54	2.2	0.0	82.0	20.3	0.968	44.2	151	2079	320	22
	s	—	0.51	0.7	0.0	15.3	9.6	0.412	13.7	54	1068	243	16
Kastanozems (n = 64)	x̄	52	7.43	3.2	11.2	34.7	0.9	0.105	12.3	530	7053	674	135
	s	15	0.50	3.6	16.5	15.9	0.3	0.032	12.4	267	3535	393	214
Lithosols (n = 23)	x̄	40	7.25	1.3	10.4	26.6	0.8	0.088	8.2	356	5786	305	8
	s	18	0.64	0.4	11.6	12.7	0.4	0.046	4.4	158	3231	155	3
Luvisols (n = 217)	x̄	43	6.79	1.9	7.7	24.1	1.0	0.105	24.0	244	4127	404	53
	s	14	0.98	1.3	12.6	12.4	0.5	0.047	29.4	186	2814	382	115
Nitosols (n = 106)	x̄	28	5.60	0.9	0.1	13.9	1.1	0.104	14.9	152	1008	187	5
	s	9	0.85	0.6	0.1	6.8	0.5	0.054	17.6	151	687	168	6
Phaeozems (n = 307)	x̄	44	6.08	1.8	1.3	29.1	1.9	0.198	26.4	581	3280	355	34
	s	9	0.88	1.6	3.4	8.1	0.7	0.067	21.4	324	1893	178	112
Planosols (n = 117)	x̄	52	6.32	3.3	0.6	37.3	2.2	0.210	24.6	545	4220	625	74
	s	14	1.07	2.2	1.0	8.6	0.8	0.087	12.2	277	2107	343	36
Podzols (n = 11)	x̄	23	5.23	1.9	0.0	24.0	3.1	0.221	80.3	182	1460	87	14
	s	8	0.46	0.8	0.0	7.1	1.3	0.079	40.2	91	1619	34	16
Regosols (n = 42)	x̄	42	6.62	2.6	5.0	21.8	1.1	0.114	33.8	348	4037	361	151
	s	18	0.97	2.9	10.5	11.9	0.5	0.047	31.2	283	3524	363	484
Rendzinas (n = 48)	x̄	60	7.40	1.6	12.4	57.8	1.8	0.168	11.9	264	10730	573	22
	s	14	0.29	0.4	14.3	11.9	0.5	0.053	10.3	169	2869	315	25
Vertisols (n = 135)	x̄	58	7.16	2.8	8.9	41.1	1.1	0.112	12.0	352	7256	681	104
	s	12	0.72	5.3	12.1	14.4	0.6	0.059	14.2	237	2854	451	278
Yermosols (n = 92)	x̄	44	7.83	6.2	6.1	20.3	0.6	0.082	7.3	438	5268	480	447
	s	10	0.24	9.7	4.3	10.6	0.2	0.020	7.8	307	1770	270	591
Xerosols (n = 101)	x̄	48	7.62	4.4	12.4	29.0	0.8	0.096	13.5	601	6207	659	227
	s	13	0.27	5.9	13.7	10.5	0.3	0.032	15.1	344	2300	362	301

APPENDIX 7. Average micronutrient (B, Cu, Fe, Mn, Mo and Zn) contents of soils and respective pot-grown wheats for FAO/Unesco soil units: Acr = Acrisol (n = 73), And = Andosol (8), Are = Arenosol (22), Cam = Cambisol (246), Che = Chernozem (48), Fer = Ferralsol (127), Flu = Fluvisol (470), Gle = Gleysol (31), Hal = Halosol (60), His = Histosol (7), Kas = Kastanozem (64), Lit = Lithosol (23), Luv = Luvisol (217), Nit = Nitosol (106), Pha = Phaeozem (307), Pla = Planosol (11), Pod = Podzol (27), Reg = Regosol (42), Ren = Rendzina (48), Ver = Vertisol (135), Yer = Yermosol (92), Xer = Xerosol (101). See also Section 2.8.

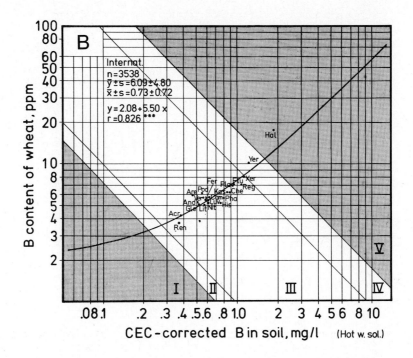

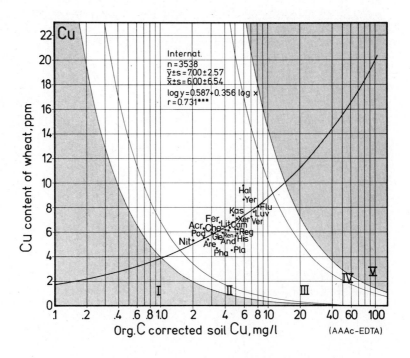

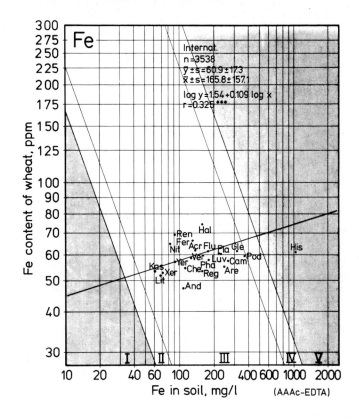
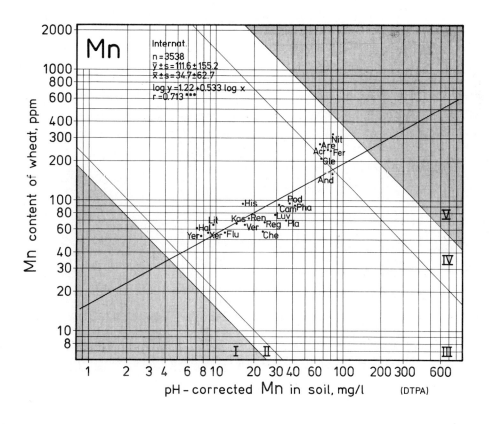

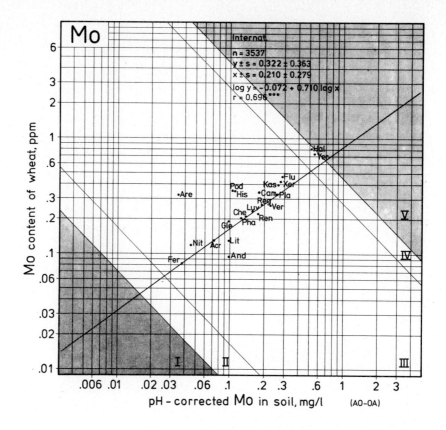

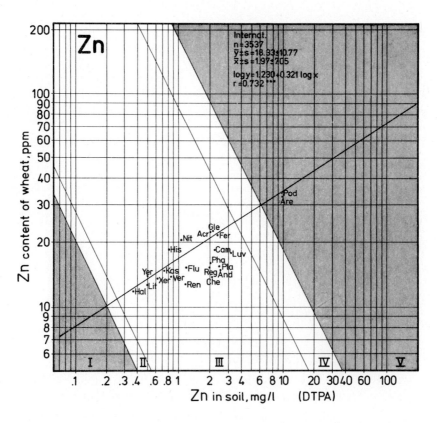

FAO SALES AGENTS AND BOOKSELLERS

Algeria	Société nationale d'édition et de diffusion, 92, rue Didouche Mourad, Algiers.
Argentina	Editorial Hemisferio Sur S.A., Librería Agropecuaria, Pasteur 743, 1028 Buenos Aires.
Australia	Hunter Publications, 58A Gipps Street, Collingwood, Vic. 3066; Australian Government Publishing Service, P.O. Box 84, Canberra, A.C.T. 2600; and Australian Government Service Bookshops at 12 Pirie Street, Adelaiden S.A.; 70 Alinga Street, Canberra, A.C.T.; 162 Macquarie Street Hobart, Tas.; 347 Swanson Street, Melbourne, Vic.; 200 St. Georges Terrace, Perth, W.A.; 309 Pitt Street, Sydney, N.S.W.; 294 Adelaide Street, Brisbane, Qld.
Austria	Gerold & Co., Buchhandlung und Verlag, Graben 31, 1011 Vienna.
Bangladesh	ADAB, 79 Road 11A, P.O. Box 5045, Dhanmondi, Dacca.
Belgium	Service des publications de la FAO, M.J. de Lannoy, 202, avenue du Roi, 1060 Brussels. CCP 000-0808993-13.
Bolivia	Los Amigos del Libro, Perú 3712, Casilla 450, Cochabamba; Mercado 1315, La Paz; René Moreno 26, Santa Cruz; Junín esq. 6 de Octubre, Oruro.
Brazil	Livraria Mestre Jou, Rua Guaipá 518, São Paulo 05089; Rua Senador Dantas 19-S205/206, 20.031 Rio de Janeiro; PRODIL, Promoção e Dist. de Livros Ltda., Av. Venâncio Aires 196, Caixa Postal 4005, 90.000 Porto Alegre; A NOSSA LIVRARIA, CLS 104, Bloco C, Lojas 18/19, 70.000 Brasilia, D.F.
Brunei	SST Trading Sdn. Bhd., Bangunan Tekno No. 385, Jln 5/59, P.O. Box 227, Petaling Jaya, Selangor.
Canada	Renouf Publishing Co. Ltd, 2182 St Catherine West, Montreal, Que. H3H 1M7.
Chile	Tecnolibro S.A., Merced 753, entrepiso 15, Santiago.
China	China National Publications Import Corporation, P.O. Box 88, Beijing.
Colombia	Editorial Blume de Colombia Ltda., Calle 65 N° 16-65, Apartado Aéreo 51340, Bogotá D.E.
Costa Rica	Librería, Imprenta y Litografía Lehmann S.A., Apartado 10011, San José.
Cuba	Empresa de Comercio Exterior de Publicaciones, O'Reilly 407 Bajos entre Aguacate y Compostela, Havana.
Cyprus	MAM, P.O. Box 1722, Nicosia.
Czechoslovakia	ARTIA, Ve Smeckach 30, P.O. Box 790, 111 27 Praha 1.
Denmark	Munksgaard Export and Subscription Service, 35 Nørre Søgade, DK 1370 Copenhagen K.
Dominican Rep.	Fundación Dominicana de Desarrollo, Casa de las Gárgolas, Mercedes 4, Apartado 857, Zona Postal 1, Santo Domingo.
Ecuador	Su Libreria Cía. Ltda., García Moreno 1172 y Mejía, Apartado 2556, Quito; Chimborazo 416, Apartado 3565, Guayaquil.
El Salvador	Librería Cultural Salvadoreña S.A. de C.V., Calle Arce 423, Apartado Postal 2296, San Salvador.
Finland	Akateeminen Kirjakauppa, 1 Keskuskatu, P.O. Box 128, 00101 Helsinki 10.
France	Editions A. Pedone, 13, rue Soufflot, 75005 Paris.
Germany, F.R.	Alexander Horn Internationale Buchhandlung, Spiegelgasse 9, Postfach 3340, 6200 Wiesbaden.
Ghana	Fides Enterprises, P.O. Box 14129, Accra; Ghana Publishing Corporation, P.O. Box 3632, Accra.
Greece	G.C. Eleftheroudakis S.A., International Bookstore, 4 Nikis Street, Athens (T-126); John Mihalopoulos & Son S.A., International Booksellers, 75 Hermou Street, P.O. Box 73, Thessaloniki.
Guatemala	Distribuciones Culturales y Técnicas "Artemis", 5a. Avenida 12-11, Zona 1, Apartado Postal 2923, Guatemala.
Guinea-Bissau	Conselho Nacional da Cultura, Avenida da Unidade Africana, C.P. 294, Bissau.
Guyana	Guyana National Trading Corporation Ltd, 45-47 Water Street, P.O. Box 308, Georgetown.
Haiti	Librairie "A la Caravelle", 26, rue Bonne Foi, B.P. 111, Port-au-Prince.
Hong Kong	Swindon Book Co., 13-15 Lock Road, Kowloon.
Hungary	Kultura, P.O. Box 149, 1389 Budapest 62.
Iceland	Snaebjörn Jónsson and Co. h.f., Hafnarstraeti 9, P.O. Box 1131, 101 Reykjavik.
India	Oxford Book and Stationery Co., Scindia House, New Delhi 110001; 17 Park Street, Calcutta 700016.
Indonesia	P.T. Sari Agung, 94 Kebon Sirih, P.O. Box 411, Djakarta.
Iraq	National House for Publishing, Distributing and Advertising, Jamhuria Street, Baghdad.
Ireland	The Controller, Stationery Office, Dublin 4.
Italy	Distribution and Sales Section, Food and Agriculture Organization of the United Nations, Via delle Terme di Caracalla, 00100 Rome; Libreria Scientifica Dott. Lucio de Biasio "Aeiou", Via Meravigli 16, 20123 Milan; Libreria Commissionaria Sansoni S.p.A. "Licosa", Via Lamarmora 45, C.P. 552, 50121 Florence.
Japan	Maruzen Company Ltd, P.O. Box 5050, Tokyo International 100-31.
Kenya	Text Book Centre Ltd, Kijabe Street, P.O. Box 47540, Nairobi.
Kuwait	Saeed & Samir Bookstore Co. Ltd, P.O. Box 5445, Kuwait.
Luxembourg	Service des publications de la FAO, M.J. de Lannoy, 202, avenue du Roi, 1060 Brussels (Belgium).
Malaysia	SST Trading Sdn. Bhd., Bangunan Tekno No. 385, Jln 5/59, P.O. Box 227, Petaling Jaya, Selangor.
Mauritius	Nalanda Company Limited, 30 Bourbon Street, Port Louis.
Mexico	Dilitsa S.A.; Puebla 182-D. Apartado 24-448, Mexico 7, D.F.
Morocco	Librairie "Aux Belles Images", 281, avenue Mohammed V, Rabat.
Netherlands	Keesing Boeken V.B., Joan Muyskenweg 22, 1096 CJ Amsterdam.
New Zealand	Government Printing Office. Government Printing Office Bookshops: Retail Bookshop, 25 Rutland Street, Mail Orders, 85 Beach Road, Private Bag C.P.O., Auckland; Retail, Ward Street, Mail Orders, P.O. Box 857, Hamilton; Retail, Mulgrave Street (Head Office), Cubacade World Trade Centre, Mail Orders, Private Bag, Wellington; Retail, 159 Hereford Street, Mail Orders, Private Bag, Christchurch; Retail, Princes Street, Mail Orders, P.O. Box 1104, Dunedin.
Nigeria	University Bookshop (Nigeria) Limited, University of Ibadan, Ibadan.
Norway	Johan Grundt Tanum Bokhandel, Karl Johansgate 41-43, P.O. Box 1177 Sentrum, Oslo 1.
Pakistan	Mirza Book Agency, 65 Shahrah-e-Quaid-e-Azam, P.O. Box 729, Lahore 3.
Panama	Distribuidora Lewis S.A., Edificio Dorasol, Calle 25 y Avenida Balboa, Apartado 1634, Panama 1.
Paraguay	Agencia de Librerías Nizza S.A., Tacuarí 144, Asunción.
Peru	Librería Distribuidora "Santa Rosa", Jirón Apurímac 375, Casilla 4937, Lima 1.
Philippines	The Modern Book Company Inc., 922 Rizal Avenue, P.O. Box 632, Manila.
Poland	Ars Polona, Krakowskie Przedmiescie 7, 00-068 Warsaw.
Portugal	Livraria Bertrand, S.A.R.L., Rua João de Deus, Venda Nova, Apartado 37, 2701 Amadora Codex; Livraria Portugal, Dias y Andrade Ltda., Rua do Carmo 70-74, Apartado 2681, 1117 Lisbon Codex; Edições ITAU, Avda. da República 46/A-r/c Esqdo., Lisbon 1.
Korea, Rep. of	Eul-Yoo Publishing Co. Ltd, 46-1 Susong-Dong, Jongro-Gu, P.O. Box Kwang-Wha-Moon 362, Seoul.
Romania	Ilexim, Calea Grivitei N° 64-66, B.P. 2001, Bucharest.
Saudi Arabia	The Modern Commercial University, P.O. Box 394, Riyadh.
Sierra Leone	Provincial Enterprises, 26 Garrison Street, P.O. Box 1228, Freetown.
Singapore	MPH Distributors (S) Pte. Ltd, 71/77 Stamford Road, Singapore 6; Select Books Pte. Ltd, 215 Tanglin Shopping Centre, 19 Tanglin Road, Singapore 1024; SST Trading Sdn. Bhd., Bangunan Tekno No. 385, Jln 5/59, P.O. Box 227, Petaling Jaya, Selangor.
Somalia	"Samater's", P.O. Box 936, Mogadishu.
Spain	Mundi Prensa Libros S.A., Castelló 37, Madrid 1; Librería Agrícola, Fernando VI 2, Madrid 4.
Sri Lanka	M.D. Gunasena & Co. Ltd, 217 Olcott Mawatha, P.O. Box 246, Colombo 11.
Sudan	University Bookshop, University of Khartoum, P.O. Box 321, Khartoum.
Suriname	VACO n.v. in Suriname, Dominee Straat 26, P.O. Box 1841, Paramaribo.
Sweden	C.E. Fritzes Kungl. Hovbokhandel, Regeringsgatan 12, P.O. Box 16356, 103 27 Stockholm.
Switzerland	Librairie Payot S.A., Lausanne et Genève; Buchhandlung und Antiquariat Heinimann & Co., Kirchgasse 17, 8001 Zurich.
Thailand	Suksapan Panit, Mansion 9, Rajadamnern Avenue, Bangkok.
Togo	Librairie du Bon Pasteur, B.P. 1164, Lomé.
Tunisia	Société tunisienne de diffusion, 5, avenue de Carthage, Tunis.
United Kingdom	Her Majesty's Stationery Office, 49 High Holborn, London WC1V 6HB (callers only); P.O. Box 569, London SE1 9NH (trade and London area mail orders); 13a Castle Street, Edinburgh EH2 3AR; 41 The Hayes, Cardiff CF1 1JW; 80 Chichester Street, Belfast BT1 4JY; Brazennose Street, Manchester M60 8AS; 258 Broad Street, Birmingham B1 2HE; Southey House, Wine Street, Bristol BS1 2BQ.
Tanzania, United Rep. of	Dar es-Salaam Bookshop, P.O. Box 9030, Dar es-Salaam; Bookshop, University of Dar es-Salaam, P.O. Box 893, Morogoro.
United States of America	UNIPUB, 345 Park Avenue South, New York, N.Y. 10010.
Uruguay	Librería Agropecuaria S.R.L., Alzaibar 1328, Casilla de Correos 1755, Montevideo.
Venezuela	Blume Distribuidora S.A., Gran Avenida de Sabana Grande, Residencias Caroni, Local 5, Apartado 50.339, 1050-A Caracas.
Yugoslavia	Jugoslovenska Knjiga, Trg. Republike 5/8, P.O. Box 36, 11001 Belgrade; Cankarjeva Zalozba, P.O. Box 201-IV, 61000 Ljubljana; Prosveta, Terazije 16, P.O. Box 555, 11001 Belgrade.
Zambia	Kingstons (Zambia) Ltd, Kingstons Building, President Avenue, P.O. Box 139, Ndola.
Other countries	Requests from countries where sales agents have not yet been appointed may be sent to: Distribution and Sales Section, Food and Agriculture Organization of the United Nations, Via delle Terme di Caracalla, 00100 Rome, Italy.

October 1981